Challenges to the Second Law of Thermodynamics

Fundamental Theories of Physics

An International Book Series on The Fundamental Theories of Physics:
Their Clarification, Development and Application

Volume 146

Challenges to the Second Law of Thermodynamics

Theory and Experiment

By

Vladislav Čápek

Charles University,
Prague, Czech Republic

and

Daniel P. Sheehan

University of San Diego,
San Diego, California, U.S.A.

 Springer

A C.I.P. Catalogue record for this book is available from the Library of Congress.

ISBN 1-4020-3015-0 (HB)
ISBN 1-4020-3016-9 (e-book)

Published by Springer,
P.O. Box 17, 3300 AA Dordrecht, The Netherlands.

Sold and distributed in North, Central and South America
by Springer,
101 Philip Drive, Norwell, MA 02061, U.S.A.

In all other countries, sold and distributed
by Springer,
P.O. Box 322, 3300 AH Dordrecht, The Netherlands.

Printed on acid-free paper

To our wives, Jana and Annie

In Memoriam
Vláda
(1943-2002)

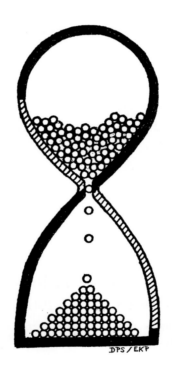

DPS/EKP

Contents

Preface

The advance of scientific thought in ways resembles biological and geologic transformation: long periods of gradual change punctuated by episodes of radical upheaval. Twentieth century physics witnessed at least three major shifts — relativity, quantum mechanics and chaos theory — as well many lesser ones. Now, early in the 21st, another shift appears imminent, this one involving the second law of thermodynamics.

Over the last 20 years the absolute status of the second law has come under increased scrutiny, more than during any other period its 180-year history. Since the early 1980's, roughly 50 papers representing over 20 challenges have appeared in the refereed scientific literature. In July 2002, the first conference on its status was convened at the University of San Diego, attended by 120 researchers from 25 countries (QLSL2002) [1]. In 2003, the second edition of Leff's and Rex's classic anthology on Maxwell demons appeared [2], further raising interest in this emerging field. In 2004, the mainstream scientific journal *Entropy* published a special edition devoted to second law challenges [3]. And, in July 2004, an echo of QLSL2002 was held in Prague, Czech Republic [4].

Modern second law challenges began in the early 1980's with the theoretical proposals of Gordon and Denur. Starting in the mid-1990's, several proposals for experimentally testable challenges were advanced by Sheehan, et al. By the late 1990's and early 2000's, a rapid succession of theoretical quantum mechanical challenges were being advanced by Čápek, et al., Allahverdyan, Nieuwenhuizen, et al., classical challenges by Liboff, Crosignani and Di Porto, as well as more experimentally-based proposals by Nikulov, Keefe, Trupp, Gräff, and others.

The breadth and depth of recent challenges are remarkable. They span three orders of magnitude in temperature, twelve orders of magnitude in size; they are manifest in condensed matter, plasma, gravitational, chemical, and biological physics; they cross classical and quantum mechanical boundaries. Several have strong corroborative experimental support and laboratory tests attempting bona fide violation are on the horizon. Considered en masse, the second law's absolute status can no longer be taken for granted, nor can challenges to it be casually dismissed.

This monograph is the first to examine modern challenges to the second law. For more than a century this field has lain fallow and beyond the pale of legitimate scientific inquiry due both to a dearth of scientific results and to a surfeit of peer pressure against such inquiry. It is remarkable that 20th century physics, which embraced several radical paradigm shifts, was unwilling to wrestle with this remnant of 19th century physics, whose foundations were admittedly suspect and largely unmodified by the discoveries of the succeeding century. This failure is due in part to the many strong imprimaturs placed on it by prominent scientists like Planck, Eddington, and Einstein. There grew around the second law a nearly inpenetrable mystique which only now is being pierced.

The second law has no general theoretical proof and, like all physical laws, its status is tied ultimately to experiment. Although many theoretical challenges to it have been advanced and several corroborative experiments have been conducted, no experimental violation has been claimed and confirmed. In this volume we will attempt to remain clear on this point; that is, while the second law might be *potentially* violable, it has not been violated *in practice*. This being the case, it is our position that the second law should be considered absolute unless experiment demonstrates otherwise. It is also our position, however, given the strong evidence for its potential violability, that inquiry into its status should not be stifled by certain unscientific attitudes and practices that have operated thus far.

This volume should be of interest to researchers in any field to which the second law pertains, especially to physicists, chemists and engineers involved with thermodynamics and statistical physics. Individual chapters should be valuable to more select readers. Chapters 1-2, which give an overview of entropy, the second law, early challenges, and classical arguments for second law inviolability, should interest historians and philosophers of science. Chapter 3, which develops quantum mechanical formalism, should interest theorists in quantum statistical mechanics, decoherence, and entanglement. Chapters 4-9 unpack individual, experimentally-testable challenges and can be profitably read by researchers in the various subfields in which they arise, *e.g.*, solid state, plasma, superconductivity, biochemistry. The final chapter explores two topics at the forefront of second law research: thermosynthetic life and physical eschatology. The former is a proposed third branch of life — beyond the traditional two (chemosynthetic and photosynthetic) — and is relevant to evolutionary and extremophile biology, biochemistry, and origin-of-life studies. The latter topic explores the fate of life in the cosmos in light of the second law and its possible violation. Roughly 80% of this volume covers research currently in the literature, rearranged and interpreted; the remaining 20% represents new, unpublished work. Chapter 3 was written exclusively by Čápek (with editing by d.p.s.), Chapters 4-10 exclusively by Sheehan, Chapter 1 primarily by Sheehan, and Chapter 2 jointly. As much as possible, each chapter is self-contained and understandable without significant reference to other chapters. Whenever possible, the mathematical notation is identical to that employed in the original research.

It is likely that many of the challenges in this book will fall short of their marks, but such is the nature of exploratory research, particularly when the quarry is as formidable as the second law. It has 180 years of historical inertia behind it and the adamantine support of the scientific community. It has been confirmed by countless experiments and has survived scores of challenges unscathed. Arguably, it is the best tested, most central and profound physical principle crosscutting the sciences, engineering, and humanities. For good reasons, its absolute status is unquestioned.

However, as the second law itself teaches: *Things change.*

Daniel P. Sheehan
San Diego, California
August 4, 2004

References

[1] Sheehan, D.P., Editor, *First International Conference on Quantum Limits to the Second Law*, AIP Conference Proceedings, Volume 643 (AIP Press, Melville, NY, 2002).

[2] Leff, H.S. and Rex, A.F., *Maxwell's Demon 2: Entropy, Classical and Quantum Information, Computing* (Institute of Physics, Bristol, 2003).

[3] Special Edition: Quantum Limits to the Second Law of Thermodynamics; Nikulov, A.V. and Sheehan, D.P., Guest Editors, Entropy **6** 1-232 (2004).

[4] *Frontiers of Quantum and Mesoscopic Thermodynamics*, Satellite conference of 20th CMD/EPS, Prague, Czech Republic, July 26-29, 2004.

Acknowledgements

It is a pleasure to acknowledge a number of colleagues, associates, and staff who assisted in the completion of this book. We gratefully thank Emily Perttu for her splendid artwork and Amy Besnoy for her library research support. The following colleagues are acknowledged for their review of sections of the book, particularly as they pertain to their work: Lyndsay Gordon, Jack Denur, Peter Keefe, Armen Allahverdyan, Theo Nieuwenhuizen, Andreas Trupp, Bruno Crosignani, Jeremy Fields, Anne Sturz, Václav Špička, and William Sheehan. Thank you all!

Special thanks are extended to USD Provost Frank Lazarus, USD President-Emeritus Alice B. Hayes, and Dean Patrick Drinan for their financial support of much of the research at USD. This work was also tangentially supported by the Research Corporation and by the United States Department of Energy.

We are especially indebted to Alwyn van der Merwe for his encouragement and support of this project. We are also grateful to Sabine Freisem and Kirsten Theunissen for their patience and resolve in seeing this volume to completion. I (d.p.s.) especially thank my father, William F. Sheehan, for introducing me to this ancient problem.

Lastly, we thank our lovely and abiding wives, Jana and Annie, who stood by us in darkness and in light.

D.P.S.
(V.Č.)

Postscript

Although this book is dedicated to our wives, for me (d.p.s.), it is also dedicated to Vláda, who died bravely October 28, 2002. He was a lion of a man, possessing sharp wit, keen insight, indominable spirit, and deep humanity. He gave his last measure of strength to complete his contribution to this book, just months before he died. He is sorely missed.

d.p.s.
July, 2004

1

Entropy and the Second Law

Various formulations of the second law and entropy are reviewed. Longstanding foundational issues concerned with their definition, physical applicability and meaning are discussed.

1.1 Early Thermodynamics

The origins of thermodynamic thought are lost in the furnace of time. However, they are written into flesh and bone. To some degree, all creatures have an innate 'understanding' of thermodynamics — as well they should since they are bound by it. Organisms that display thermotaxis, for example, have a somatic familiarity with thermometry: zeroth law. Trees grow tall to dominate solar energy reserves: first law. Animals move with a high degree of energy efficiency because it is 'understood' at an evolutionary level that energy wasted cannot be recovered: second law. Nature culls the inefficient.

Human history and civilization have been indelibly shaped by thermodynamics. Survival and success depended on such things as choosing the warmest cave for winter and the coolest for summer, tailoring the most thermally insulating furs, rationing food, greasing wheels against friction, finding a southern exposure for a home (in the northern hemisphere), tidying up occasionally to resist the tendencies of entropy. Human existence and civilization have always depended implicitly on an understanding of thermodynamics, but it has only been in the last 150 years that this understanding has been codified. Even today it is not complete.

Were one to be definite, the first modern strides in thermodynamics began perhaps with James Watt's (1736-1819) steam engine, which gave impetus to what we now know as the Carnot cycle. In 1824 Sadi Nicolas Carnot (1796-1832), published his only scientific work, a treatise on the theory of heat (*Réflexions sur la Puissance Motice du Feu*) [1]. At the time, it was not realized that a portion of the heat used to drive steam engines was converted into work. This contributed to the initial disinterest in Carnot's research.

Carnot turned his attention to the connection between heat and work, abandoning his previous opinion about heat as a fluidum, and almost surmised correctly the mechanical equivalent of heat[1]. In 1846, James Prescott Joule (1818-1889) published a paper on thermal and chemical effects of the electric current and in another (1849) he reported mechanical equivalent of heat, thus erasing the sharp boundary between mechanical and thermal energies. There were also others who, independently of Joule, contributed to this change of thinking, notably Hermann von Helmholtz (1821-1894).

Much of the groundwork for these discoveries was laid by Benjamin Thompson (Count of Rumford 1753-1814). In 1798, he took part in boring artillery gun barrels. Having ordered the use of blunt borers – driven by draught horses – he noticed that substantial heat was evolved, in fact, in quantities sufficient to boil appreciable quantities of water. At roughly the same time, Sir Humphry Davy (1778-1829) observed that heat developed upon rubbing two pieces of metal or ice, even under vacuum conditions. These observations strongly contradicted the older fluid theories of heat.

The law of energy conservation as we now know it in thermodynamics is usually ascribed to Julius Robert von Mayer (1814-1878). In classical mechanics, however, this law was known intuitively at least as far back as Galileo Galilei (1564-1642). In fact, about a dozen scientists could legitimately lay claim to discovering energy conservation. Fuller accounts can be found in books by Brush [2] and von Baeyer [3]. The early belief in energy conservation was so strong that, since 1775, the French Academy has forbidden consideration of any process or apparatus that purports to produce energy *ex nihilo*: a *perpetuum mobile* of the first kind.

With acceptance of energy conservation, one arrives at the first law of thermodynamics. Rudolph Clausius (1822-1888) summarized it in 1850 thus: "In any process, energy may be changed from one to another form (including heat and work), but can never be produced or annihilated." With this law, any possibility of realizing a *perpetuum mobile* of the first kind becomes illusory.

Clausius' formulation still stands in good stead over 150 years later, despite unanticipated discoveries of new forms of energy — *e.g.*, nuclear energy, rest mass energy, vacuum energy, dark energy. Because the definition of energy is malleable, in a practical sense, the first law probably need not ever be violated because, were one to propose a violation, energy could be redefined so as to correct it. Thus, conservation of energy is reduced to a tautology and the first law to a powerfully convenient accounting tool for the two general forms of energy: heat and work.

---

[1]Unfortunately, this tract was not published, but was found in his inheritance in 1878.

In equilibrium thermodynamics, the first law is written in terms of an additive state function, the internal energy U, whose exact differential dU fulfills

$$dU = \delta Q + \delta W. \tag{1.1}$$

Here δQ and δW are the inexact differentials of heat and work added to the system. (In nonequilibrium thermodynamics, there are problems with introducing these quantities rigorously.) As inexact differentials, the integrals of δQ and δW are path dependent, while dU, an exact differential is path independent; thus, U is a state function. Other state functions include enthalpy, Gibbs free energy, Helmholtz free energy and, of course, entropy.

1.2 The Second Law: Twenty-One Formulations

The second law of thermodynamics was first enunciated by Clausius (1850) [4] and Kelvin (1851) [5], largely based on the work of Carnot 25 years earlier [1]. Once established, it settled in and multiplied wantonly; the second law has more common formulations than any other physical law. Most make use of one or more of the following terms — entropy, heat, work, temperature, equilibrium, *perpetuum mobile* — but none employs all, and some employ none. Not all formulations are equivalent, such that to satisfy one is not necessarily to satisfy another. Some versions overlap, while others appear to be entirely distinct laws. Perhaps this is what inspired Truesdell to write, "Every physicist knows exactly what the first and second laws mean, but it is my experience that no two physicists agree on them."

Despite — or perhaps because of — its fundamental importance, no single formulation has risen to dominance. This is a reflection of its many facets and applications, its protean nature, its colorful and confused history, but also its many unresolved foundational issues. There are several fine accounts of its history [2, 3, 6, 7]; here we will give only a sketch to bridge the many versions we introduce. Formulations can be catagorized roughly into five catagories, depending on whether they involve: 1) device and process impossibilities; 2) engines; 3) equilibrium; 4) entropy; or 5) mathematical sets and spaces. We will now consider twenty-one standard (and non-standard) formulations of the second law. This survey is by no means exhaustive.

The first explicit and most widely cited form is due to Kelvin[2] [5, 8].

> **(1) Kelvin-Planck** No device, operating in a cycle, can produce the sole effect of extraction a quantity of heat from a heat reservoir and the performance of an equal quantity of work.

---

[2]William Thomson (1824-1907) was known from 1866-92 as Sir William Thomson and after 1892 as Lord Kelvin of Largs.

In this, its most primordial form, the second law is an injunction against *perpetuum mobile* of the second type (PM2). Such a device would transform heat from a heat bath into useful work, in principle, indefinitely. It formalizes the reasoning undergirding Carnot's theorem, proposed over 25 years earlier.

The second most cited version, and perhaps the most natural and experientially obvious, is due to Clausius (1854) [4]:

> **(2) Clausius-Heat** No process is possible for which the sole effect is that heat flows from a reservoir at a given temperature to a reservoir at higher temperature.

In the vernacular: Heat flows from hot to cold. In contradistinction to some formulations that follow, these two statements make claims about strictly *nonequilibrium* systems; as such, they cannot be considered equivalent to later equilibrium formulations. Also, both versions turn on the key term, *sole effect*, which specifies that the heat flow must not be aided by external agents or processes. Thus, for example, heat pumps and refrigerators, which do transfer heat from a cold reservoir to a hot reservoir, do so without violating the second law since they require work input from an external source that inevitably satisfies the law.

Other common (and equivalent) statements to these two include:

> **(3) Perpetual Motion** *Perpetuum mobile* of the second type are impossible.

and

> **(4) Refrigerators** Perfectly efficient refrigerators are impossible.

The primary result of Carnot's work and the root of many second law formulations is Carnot's theorem [1]:

> **(5) Carnot theorem** All Carnot engines operating between the same two temperatures have the same efficiency.

Carnot's theorem is occasionally but not widely cited as the second law. Usually it is deduced from the Kelvin-Planck or Clausius statements. Analysis of the Carnot cycle shows that a portion of the heat flowing through a heat engine must always be lost as waste heat, not to contribute to the overall useful heat output[3]. The maximum efficiency of heat engines is given by the Carnot efficiency: $\eta = 1 - \frac{T_c}{T_h}$, where $T_{c,h}$ are the temperatures of the colder and hotter heat reservoirs between which the heat engine operates. Since absolute zero ($T_c = 0$) is unattainable (by one version of the third law) and since $T_h \neq \infty$ for any realistic system, the Carnot efficiency forbids perfect conversion of heat into work (*i.e.*, $\eta = 1$). Equivalent second law formulations embody this observation:

---

[3]One could say that the second law is Nature's tax on the first.

(6) Efficiency All Carnot engines have efficiencies satisfying:
$0 < \eta < 1$.

and,

(7) Heat Engines Perfectly efficient heat engines ($\eta = 1$) are impossible.

The *efficiency* form is not cited in textbooks, but is suggested as valid by Koenig [9]. There is disagreement over whether Carnot should be credited with the discovery of the second law [10]. Certainly, he did not enunciate it explicitly, but he seems to have understood it in spirit and his work was surely a catalyst for later, explicit statements of it.

Throughout this discussion it is presumed that realizable heat engines must operate between two reservoirs at different temperatures. (T_c and T_h). This condition is considered so stringent that it is often invoked as a litmus test for second law violators; that is, if a heat engine purports to operate at a single temperature, it violates the second law. Of course, mathematically this is no more than asserting $\eta = 1$, which is already forbidden.

Since thermodynamics was initially motivated by the exigencies of the industrial revolution, it is unsurprising that many of its formulations involve engines and cycles.

(8) Cycle Theorem Any physically allowed heat engine, when operated in a cycle, satisfies the condition

$$\oint \frac{\delta Q}{T} = 0 \qquad\qquad (1.2)$$

if the cycle is reversible; and

$$\oint \frac{\delta Q}{T} < 0 \qquad\qquad (1.3)$$

if the cycle is irreversible.

Again, δQ is the inexact differential of heat. This theorem is widely cited in the thermodynamic literature, but is infrequently forwarded as a statement of the second law. In discrete summation form for reversible cycles ($\sum_i Q_i/T_i = 0$), it was proposed early on by Kelvin [5] as a statement of the second law.

(9) Irreversibility All natural processes are irreversible.

Irreversibility is an essential feature of natural processes and it is the essential thermodynamic characteristic defining the direction of time[4] — *e.g.*, omelettes do

---

[4]It is often said that irreversibility gives direction to time's arrow. Perhaps one should say irreversibility *is* time's arrow [11-17].

not spontaneously unscramble; redwood trees do not 'ungrow'; broken Ming vases do not reassemble; the dead to not come back to life. An irreversible process is, by definition, not quasi-static (reversible); it cannot be undone without additional irreversible changes to the universe. Irreversibility is so undeniably observed as an essential behavior of the physical world that it is put forward by numerous authors in second law statements.

In many thermodynamic texts, *natural* and *irreversible* are equated, in which case this formulation is tautological; however, as a reminder of the essential content of the law, it is unsurpassed. In fact, it is so deeply understood by most scientists as to be superfluous.

A related formulation, advanced by Koenig [9] reads:

>   **(10) Reversibility** All normal quasi-static processes are reversible, and conversely.

Koenig claims, "this statement goes further than [the irreversibility statement] in that it supplies a necessary and sufficient condition for reversibility (and irreversibility)." This may be true, but it is also sufficiently obtuse to be forgettable; it does not appear in the literature beyond Koenig.

Koenig also offers the following orphan version [9]:

>   **(11) Free Expansion** Adiabatic free expansion of a perfect gas is an irreversible process.

He demonstrates that, within his thermodynamic framework, this proposition is equivalent to the statement, "If a [PM2] is possible, then free expansion of a gas is a reversible process; and conversely." Of course, since adiabatic free expansion is irreversible, it follows *perpetuum mobile* are logically impossible — a standard statement of the second law. By posing the second law in terms of a particular physical process (adiabatic expansion), the door is opened to use any natural (irreversible) process as the basis of a second law statement. It also serves as a reminder that the second law is not only *of* the world and *in* the world, but, in an operational sense, it *is* the world. This formulation also does not enjoy citation outside Koenig [9].

A relatively recent statement is proposed by Macdonald [18]. Consider a system Z, which is closed with respect to material transfers, but to which heat and work can be added or subtracted so as to change its state from A to B by an arbitrary process $\mathcal{P}$ that is not necessarily quasi-static. Heat ($H_{\mathcal{P}}$) is added by a *standard heat source*, taken by Macdonald to be a reservoir of water at its triple point. The second law is stated:

>   **(12) Macdonald** [18] It is impossible to transfer an arbitrarily large amount of heat from a standard heat source with processes terminating at a fixed state of Z. In other words, for every state B of Z,

$$\text{Sup}[H_{\mathcal{P}} : \mathcal{P} \text{ terminates at } B] < \infty,$$

where Sup[...] is the supremum of heat for the process $\mathcal{P}$.

Absolute entropy is defined easily from here as the supremum of the heat $H_{\mathcal{P}}$ divided by a fiduciary temperature T_o, here taken to be the triple point of water (273.16 K); that is, $S(B) = \text{Sup}[H_{\mathcal{P}}/T_o : P \text{ terminates at } B]$. Like most formulations of entropy and the second law, these apply strictly to closed equilibrium systems.

Many researchers take equilibrium as the *sine qua non* for the second law.

(13) Equilibrium The macroscopic properties of an isolated nonstatic system eventually assume static values.

Note that here, as with many equivalent versions, the term *equilibrium* is purposefully avoided. A related statement is given by Gyftopolous and Beretta [19]:

(14) Gyftopolous and Beretta Among all the states of a system with given values of energy, the amounts of constituents and the parameters, there is one and only one stable equilibrium state. Moreover, starting from any state of a system it is always possible to reach a stable equilibrium state with arbitrary specified values of amounts of constituents and parameters by means of a reversible weight process.

(Details of nomenclature (*e.g.*, weight process) can be found in §1.3.) Several aspects of these two equilibrium statements merit unpacking.

- *Macroscopic* properties (*e.g.*, temperature, number density, pressure) are ones that exhibit statistically smooth behavior at equilibrium. Scale lengths are critical; for example, one expects macroscopic properties for typical liquids at scale lengths greater than about 10^{-6}m. At shorter scale lengths statistical fluctuations become important and can undermine the second law. This was understood as far back as Maxwell [20, 21, 22, 23].

- There are no truly *isolated* systems in nature; all are connected by long-range gravitational and perhaps electromagnetic forces; all are likely affected by other uncontrollable interactions, such as by neutrinos, dark matter, dark energy and perhaps local cosmological expansion; and all are inevitably coupled thermally to their surroundings to some degree. Straightforward calculations show, for instance, that the gravitational influence of a minor asteroid in the Asteroid Belt is sufficient to instigate chaotic trajectories of molecules in a parcel of air on Earth in less than a microsecond. Since gravity cannot be screened, the exact molecular dynamics of all realistic systems are constantly affected in essentially unknown and uncontrollable ways. Unless one is able to model the entire universe, one probably cannot exactly model any subset of it[5]. Fortunately, statistical arguments (*e.g.*, molecular chaos, ergodicity) allow thermodynamics to proceed quite well in most cases.

---

[5]Quantum mechanical entanglement, of course, further complicates this task.

- One can distinguish between *stable* and *unstable* static (or equilibrium) states, depending on whether they "persist over time intervals significant for some particular purpose in hand." [9]. For instance, to say "*Diamonds are forever.*" is to assume much. Diamond is a metastable state of carbon under everyday conditions; at elevated temperatures ($\sim$ 2000 K), it reverts to graphite. In a large enough vacuum, graphite will evaporate into a vapor of carbon atoms and they, in turn, will thermally ionize into a plasma of electrons and ions. After 10^{33} years, the protons might decay, leaving a tenuous soup of electrons, positrons, photons, and neutrinos. Which of these is a stable equilibrium? None or each, depending on the time scale and environment of interest. By definition, a *stable* static state is one that can change only if its surroundings change, but still, time is a consideration. To a large degree, equilibrium is a matter of taste, time, and convenience.

- Gyftopoulos and Beretta emphasise *one and only one stable equilibrium state*. This is echoed by others, notably by Mackey who reserves this caveat for his *strong form* of the second law [24].

Thus far, entropy has not entered into any of these second law formulations. Although, in everyday scientific discourse the two are inextricably linked, this is clearly not the case. Entropy was defined by Clausius in 1865, nearly 15 years after the first round of explicit second law formulations. Since entropy was originally wrought in terms of heat and temperature, this allows one to recast earlier formulations easily. Naturally, the first comes from Clausius:

(15) **Clausius-Entropy** [4, 6] For an adiabatically isolated system that undergoes a change from one equilibrium state to another, if the thermodynamic process is reversible, then the entropy change is zero; if the process is irreversible, the entropy change is positive. Respectively, this is:

$$\int_i^f \frac{\delta Q}{T} = S_f - S_i \qquad (1.4)$$

and

$$\int_i^f \frac{\delta Q}{T} < S_f - S_i \qquad (1.5)$$

Planck (1858-1947), a disciple of Clausius, refines this into what he describes as "the most general expression of the second law of thermodynamics." [8, 6]

(16) **Planck** Every physical or chemical process occurring in nature proceeds in such a way that the sum of the entropies of all bodies which participate in any way in the process is increased. In the limiting case, for reversible processes, the sum remains unchanged.

Alongside the Kelvin-Planck version, these two statements have dominated the scientific landscape for nearly a century and a half. Planck's formulation implicitly cuts the original ties between entropy and heat, thereby opening the door for other versions of entropy to be used. It is noteworthy that, in commenting on the possible limitations of his formulation, Planck explicitly mentions the *perpetuum mobile*. Evidently, even as thermodynamics begins to mature, the specter of the *perpetuum mobile* lurks in the background.

Gibbs takes a different tack to the second law by avoiding thermodynamic processes, and instead conjoins entropy with equilibrium [25, 6]:

> **(17) Gibbs** For the equilibrium of an isolated system, it is necessary and sufficient that in all possible variations of the state of the system which do not alter its energy, the variation of its entropy shall either vanish or be negative.

In other words, thermodynamic equilibrium for an isolated system is the state of maximum entropy. Although Gibbs does not refer to this as a statement of the second law, *per se*, this *maximum entropy principle* conveys its essential content. The maximum entropy principle [26] has been broadly applied in the sciences, engineering economics, information theory — wherever the second law is germane, and even beyond. It has been used to reformulate classical and quantum statistical mechanics [26, 27]. For instance, starting from it one can derive on the back of an envelope the continuous or discrete Maxwell-Boltzmann distributions, the Planck blackbody radiation formula (and, with suitable approximations, the Rayleigh-Jeans and Wien radiation laws) [24].

Some recent authors have adopted more definitional entropy-based versions [9]:

> **(18) Entropy Properties** Every thermodynamic system has two properties (and perhaps others): an intensive one, absolute temperature T, that may vary spatially and temporally in the system $T(x, t)$; and an extensive one, entropy S. Together they satisfy the following three conditions:
> (i) The entropy change dS during time interval dt is the sum of: (a) entropy flow through the boundary of the system $d_e S$; and (b) entropy production within the system, $d_i S$; that is, $dS = d_e S + d_i S$.
> (ii) Heat flux (not matter flux) through a boundary at uniform temperature T results in entropy change $d_e S = \frac{\delta Q}{T}$.
> (iii) For reversible processes within the system, $d_i S = 0$, while for irreversible processes, $d_i S > 0$.

This version is a starting point for some approaches to irreversible thermodynamics.

While there is no agreement in the scientific community about how best to state the second law, there is general agreement that the current melange of statements, taken en masse, pretty well covers it. This, of course, gives fits to mathematicians, who insist on precision and parsimony. Truesdell [28, 6] leads the charge:

> Clausius' verbal statement of the second law makes no sense.... All that
> remains is a Mosaic prohibition; a century of philosophers and journal-
> ists have acclaimed this commandment; a century of mathematicians
> have shuddered and averted their eyes from the unclean.

Arnold broadens this assessment [29, 6]:

> Every mathematician knows it is impossible to understand an elemen-
> tary course in thermodynamics.

In fact, mathematicians have labored to drain this "dismal swamp of obscurity"
[28], beginning with Carathéodory [30] and culminating with the recent *tour de
force* by Lieb and Yngvason [31]. While both are exemplars of mathematical rigor
and logic, both suffer from incomplete generality and questionable applicability to
realistic physical systems; in other words, there are doubts about their empirical
content.

Carathéodory was the first to apply mathematical rigor to thermodynamics
[30]. He imagines a state space Γ of all possible equilibrium states of a generic
system. Γ is an n-dimensional manifold with continuous variables and Euclidean
topology. Given two arbitrary states s and t, if s can be transformed into t by
an adiabatic process, then they satisfy *adiabatically accessibility* condition, written
$s \prec t$, and read *s precedes t*. This is similar to Lieb and Yngvason [31], except
that Lieb and Yngvason allow sets of possibly disjoint ordered states, whereas
Carathéodory assumes continuous state space and variables. Max Born's simplified
version of Carathéodory's second law reads [32]:

> **(19a) Carathéodory (Born Version):** In every neighborhood of
> each state (s) there are states (t) that are inaccessible by means of
> adiabatic changes of state. Symbolically, this is:

$$(\forall s \in \Gamma, \forall U_s) : \exists t \in U_s s \nprec t, \tag{1.6}$$

where U_s and U_t are open neighborhoods surrounding the states s and t.

Carathéodory's originally published version is more precise [30, 6].

> **(19b) Carathéodory Principle** In every open neighborhood $U_s \subset \Gamma$
> of an arbitrarily chosen state s there are states t such that for some
> open neighborhood U_t of t: all states r within U_t cannot be reached
> adiabatically from s. Symbolically this is:

$$\forall s \in \Gamma \forall U_s \exists t \in U_s \& \exists U_t \subset U_s \forall r \in U_t : s \nprec r. \tag{1.7}$$

Lieb and Yngvason [31] proceed along similar lines, but work with an set of
distinct states, rather than a continuous space of them. For them, the second law
is a theorem arising out of the ordering of the states via adiabatic accessibility.
Details can be found in §1.3.

In connection with analytical microscopic formulations of the second law, the recent work by Allahverdyan and Nieuwenhuizen [33] is noteworthy. They rederive and extend the results of Pusz, Woronowicz [34] and Lenard [35], and provide an analytical proof of the following equilibrium formulation of the Thomson (Kelvin) statement:

> **(20) Thomson (Equilibrium)** No work can be extracted from a closed equilibrium system during a cyclic variation of a parameter by an external source.

The Allahverdyan-Niewenhuizen (A-N) theorem is proved by rigorous quantum mechanical methods without invoking the time-invariance principle. This makes it superior to previous treatments of the problem. Although significant, it is insufficient to resolve most types of second law challenges, for multiple reasons. First, the A-N theorem applies to equilibrium systems only, whereas the original forms of the second law (Kelvin and Clausius) are strictly nonequilibrium in character and most second law challenges are inherently nonequilibrium in character; thus, the pertinence of the A-N theorem is limited. Second, it assumes that the system considered is isolated, but realistically, no such system exists in Nature. Third, it assumes the Gibbs form of the initial density matrix. While this assumption is natural when temperature is well defined, once *finite* coupling of the system to a bath is introduced, this assumption can be violated appreciably, especially for systems which purport second law violation (*e.g.*, [36]).

The relationships between these various second law formulations are complex, tangled and perhaps impossible to delineate completely, especially given the muzziness with which many of them and their underlying assumptions and definitions are stated. Still, attempts have been made along these lines [2, 6, 7, 9] [6]. This exercise of tracing the connections between the various formulations has historical, philosophical and scientific value; hopefully, it will help render a more inclusive formulation of the second law in the future.

In addition to academic formulations there are also many folksy aphorisms that capture aspects of the law. Many are catchphrases for more formal statements. Although loathe to admit it, most of these are used as primary rules of thumb by working scientists. Most are anonymous; when possible, we try to identify them with academic forms. Among these are:

- Disorder tends to increase.   (*Clausius, Planck*)

- Heat goes from hot to cold.   (*Clausius*)

- There are no perfect heat engines.   (*Carnot*)

- There are no perfect refrigerators.   (*Clausius*)

- Murphy's Law (and corollary)   (*Murphy* ~ 1947)

---

[6] See, Table I in Uffink [6] and Table II (Appendix A) in Koenig [9]

1. If anything can go wrong it will.

2. Situations tend to progress from bad to worse.

- A mess expands to fill the space available.

- The only way to deal with a can of worms is to find a bigger can.

- Laws of Poker in Hell:

    1. Poker exists in Hell.   (*Zeroth Law*)

    2. You can't win.   (*First Law*)

    3. You can't break even.   (*Second Law*)

    4. You can't leave the game.   (*Third Law*)

- Messes don't go away by themselves.   (*Mom*)

- Perpetual motion machines are impossible.   (*Nearly everyone*)

Interestingly, in number, second law aphorisms rival formal statements. Perhaps this is not surprising since the second law began with Carnot and Kelvin as an injunction against perpetual motions machines, which have been scorned publically back to times even before Leonardo da Vinci ($\sim$ 1500). Arguably, most versions of the second law add little to what we already understand intuitively about the dissipative nature of the world; they only confirm and quantify it. As noted by Pirrucello [37]:

> Perhaps we'll find that the second law is rooted in folk wisdom, platitudes about life. The second law is ultimately an expression of human disappointment and frustration.

For many, the first and best summary of thermodynamics was stated by Clausius 150 years ago [4]:

1. Die Energie der Welt ist konstant.
2. Die Entropie der Welt strebt einem Maximum zu.

or, in English,

1. The energy of the universe is constant.
2. The entropy of the universe strives toward a maximum.

Although our conceptions of energy, entropy and the universe have undergone tremendous change since his time, remarkably, Clausius' summary still rings true today — and perhaps even more so now for having weathered so much.

In surveying these many statements, one can get the impression of having stumbled upon a scientific Rorschauch test, wherein the second law becomes a reflection of one's own circumstances, interests and psyche. However, although there is much disagreement on how best to state it, its primordial injunction against *perpetuum mobile* of the second type generally receives the most support and the least dissention. It is the gold standard of second law formulations. If the second law is the flesh of thermodynamics, this injunction is its heart.

If the second law should be shown to be violable, it would nonetheless remain valid for the vast majority of natural and technological processes. In this case, we propose the following tongue-in-cheek formulation for a post-violation era, should it come to pass:

(21) **Post-Violation** For any spontaneous process the entropy of the universe does not decrease — except when it does.

1.3 Entropy: Twenty-One Varieties

The discovery of thermodynamic entropy as a state function is one of the triumphs of nineteenth-century theoretical physics. Inasmuch as the second law is one of the central laws of nature, its handmaiden — entropy — is one of the most central physical concepts. It can pertain to almost any system with more than a few particles, thereby subsuming nearly everything in the universe from nuclei to superclusters of galaxies [38]. It is protean, having scores of definitions, not all of which are equivalent or even mutually compatible[7]. To make matters worse, "perhaps every month someone invents a new one," [39]. Thus, it is not surprising there is considerable controversy surrounding its nature, utility, and meaning. It is fair to say that no one *really* knows what entropy is.

Roughly, entropy is a quantitative macroscopic measure of microscopic disorder. It is the only major physical quantity predicated and reliant upon wholesale ignorance of the system it describes. This approach is simultaneously its greatest strength and its Achilles heel. On one hand, the computational complexities of even simple dynamical systems often mock the most sophisticated analytic and numerical techniques. In general, the dynamics of n-body systems (n > 2) cannot be solved exactly; thus, thermodynamic systems with on the order of a mole of particles (10^{23}) are clearly hopeless, even in a perfectly deterministic Laplacian world, sans chaos. Thus, it is both convenient and wise to employ powerful physical assumptions to simplify entropy calculations — *e.g.*, equal *a priori* probability, ergodicity, strong mixing, extensivity, random phases, thermodynamic limit. On the other hand, although they have been spectacularly predictive and can be shown to be reasonable for large classes of physical systems, these assumptions are known not to be universally valid. Thus, it is not surprising that no completely satisfactory definition of entropy has been discovered, despite 150 years of effort. Instead, there has emerged a menagerie of different types which, over the decades, have grown increasingly sophisticated both in response to science's deepening understanding of nature's complexity, but also in recognition of entropy's inadequate expression.

This section provides a working man's overview of entropy; it focuses on the most pertinent and representative varieties. It will not be exhaustive, nor will

---

[7]P. Hänggi claims to have compiled a list of 55 different varieties; here we present roughly 21.

it respect many of the nuances of the subject; for these, the interested reader is directed to the many fine treatises on the subject.

Most entropies possess a number of important physical and mathematical properties whose adequate discussion extends beyond the aims of this volume; these include additivity, subadditivity, concavity, invariance, insensitivity, continuity conditions, and monotonicity [24, 31, 39]. Briefly, for a system A composed of two subsystems A_1 and A_2 such that $A = A_1 + A_2$, the entropy is additive if $S(A) = S(A_1) + S(A_2)$. For two independent systems A and B, the entropy is subadditive if their entropy when joined (composite entropy) is never less than the sum of their individual entropies; i.e., $S(A + B) \geq S(A) + S(B)$. (Note that for additivity the subsystems (A_1, A_2) retain their individual identities, while for subadditivity the systems (A, B) lose their individual identities.) For systems A and B, entropy demonstrates concavity if $S(\lambda A + (\lambda - 1)B) \geq \lambda S(A) + (1 - \lambda)S(B)$; $0 \leq \lambda \leq 1$.

A workingman's summary of standard properties can be extracted from Gyftopoulous and Beretta [19]. Classical entropy must[8]:

  a) be well defined for every system and state;
  b) be invariant for any reversible adiabatic process ($dS = 0$) and increase for any irreversible adiabatic process ($dS > 0$);
  c) be additive and subadditive for all systems, subsystems and states.
  d) be non-negative, and vanish for all states described by classical mechanics;
  e) have one and only one state corresponding to the largest value of entropy;
  f) be such that graphs of entropy versus energy for stable equilibria are smooth and concave; and
  g) reduce to relations that have been established experimentally.

The following are summaries of the most common and salient formulations of entropy, spiced with a few distinctive ones. There are many more.

(1) Clausius [4] The word *entropy* was coined by Rudolf Clausius (1865) as a thermodynamic complement to energy. The *en* draws parallels to *energy*, while *tropy* derives from the Greek word $\tau\rho o\pi\eta$, meaning *change*. Together *en-tropy* evokes quantitative measure for thermodynamic change[9].

Entropy is a macroscopic measure of the microscopic state of disorder or chaos in a system. Since heat is a macroscopic measure of microscopic random kinetic energy, it is not surprising that early definitions of entropy involve it. In its original and most utilitarian form, entropy (or, rather, entropy *change*) is expressed in terms of heat Q and temperature T. For reversible thermodynamic processes, it is

$$dS = \frac{\delta Q}{T}, \tag{1.8}$$

---

[8]Many physical systems in this volume do not abide these restrictions, most notably, additivity.

[9]Strictly speaking, Clausius coined *entropy* to mean *in transformation*.

while for irreversible processes, it is

$$dS > \frac{\delta Q}{T} \tag{1.9}$$

These presume that T is well defined in the surroundings, thus foreshadowing the zeroth law. To establish fiduciary entropies the third law is invoked. For systems "far" from equilibrium, neither entropy nor temperature is well defined.

(2) Boltzmann-Gibbs [40, 41] The most famous classical formulation of entropy is due to Boltzmann:

$$S_{BG,\mu} = S(E,N,V) = k\ln\Omega(E,N,V) \tag{1.10}$$

Here $\Omega(E,N,V)$ is the total number of distinct microstates (complexions) accessible to a system of energy E, particle number N in volume V. The Boltzmann relation provides the first and most important bridge between microscopic physics and equilibrium thermodynamics. It carries with it a minimum number of assumptions and, therefore, is quite general. It applies directly to the microcanonical ensemble (fixed E, N, V), but, with appropriate inclusion of heat and particle reservoirs, also to the canonical and grand canonical ensembles. In principle, it applies to both extensive and nonextensive systems and does not presume the standard thermodynamic limit (*i.e.*, infinite particle number and volume [$N \to \infty$, $V \to \infty$], finite density [$\frac{N}{V} = C < \infty$]) [38]; it can be used with boundary conditions, which often handicap other formalisms; it does not presume temperature. However, ergodicity (or quasi-ergodicity) is presumed in that the system's phase space trajectory is assumed to visit smoothly and uniformly all neighborhoods of the (6N-1)-dimensional constant-energy manifold consistent with $\Omega(E,N,V)$ [10].

The Gibbs entropy is similar to Boltzmann's except that it is defined via ensembles, distributions of points in classical phase space consistent with the macroscopic thermodynamic state of the system. Hereafter, it is called the Boltzmann-Gibbs (BG) entropy. Like other standard forms of entropy, $S_{BG,\mu}$ applies strictly to equilibrium systems.

Note that Ω is not well defined for classical systems since phase space variables are continuous. To remedy this, the phase space can be measured in unit volumes, often in units of $\hbar$. This motivates *coarse-grained* entropy. Coarse-graining reduces the information contained in Ω and may be best described as a kind of phase space averaging procedure for a distribution function. The coarse-grained distribution leads to a proper increase of the corresponding statistical (information) entropy. A perennial problem with this, however, is that the averaging procedure is not unique so that the rate of entropy increase is likewise not unique, in contrast to presumably uniquely defined increase of the thermodynamic entropy.

Starting from $S_{BG,\mu}$, primary intensive parameters (temperature T, pressure P, and chemical potential μ) can be calculated [42-46]:

---

[10]Alternatively, ergodicity is defined as the condition that the ensemble-averaged and time-averaged thermodynamic properties of a system be the same.

$$\left(\frac{\partial S}{\partial E}\right)_{N,V} \equiv \frac{1}{T} \tag{1.11}$$

$$\left(\frac{\partial S}{\partial V}\right)_{E,N} \equiv \frac{P}{T} \tag{1.12}$$

$$\left(\frac{\partial S}{\partial N}\right)_{V,E} \equiv -\frac{\mu}{T}. \tag{1.13}$$

If one drops the condition of fixed E and couples the system to a heat reservoir at fixed temperature T, allowing free exchange of energy between the system and reservoir, allowing E to vary as $(0 \leq E \leq \infty)$, then one passes from the microcanonical to the canonical ensemble [41-46].

For the canonical ensemble, entropy is defined as

$$S_{BG,c} \equiv k \left[\ln(Z) + \beta \overline{E}\right] = k \left[\frac{\partial}{\partial T}(T \ln(Z))\right]. \tag{1.14}$$

Here $\beta \equiv \frac{1}{kT}$ and Z is the partition function (*Zustandsumme* or "sum over states") upon which most of classical equilibrium thermodynamic quantities can be founded:

$$Z \equiv \sum_i e^{-\beta E_i}, \tag{1.15}$$

where E_i are the constant individual system energies and $\overline{E}$ is the mean (average) system energy:

$$\overline{E} \equiv \frac{\sum_i E_i e^{-\beta E_i}}{\sum_i e^{-\beta E_i}} = \sum_i E_i p_i. \tag{1.16}$$

The probability p_i is the Boltzmann factor $\exp[-E_i/kT]$. One can define entropy through the probability sum

$$S_{BG} = -k \sum_i p_i \ln p_i, \tag{1.17}$$

or in the continuum limit

$$S_{BG} = -k \int f \ln f \, dv, \tag{1.18}$$

where f is a distribution function over a variable v. This latter expression is apropos to particle velocity distributions.

If, in addition to energy exchange, one allows particle exchange between a system and a heat-particle reservoir, one passes from the canonical ensemble (fixed T, N, V) to the grand canonical ensemble (fixed T, μ, V), for which entropy is defined [41-46]:

$$S_{BG,gc} \equiv \frac{1}{\beta}\left(\frac{\partial q}{\partial T}\right)_{z,V} - Nk \ln(z) + kq = k\left[\frac{\partial(T \ln(\mathcal{Z}))}{\partial T}\right]_{\mu,V}. \tag{1.19}$$

Here q is the q-potential:

$$q = q(z, V, T) \equiv \ln[\mathcal{Z}(z, V, T)], \qquad (1.20)$$

defined in terms of the grand partition function:

$$\mathcal{Z}(z, V, T) \equiv \sum_{i,j} \exp(-\beta E_i - \alpha N_j) = \sum_{N_j=0}^{\infty} z^{N_j} Z_{N_j}(V, T). \qquad (1.21)$$

Here $z \equiv e^{-\beta\mu}$ is the fugacity, Z_{N_j} is the regular partition function for fixed particle number N_j, and $\alpha = -\frac{\mu}{kT}$. The sum is over all possible values of particle number and energy, exponentially weighted by temperature. It is remarkable that such a simple rule is able to predict successfully particle number and energy occupancy and, therefrom, the bulk of equilibrium thermodynamics. This evidences the power of the physical assumptions underlying the theory.

(3) von Neumann [47] In quantum mechanics, entropy is not an *observable*, but a *state* defined through the density matrix, ρ:

$$S_{vN}(\rho) = -kTr[\rho \ln(\rho)]. \qquad (1.22)$$

(Recall the expectation value of an observable is $\langle A \rangle = Tr(\rho A)$.) Roughly, $S_{vN}(\rho)$ is a measure of the quantity of chaos in a quantum mechanical mixed state. The von Neumann entropy has advantage over the Boltzmann formulation in that, presumably, it is a more basic and faithful description of nature in that the number of microstates for a system is well defined in terms of pure states, unlike the case of the classical continuum. On the other hand, unlike the Boltzmann microcanonical entropy, for the von Neumann formulation, important properties like ergodicity, mixing and stability strictly hold only for infinite systems.

The time development of ρ for an isolated system is governed by the Liouville equation

$$i\frac{d}{dt}\rho(t) = \frac{1}{\hbar}[H, \rho(t)] \equiv \mathcal{L}\rho(t). \qquad (1.23)$$

Here H is the Hamiltonian of the system and $\mathcal{L} \dots = \frac{1}{\hbar}[H, \dots]$ is the Liouville superoperator. It follows that the entropy is constant in time. As noted by Wehrl [39],

> ... the entropy of a system obeying the Schrödinger equation (with a time-independent Hamiltonian) *always remains constant* [because the density matrix time evolves as] $\rho(t) = e^{-iHt}\rho e^{iHt}$. Since e^{iHt} is a unitary operator, the eigenvalues of $\rho(t)$ are the same eigenvalues of ρ. But the expression for the entropy only involves the eigenvalues of the density matrix, hence $S(\rho(t)) = S(\rho)$. (In the classical case, the analogous statement is a consequence of Liouville's theorem.)[11]

---

[11]This statement holds if H is a function of time; *i.e.*, $\rho(t) = \mathcal{U}\rho(0)\mathcal{U}^\dagger$, where $\mathcal{U} = T\exp(-\frac{i}{\hbar}\int_0^t H dt)$.

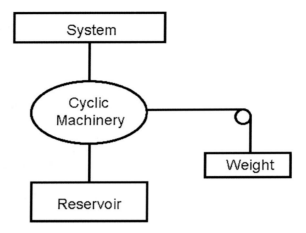

Figure 1.1: S_{GHB} is based on *weight processes.*

Since the Schrödinger equation alone is not sufficient to motivate the time evolution of entropy as normally observed in the real world, one usually turns to the
Boltzmann equation, the master equation, or other time-asymmetric formalisms
to achieve this end [43, 48, 49, 50]. Finally, the von Neumann entropy depends
on time iff ρ is coarse-grained; in contrast, the fine-grained entropy is constant.
(This, of course, ignores the problematic issues surrounding the non-uniqueness of
the coarse graining process.)

(4) Gyftopoulous, et al. [19, 51] A utilitarian approach to entropy is advanced
by Gyftopoulos, Hatsopoulos, and Beretta. Entropy S_{GHB} is taken to be an intrinsic, non-probabilistic property of any system whether microscopic, macroscopic,
equilibrium, or nonequilibrium. Its development is based on *weight processes* in
which a system A interacts with a reservoir R via cyclic machinery to raise or
lower a weight (Figure 1.1). Of course, the weight process is only emblematic of
any process of pure work. S_{GHB} is defined in terms of energy E, a constant that
depends on a reservoir c_R, and *generalized available energy* Ω^R as:

$$S_{GHB} = S_0 + \frac{1}{c_R}[(E - E_0) - (\Omega^R - \Omega_0^R)], \qquad (1.24)$$

for a system A that evolves from state A_1 to state A_0. E_0 and Ω_0^R are values
of a reference state and S_0 is a constant fixed value for the system at all times.
Temperature is not ostensibly defined for this system; rather, c_R is a carefully
defined reservoir property (which ultimately can be identified with temperature).
Available energy Ω^R is the largest amount of energy that can be extracted from
the system A-reservoir combination by weight processes. Like S_{GHB}, it applies to
all system sizes and types of equilibria.

At first meeting, S_{GHB} may seem contrived and circular, but its method of
weight processes is similar to and no more contrived than that employed by Planck

and others; its theoretical development is no more circular than that of Lieb and Yngvason [31]; furthermore, it claims to encompass broader territory than either by applying both to equilibrium and nonequilibrium systems. It does not, however, provide a microscopic picture of entropy and so is not well-suited to statistical mechanics.

(5) Lieb-Yngvason [31] The Lieb-Yngvason entropy S_{LY} is defined through the mathematical ordering of sets of equilibrium states, subject to the constraints of monotonicity, additivity and extensivity. The second law is revealed as a mathematical theorem on the ordering of these sets. This formalism owes significant debt to work by Carathéodory [30], Giles [52], Buchdahl [53] and others.

Starting with a space Γ of equilibrium states X,Y,Z ..., one defines an ordering of this set via the operation denoted $\prec$, pronounced *precedes*. The various set elements of Γ can be ordered by a comparison procedure involving the criterion of *adiabatic accessibility*. For elements X and Y, [31]

> A state Y is adiabatically accessible from a state X, in symbols X $\prec$ Y, if it is possible to change the state X to Y by means of an interaction with some device (which may consist of mechanical and electrical parts as well as auxiliary thermodynamic systems) and a weight, in such a way that the device returns to its initial state at the end of the process whereas the weight may have changed its position in a gravitation field.

This bears resemblance to the GHB weight process above (Figure 1.1). Although superficially this definition seems limited, it is quite general for equilibrium states. It is equivalent to requiring that state X can proceed to state Y by *any* natural process, from as gentle and mundane as the unfolding of a Double Delight rose in a quiet garden, to as violent and ultramundane as the detonation of a supernova.

If X proceeds to Y by an irreversible adiabatic process, this is denoted X $\prec\prec$ Y, and if X $\prec$ Y and Y $\prec$ X, then X and Y are called *adiabatically equivalent*, written X $\overset{A}{\sim}$ Y. If X $\prec$ Y or Y $\prec$ X (or both), they are called *comparable*.

The Lieb-Yngvason entropy S_{LY} is defined as [31]:

> There is a real-valued function on all states of all systems (including compound systems), called **entropy** and denoted by S such that
> a) *Monotonicity*: When X and Y are comparable states then

$$\text{X} \prec \text{Y} \text{ if and only if } S(X) \leq S(Y).$$

> b) *Additivity and extensivity*: If X and Y are states of some (possibly different) systems and if (X,Y) denotes the corresponding state in the composition of the two systems, then the entropy is additive for these states, *i.e.*,

$$S(X,Y) = S(X) + S(Y)$$

S is also extensive, *i.e.*, for each $t > 0$ and each state X and its scaled copy tX,

$$S(tX) = tS(X).$$

The monotonicity clause is equivalent to the following:

$$X \overset{A}{\sim} Y \Longrightarrow S(X) = S(Y); \text{ and}$$
$$X \prec\prec Y \Longrightarrow S(X) < S(Y).$$

The second of these says that entropy increases for an irreversible adiabatic process. This is the Lieb-Yngvason formulation of the second law.

The existence and uniqueness of S_{LY} can be shown to follow from assumptions surrounding adiabatic accessibility and the comparsion process. In this formalism, temperature is not a primitive concept; rather, it is defined via S_{LY} as $\frac{1}{T} := (\frac{\partial S_{LY}}{\partial U})_V$, where U is energy and V is volume. The mathematical details of these results are beyond the scope of this discussion; the intrepid reader is directed to [31].

(6) Carathéodory Historically preceding S_{LY}, Carathéodory also defined entropy in a formal mathematical sense [30, 6].

> For simple[12] systems, Carathéodory's principle is equivalent to the proposition that the differential form $\delta Q := dU - \delta W$ possesses an integrable divisor, *i.e.*, there exists functions S and T on the state space Γ such that
>
> $$\delta Q = TdS.$$
>
> Thus, for simple systems, every equilibrum state can be assigned values for entropy and absolute temperature. Obviously these functions are not uniquely determined by the relation $[\delta Q = TdS]$.

Carathéodory's entropy was not widely accepted by working scientists during his lifetime, but it has grown in significance during the last 40 years as thermodynamic foundations have been shored up.

(7) Shannon [54] Various information-relevant entropies have been proposed over the last six decades, the most prominent of which are the Shannon entropy and algorithmic randomness [55, 56, 57]. These are especially salient in considerations of sentient Maxwell demons [21], which have helped expose the deep relationships between physics and information theory.

Let p_j be probabilities of mutually exclusive events, say for instance, the probabilties of particular letters in an unknown word. The uncertainty (entropy) of the information about this situation is the Shannon entropy:

---

[12]Consult the literature for the requirements of a *simple* system [6, 30]

$$S_{Sh} = -\sum_j p_j \log(p_j) \tag{1.25}$$

The logarithm may be taken to any fixed base, but base 2 is standard, giving entropy in bits. Shannon entropy can be seen to be a discrete form of the classical Boltzmann-Gibbs entropy, (1.17).

(8) Fisher Shannon entropy is defined over a space of unordered elements, for instance, letters. For a space of ordered elements, for example, a continuous parameter (*e.g.*, the length or brightness of meteor trails), Fisher information is appropriate. For a probability distribution $f(x; \phi)$ in the random variable x dependent on the unobservable variable ϕ, the Fisher information (entropy) is

$$S_F(\phi) = K[\frac{\partial}{\partial \phi} \log f(x; \phi)]^2 = -K[\frac{\partial^2}{\partial \phi^2} \log f(x; \phi)] \tag{1.26}$$

Clearly, the sharpness of the support curve is proportional to the expection of $S_F(\phi)$, thus high information content (low entropy) corresponds to a sharp distribution and a low information content (high entropy) to a broad distribution.

(9) Algorithmic Randomness [55, 56, 57] Algorithmic randomness (algorithmic complexity, Kolmorgorov complexity) of a string of elements is defined as the minimum size of a program (*e.g.*, in bits) executed on a universal computer that yields the string. Strings are relatively simple or complex depending on whether its program length is relatively short or long, respectively. For example, the string of 60,000 digits (121223121223121223...) is relatively simple and has relatively low algorithmic randomness since it can be programmed as 10,000 repeating blocks of (121223), whereas a completely *random* string of 60,000 digits cannot be compressed this way and thus has a relatively large algorithmic randomness. Most strings cannot be compressed and, to leading order in binary notation, their algorithmic randomness is given by their lengths in bits. By example, a random natural number N, if it can be expressed as $N \sim 2^s$, has algorithmic randomness $\sim log_2 N = s$.

Algorithmic complexity, in contrast to other standard definitions of entropy, does not rely on probabilities. However, the randomness of a string is not uniquely determined and there is no general method to discern a simple string from a complex one; this is related to Gödel's undecidability [59]. For example, the sequence (22459157718361045473442715) may appear completely random, but it is easily generated from π^e. Or, the letter sequence FPURCLK might seem random until it is unscrambled and considered in an appropriate context. Apparently, order can be in the eye of the beholder.

Zurek suggests that *physical entropy* "is the sum of (i) the missing information measured by Shannon's formula and (ii) of the [algorithmic content] in the available data about the system" [58].

(10) Tsallis [60, 61] Tsallis entropy is a controversial generalization of Boltzmann-Gibbs entropy and is an heir to the Rényi and Daróczy entropies below. It is defined as

$$S_{Ts} = \frac{1}{q-1}\left[1 - \int f^q(x)dx\right],$$ (1.27)

where q is a real number *entropic index* and $f(x)$ is a probability distribution function. For $q = 1$, S_{Ts} reduces to the Boltzmann-Gibbs entropy.

Primary virtues of the Tsallis entropy include its mathematical simplicity and descriptiveness of nonextensive systems. A physical quantity is extensive if its value scales linearly with the size of the system [13]. The extensive Boltzmann-Gibbs entropy of two independent systems A and B is $S_{BG}(A+B) = S_{BG}(A)+S_{BG}(B)$, while for the Tsallis entropy it is $S_{Ts}(A+B) = S_{Ts}+S_{Ts}(B)+(1-q)S_{Ts}(A)S_{Ts}(B)$. The parameter q can be taken as a measure of nonextensivity[14].

Tsallis entropy has been applied to numerous disparate physical phenomena that are deemed beyond the reach of equilibrium thermodynamics. Notably, these include systems with long-range nonextensive fields (*e.g.*, gravitational, electrostatic) such as plasmas and multi-particle self-gravitating systems (*e.g.*, galaxies, globular clusters). It has been applied to the behaviors of self-organizing and low-dimensional chaotic systems and processes far from equilibrium; examples include financial markets, crowds, traffic, locomotion of microorganisms, subatomic particle collisions, and tornados. Unfortunately, its underlying physical basis has not been well established, leading critics to label it *ad hoc* and its successes little more than "curve fitting." Its elegant simplicity and adaptability, however, cannot be denied.

The entropic index (nonextensivity parameter) q is taken to be a measure of the fractal nature of a system's path in phase space. Whereas under Boltzmann-Gibbs formalism, a system on average spends equal time in all accessible, equal-sized volumes of phase space (equal *a priori* probability), under the Tsallis formalism the phase space path is fractal, thereby allowing it to model chaotic, nonequilibrium systems, and display rapid and radical changes in behavior and phase.

(11-21) Other Entropies There are a number of other entropy and entropy-like quantities that are beyond the scope of this discussion. These include (with ρ the density matrix, unless otherwise noted):

Daróczy entropy [62]:

$$S_D = \frac{1}{2^{1-\alpha}-1}(Tr(\rho^\alpha)-1),$$ (1.28)

with $\alpha > 0$ and $\alpha \neq 1$.

Rényi entropy [63]:

$$S_R = \frac{k}{1-\alpha}\ln[Tr(\rho^\alpha)],$$ (1.29)

again with $\alpha > 0$ and $\alpha \neq 1$.

---

[13]Extensivity is a traditional requirement for thermodynamic quantities like energy and entropy.

[14]Notice that if $q \to 1$, then $S_{Ts} \to S_{BG}$.

Hartley entropy [64]:

$$S_H = k \ln[N(\rho)], \tag{1.30}$$

where $N(\rho)$ is the number of positive eigenvalues of ρ.

Infinite norm entropy:

$$S_{In} = -k \ln \|\rho\|_\infty, \tag{1.31}$$

where $\|\rho\|_\infty = p_{max}$ is the largest eigenvalue of ρ.

Relative entropy (classical mechanics) [65, 66]:

$$S_{Rel,c} = - \int \rho(\ln \rho - \ln \sigma)d\tau, \tag{1.32}$$

where ρ and σ are probability distributions and τ is the phase space coordinate.

Relative entropy (quantum mechanics):

$$S_{Rel,q}(\sigma|\rho) = Tr[\rho(\ln \rho - \ln \sigma)], \tag{1.33}$$

where ρ and σ are distinct density matrices. It is non-negative [67].

In addition to these, there is Segal entropy [68], which subsumes many of the quantum mechanical entropies mentioned above; Kolmogorov-Sinai (KS) entropy, which describes dynamical systems undergoing discrete time evolution; Kouch-nirenko A entropies, close relatives to KS entropy; skew entropy [69]; Ingarden-Urbanik entropy [70]; Macdonald entropy [18]. For completeness, you may add your own personal favorite here: .

1.4 Nonequilibrium Entropy

There is no completely satisfactory definition of entropy. To some degree, every definition is predicated on physical ignorance of the system it describes and, therefore, must rely on powerful *ad hoc* assumptions to close the explanatory gap. These limit their scopes of validity. Let us review a few examples. The Boltzmann-Gibbs entropy assumes equal *a priori* probability either of phase space or ensemble space. While this is a reasonable assumption for simple equilibrium systems like the ideal gas and Lorentz gas, it is known to fail for large classes of systems, especially at disequilibrium; the molecular chaos ansatz (Boltzmann's Stosszahlansatz) is similarly suspect. It is not known what the necessary conditions are for ergodicity. The thermodynamic limit, which is presumed or necessary for most quantum and classical thermodynamic formalisms, on its face cannot be completely realistic, particularly since it ignores boundary conditions that are known to be pivotal for many thermodynamic behaviors. Extensivity, also presumed for most entropies, is ostensibly violated by systems that exhibit long-range order and fields — these include systems from nuclei up to the largest scale structures of the universe [38]. Information entropies are hobbled by lack of general definitions of order, disorder

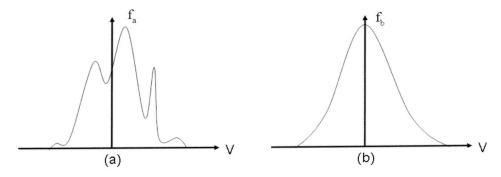

Figure 1.2: One-dimensional velocity distribution functions: (a) non-Maxwellian; (b) Maxwellian.

and complexity. Finally, as it is deduced from thermodynamics, the notion of entropy is critically dependent on the presumed validity of the second law.

Among the many foundational issues thwarting a general definition of physical entropy, none is more urgent than extending entropy into the nonequilbrium regime. After all, changes in the world *are* primarily irreversible nonequilibrium processes, but even the most basic nonequilibrium properties, like transport coefficients, cannot be reliably predicted in general[15].

The prominent classical and quantum entropies strictly apply at equilibrium only. As a simple example, consider the two one-dimensional velocity distributions in Figure 1.2. Distribution f_a is highly nonequilibrium (non-Maxwellian) and does not have a well-defined temperature, while f_b is Maxwellian and does have a well-defined temperature. Let's say we wish to add heat δQ to f_a to transform it into f_b and then calculate the entropy change for this process via $\int_i^f \frac{\delta Q}{T} = \Delta S$. This presents a problem in this formalism because T is not properly defined for f_a or any other other intermediate distribution on its way to the Maxwellian f_b [16].

While small excusions into near nonequilibrium can be made via the Onsager relations [71] or fluctuation-dissipation theorems [43, 72], in general, far nonequilibrium systems are unpredictable. Only recently has theory begun to make significant headway into these regimes. Excursions are limited to idealized systems and carry with them their own questionable baggage, but results are heartening [73]. Notable past and present exponents of nonequilibrium thermodynamics include Onsager, Prigogine, Meixner, Green, Kubo, Ruelle, Hoover, Evans, Cohen, Gallavotti, Lebowitz, Nicolis, Gaspard, Dorfmann, Maes, Jou, Eu and many others [71-89]. Notable recent advances in the microscopic descriptions of nonequilib-

---

[15]Some entropies, like S_{GHB} and S_{Ts}, are claimed to apply at nonequilibrium, but they do not have compelling microscopic descriptions.

[16]On the other hand, one might aver that, since $S = -k \int f \ln f dv$, one could calculate $\Delta S = -k \left[\int f_b \ln f_b dv - \int f_a \ln f_a dv \right]$.

rium entropy have proceded largely through study of nonequilibrium steady states (NESS), especially in fluids (gases) [73]. This formalism is apropos to many of the challenges in this volume.

For NESS, classical phase space volumes ($dx = dqdp$) are often replaced by more general measures, perhaps the best known of which is the Sinai-Ruelle-Bowen (SRB) measure. It is especially useful in describing chaotic systems whose phase space development is hyperbolic; that is, stretching in some dimensions while contracting in others. Phase space stretching gives rise to the hallmark of chaos: sensitivity to initial conditions. The separation rate of initially proximate phase space trajectories is given by Lyapounov exponents λ, one for each dimension. Negative λ indicates convergence of trajectories, while positive λ indicates exponential separation of nearby trajectories — and chaos.

Although a general definition of entropy in NESS is lacking, entropy production can be expressed as

$$\dot{S}(\rho) = \int (-\nabla_x \mathcal{X})\rho(dx), \qquad (1.34)$$

where divergence is with respect to the phase space measure coordinate and the nonequilibrium time development of x is determined via

$$\frac{dx}{dt} = \mathcal{X}(x), \qquad (1.35)$$

where $\mathcal{X}(x)$ is a vector field denoting physical forces. Using SRB measures, the second law demands that $\dot{S}(t) \geq 0$; for dissipative systems (those producing heat) $\dot{S}(t) > 0$. This is possible because SRB measures break time reversal symmetry, rendering the system non-Hamiltonian, thus allowing $\nabla_x \mathcal{X} \neq 0$.

Within the chaotic dynamics paradigm, NESS exist at nonequilibrium attractors in phase space. An example of NESS attractors among second law challenges can be inferred from Figure 6.6 in §6.2.4.3, pertaining to a gravitator that circulates at a steady-state angular velocity within a gas-filled cavity, driven by spontaneous pressure gradients. The primary difference between this and standard NESS is that, while traditional NESS are dissipative (turn work into heat), second law challenges are regenerative (turn heat into work), thus admitting $\dot{S}(t) < 0$.

Nonequilibrium, irreversibility and dissipation are the triumvirate that rules the natural thermodynamic world. Second law challenges obey the former two, but not the third. As such, much of the formalism already developed for nonequilibrium thermodynamics should be directly applicable to the challenges in this volume, the chief proviso being sign reversal for heat and entropy production. By turning this considerable theoretical machinery on the challenges, they may be either further supported or resolved in favor of the second law.

It is now commonly held that the second law arises as a consequence of the interaction between a quantum system and its thermal environment [90, 91, 92]. While this might be true, it should be noted that system-bath interactions can also take an active role in *violations* of specific formulations of this law in specific situations, as will be shown in Chapter 3.

1.5 Entropy and the Second Law: Discussion

Entropy and the second law are commonly conflated — for example, the non-decrease of entropy for a closed system is an oft-cited version — but many formulations of the second law do not involve entropy at all; consider, for instance, the Clausius and Kelvin-Planck forms. Entropy is surely handy, but it is not essential to thermodynamics — one could hobble along without it. It is more critical to statistical mechanics, which grapples with underlying dynamics and microstates, but even there its utility must be tempered by its underlying assumptions and limitations, especially when treating chaotic, nonlinear, and nonequilibrium systems (See §2.3.2.).

The majority of second law challenges are phrased in terms of heat and work, rather than in terms of entropy. This is largely because entropy *per se* is difficult to measure experimentally. Heat, temperature, pressure, and work are measured quantities, while entropy is usually inferred. Thus, entropy, the second law, and its challenges are not as intimate as is often assumed. Entropy is a handmaiden of the second law, not its peer.

At the microscopic level an individual molecule doesn't know what entropy is and it couldn't care less about the second law. A classical system of N particles is also oblivious to them insofar as its temporal trajectory in a (6N-1)-dimensional phase space is simply a moving point to which an entropy cannot be ascribed and to which entropy increases are meaningless. (In this context, for ensemble theory, entropy cannot be strictly defined since f is singular.) Entropy is a global property of a system, measurable in terms of the surface area of the constant energy manifold on which the system's phase space point wanders, but this assumes conditions on the motion of the phase space point that, by definition, are either not measured or not measurable and, hence, might not be valid.

In its very conception, entropy presumes ignorance of the microscopic details of the system it attempts to describe. In order to close the explanatory gap, one or more far-reaching assumptions about the microscopic behavior or nature of that system must be made. Many of these provisos — *e.g.*, ergodicity, strong mixing, equal *a priori* probability, extensivity, thermodynamic limit, equilibrium — allow accurate predictions for large and important classes of thermodynamic phenomena; however, *every* formulation of entropy makes assumptions that limit the parameter space in which it is valid, such that *no* known formulation applies to *all* possible thermodynamic regimes[17]. It is doubtful that any formulation of entropy can be completely inclusive since there will probably always be special cases outside the range of validity of any proviso powerful enough to close the explanatory gap. The best one can hope to do is to identify when a particular type of entropy will or will not apply to a particular case — and even the criteria for this hope are not known. Insofar as complex systems — and most realistic thermodynamic systems are complex — can display chaotic and unpredictable behavior (unpredictable to the experimenter and perhaps even to the system itself), it seems unlikely that any single form of entropy will be able to capture all the novelty Nature can produce.

---

[17]Systems are known for which one, many, or all the above provisos fail.

Entropy formulations vary across disciplines, from physics to engineering, from chaos theory to economics, from biology to information theory. Even within a single discipline (physics) there are numerous versions between classical and quantum regimes, between utilitarian and formal approaches. Not all are equivalent, or even compatible. Most become problematic at nonequilibrium, but this is where physics becomes the most interesting. Most entropies blend seemlessly into others, making clear distinctions nearly impossible. One could say the subject of entropy is well-mixed and somewhat disordered. This state of affairs is intellectually unsatisfying and epistemologically unacceptable.

It is the opinion of one of the authors (d.p.s.) that, despite its singular importance to thermodynamics and statistical mechanics, entropy will never have a completely satisfactory and general definition, nor will its sovereign status necessarily endure. Rather, like the *calorique*, which was useful but not intellectually persuasive enough to survive the 19th century, entropy could well fade into history[18]. In the end, each thermodynamic system (particularly nonequilibrium ones) should be considered individually and microscopically with respect to its boundary conditions, constraints, and composition to determine its behavior[19]. Considered classically, it is the 6N-dimensional phase space trajectory that truly matters and the various approximations that currently expedite calculations are too simplistic to capture the true richness of dynamic behaviors. Thus, each system should be considered on a case by case basis. If entropy is defined at the microscopic level of detail necessary to make completely accurate predictions about phase space trajectories, however, it loses its utility — and meaning [20].

Entropy remains enigmatic. The more closely one studies it, the less clear it becomes. Like a pointillisme painting whose meaning dissolves into a collection of meaningless points when observed too closely, so too entropy begins to lose meaning when one contemplates it at a microscopic level. Insofar as our definition of entropy is predicated on what is presumed unknown or unknowable about a system, it is epistemologically unsatisfactory and must ultimately be surpassed. As our understanding of the underlying dynamics of complex systems brightens, so must the utility of entropy dim and, perhaps, entirely disappear. Fortunately, the second law can survive without its handmaiden.

1.6 Zeroth and Third Laws of Thermodynamics

The first law is the skeleton of thermodynamics; the second law is its flesh. The first gives structure; the second gives life. By comparison, the zeroth and

---

[18] In the near term, however, this will surely not be the case.

[19] A few simple cases, like the ideal gas, will be predictable due to their thermodynamic simplicity, but realistically complex nonequilibrium systems that possess significant thermodynamic depth — like life — will defy tidy description in terms of entropy, or easy prediction in terms of behavior. In the most interesting cases, *chaos rules*.

[20] On the other hand, perhaps if a completely general definition of *order* and *complexity* is discovered, this will lead to a general definition of physical entropy.

third laws are mere hat and slippers. Since one should not go about undressed, let us briefly consider the latter two.

Zeroth Law The zeroth law pertains to the transitivity of equilibrium. It can be stated:

> If system A is in equilibrium with systems B and C, then system B is in equilibrium with system C.

More commonly, it is expressed in terms of temperature, because temperature is the easiest equilibrium property to measure experimentally:

> If the temperature of system A is equal to the temperature of system B, and the temperature of system B is equal to the temperature of system C, then the temperature of system A is equal to the temperature of system C. (If $T_A = T_B$ and $T_B = T_C$, then $T_A = T_C$.)

Or, to put it succinctly:

> Thermometers exist.

The role of this law is far-reaching since it allows one to introduce, within axiomatic thermodynamics, integral intensive characteristics of mutually equilibrium systems, such as temperature, pressure, or chemical potential. It is therefore unsettling that quantum mechanical models exist that predict its violation (§3.6.7).

Third Law[21] The third law of thermodynamics pertains primarily to establishing fiduciary entropies. Like the second and zeroth, it can be stated in various ways. The first, the Nernst-Planck form states:

> **Nernst-Planck** Any change in condensed matter is, in the limit of the zero absolute temperature, performed without change in entropy. (Nernst 1906)

Planck supplemented this in 1912 (in modern form):

> **Planck** The entropy of any pure substance at $T = 0$ is finite, and, therefore, can be taken to be zero.

The third law says that any substance that has a unique stable or metastable state as its temperature is reduced toward absolute zero can be taken to have zero entropy at absolute zero [93]. In fact, at $T = 0$ most substances will have residual *zero point entropies* associated with such things as mixed isotopic composition, randomly oriented nuclear spins, minor chemical impurities, or crystal defects, but if these do not affect the thermodynamic process for which entropy is pertinent, they can be safely ignored since "they just go along for the ride." In some sense, the entropy depends on the knowledge or opinion of the observer. Ideally, if the

---

[21]M.O. Scully maintains, "The third law has all the weight of an Italian traffic advisory."

number of microstates that describes this perfect substance at $T = 0$ is $\Omega = 1$, then its entropy via the Boltzmann formula is $S = k \ln[\Omega = 1] = 0$ exactly. This law has far reaching consequences like the zero-temperature vanishing of specific heats, thermal expansion coefficients, latent heats of phase transitions.

The third law can also be stated as impossibilities, for instance:

It is impossible to reduce the temperature of a system to absolute zero via any finite sequence of steps.

or,

Perpetuum mobile of the third type are impossible.

The first of these can be argued formally [93] and has not been violated experimentally; the current lower limit of experimentally achieved temperatures is about 10^{-9}K. The second of these has been effectively violated by a number of non-dissipative systems, notably superfluids in motion and supercurrents, whose theoretical decay time exceeds 10^5 years.

In summary, the laws of thermodynamics are not as sacrosanct as one might hope. The third law has been violated experimentally (in at least one form); the zeroth law has a warrant out for its arrest; and the first law can't be violated because it's effectively tautological. The second law is intact (for now), but as we will discuss, it is under heavy attack both experimentally and theoretically.

References

[1] Carnot, S., *Réflexions sur la Puissance Motrice due Feu*, Édition Critique par Robert Fox (J.Vrin, Paris, 1978).

[2] Brush, S., *The Kind of Motion We Call Heat* (North-Holland, Amsterdam, 1976).

[3] von Baeyer, H.C., *Warmth Disperses and Time Passes — The History of Heat* (The Modern Library, New York, 1998).

[4] Clausius, R., *Abhandlungungen über die mechanische Wärmetheorie*, Vol. 1, (F. Vieweg, Braunschweig, 1864); Vol. 2 (1867); *The Mechanical Theory of Heat* (Macmillan, London, 1879); Phil. Mag. **2** 1, 102 (1851); **24**, 201, 801 (1862).

[5] Kelvin, Lord (Thomson, W.), *Mathematical and Physical Papers*, Vol. I (Cambridge University Press, Cambridge, 1882).

[6] Uffink, J., Studies Hist. Phil. Mod. Phys. **32** 305 (2001).

[7] Kestin, J., *The Second Law of Thermodynamics* (Dowden, Hutchinson, and Ross, Stroutsburg, PA, 1976).

[8] Planck, M., *Vorlesungen über die Theorie der Wärmestrahlung* (Barth, Leipzig, 1906); *Treatise on Thermodynamics* 7th ed., Translated by Ogg, A. (Dover, New York, 1945).

[9] König, F.O., Surv. Prog. Chem. **7** 149 (1976).

[10] Langton, S., in *First International Conference on Quantum Limits to the Second Law*, AIP Conf. Proc., Vol. 643, Sheehan, D.P., Editor, (AIP Press, Melville, NY, 2002) pg. 448.

[11] Davies, P.C.W., *The Physics of Time Asymmetry* (University of California Press, Berkeley, 1974).

[12] Reichenbach, H., *The Direction of Time* (University of California Press, Berkeley, 1957).

[13] Sachs, R.G., *The Physics of Time Reversal* (University of Chicago Press, Chicago, 1987).

[14] Halliwell, J.J., Pérez-Mercader, J. and Zurek, W.H., Editors, *Physical Origins of Time Asymmetry* (Cambridge University Press, Cambridge, 1994).

[15] Zeh, H.D., *The Physical Basis of the Direction of Time* (Springer-Verlag, Heidelberg, 1989).

[16] Savitt, S.F., Editor, *Time's Arrow Today* (Cambridge University Press, Cambridge, 1995).

[17] Price, H., *Time's Arrow and Archimedes' Point* (Oxford University Press, Oxford, 1996).

[18] Macdonald, A., Am. J. Phys. **63** 1122 (1995).

[19] Gyftopoulos, E.P. and Beretta, G.P., *Thermodynamics: Foundations and Applications* (Macmillan, New York, 1991).

[20] Maxwell, J.C., *Theory of Heat* (Longmans, Green, and Co., London, 1871).

[21] Leff, H.S. and Rex, A.F., *Maxwell's Demon 2: Entropy, Classical and Quantum Information, Computing*, (Institute of Physics, Bristol, 2003); *Maxwell's Demon. Entropy, Information, Computing* (Hilger & IOP Publishing, Bristol, 1990).

[22] Smoluchowsky, M., Phys. Zeit. **12** 1069 (1912).

[23] Wang, G.M., Sevick, E.M., Mittag, E., Searles, D.J., and Evans, D.J., Phys. Rev. Lett. **89** 50601 (2002).

[24] Mackey, M.C., *Time's Arrow: The Origins of Thermodynamic Behavior* (Springer-Verlag, New York, 1992); Rev. Mod. Phys. **61** 981 (1989).

[25] Gibbs, J.W., *The Scientific Papers of J. Willard Gibbs*, Vol. 1 Thermodynamics, (Longmans, London, 1906).

[26] Jaynes, E.T., Phys. Rev. **106** 620 (1957).

[27] Katz, A. *Principles of Statistical Mechanics* (Freeman, San Francisco, 1967).

[28] Truesdell, C., *The Tragicomical History of Thermodynamics 1822-54* (Springer-Verlag, New York, 1980).

[29] Arnold, V., *Proceedings of the Gibbs Symposium* (American Mathematical Society, Providence, 1990) pg. 163.

[30] Carathéodory, C., Math. Annalen **67** 355 (1909).

[31] Lieb, E.H. and Yngvason, J., Physics Reports **310** 1 (1999); Phys. Today **53** 32 (2000).

[32] Born, M., Phys. Zeit. **22** 218, 249, 282 (1921).

[33] Allahverdyan, A.E. and Nieuwenhuizen, Th. M., http://arxiv.org/abs/cond-mat/0110422.

[34] Pusz, W. and Woronowicz, L, Commun. Math. Phys. **58** 273 (1978).

[35] Lenard, A., J. Stat. Phys. **19** 575 (1978).

[36] Čápek, V. and Mančal, T., Europhys. Letters **48** 365 (1999).

[37] Pirruccullo, A., private communications (2004).

[38] Gross, D.H.E., *Microcanonical Thermodynamics*, World Scientific Lecture Notes in Physics, Volume 66, (World Scientific, Singapore, 2001).

[39] Wehrl, A., Rev. Mod. Phys. **50** 221 (1978).

[40] Boltzmann, L., Wiener Ber. **75** 67; **76** 373 (1877).

[41] Gibbs, J.W., *Elementary Principles in Statistical Mechanics* (Yale University Press, Boston, 1902).

[42] Tolman, R.C., *The Principles of Statistical Mechanics* (Oxford University Press, Oxford, 1938).

[43] Reif, F., *Fundamentals of Statistical and Thermal Physics* (McGraw-Hill, New York, 1965).

[44] Penrose, O., *Foundations of Statistical Mechanics* (Pergamon Press, Oxford, 1970).

[45] Reichl, L.E., *A Modern Course in Statistical Physics* (Unversity of Texas Press, Austin, 1980).

[46] Pathria, R.K., *Statistical Mechanics* (Pergamon Press, Oxford, 1985).

[47] von Neumann, J., Z. Phys. **57** 30 (1929); Gött. Nachr. 273 (1927).

[48] Boltzmann, L., Wiener Ber. **66** 275 (1872).

[49] Prigogine, I., *Introduction to Thermodynamics of Irreversible Processes* (Interscience/Wiley, New York, 1968).

[50] Beretta, G.P., Gyftopoulos, E.P., and Hatsopoulos, G.N., Nuovo Cimento Soc. Ital. Fis. B. **82B** 169 (1984).

[51] Gyftopoulos, E.P. and Çubukçu, E., Phys. Rev. E **55** 3851 (1997).

[52] Giles, R., *Mathematical Foundations of Thermodynamics* (Pergamon, Oxford, 1964).

[53] Buchdahl, H.A., *The Concepts of Classical Thermodynamics* (Cambridge University Press, Cambridge, 1966).

[54] Shannon, C.E., Bell System Tech. J. **27** 379 (1948); Shannon, C.E. and Weaver, W., *The Mathematical Theory of Communication* (University of Illinois, Urbana, 1949).

[55] Kolmogorov, A.N., Inf. Transmission **1** 3 (1965)

[56] Bennett, C.H., Int. J. Theor. Phys. **21** 905 (1982).

[57] Zurek, W.H., Editor, *Complexity, Entropy, and the Physics of Information* Vol. VIII, Santa Fe Institute (Perseus Books, Cambridge, 1990).

[58] Zurek, W.H., Nature **341** 119 (1989); Phys. Rev. A **40** 4731 (1989).

[59] Chaitin, G.J., *Algorithmic Information Theory* (Cambridge University Press, Cambridge, 1987).

[60] Tsallis, C., J. Stat. Phys. **52** 479 (1988).

[61] Gell-Mann, M. and Tsallis, C., Editors, *Nonextensive Entropy* (Oxford University Press, Oxford, 2004).

[62] Daróczy, Z., Inf. Control **16** 36,74 (1970).

[63] Rényi, A., *Wahrscheinlichkeitsrechnung* (VEB Deutcher Verlag der Wissenschaften, Berlin, 1966).

[64] Hartley, R.V., Bell Syst. Tech. J. **7** 535 (1928).

[65] Umegaki, H., Kodai Math. Sem. Rep. **14** 59 (1962).

[66] Lindblad, G., Commun. Math. Phys. **33** 305 (1973).

[67] Balian, R., *From Microphysics to Macrophysics; Methods and Applications of Statistical Mechanics, II* (Springer-Verlag, Berlin, 1992).

[68] Segal, I.E., J. Math. Mech. **9** 623 (1960).

[69] Wigner, E.P. and Yanase, M.M., Proc. Natl. Acad. Sci. USA, **49** 910 (1963).

[70] Ingarden, R.S. and Urbanik, K., Acta Phys. Pol. **21** 281 (1962).

[71] Onsager, L., Phys. Rev. **37**, 405 (1931); Phys. Rev. **38**, 2265 (1931).

[72] Gallavotti, G. and Cohen, E.G.D., Phys. Rev. Lett. **74**, 2694 (1995); J. Stat. Phys. **80**, 931 (1995).

[73] Ruelle, D., Phys. Today **57**, 48 (2004).

[74] Prigogine, I., *Non Equilibrium Statistical Mechanics* (Wiley and Sons, New York, 1962).

[75] Evans, D.J. and Morriss, G.P., *Statistical Mechanics of Nonequilibrium Liquids* (Academic Press, London, 1990).

[76] Nicolis, G., *Introduction to Nonlinear Science* (Cambridge University Press, New York, 1995).

[77] Gaspard, P., *Chaos, Scattering, and Statistical Mechanics* (Cambridge University Press, New York, 1998).

[78] Dorfmann, J.R., *An Introduction to Chaos in Nonequilibrium Statistical Mechanics* (Cambridge University Press, New York, 1999).

[79] Jou, D., Casas-Vázquez, J., and Lebon, G., Rep. Prog. Phys. **62** 1035 (1999); Jou, D., Casas-Vázquez, J., and Lebon, G., *Extended Irreversible Thermodynamics* (Springer-Verlag, Berlin, 2001).

[80] de Groot, S.R. and Mazur, P., *Non-Equilibrium Thermodynamics* (Dover, New York, 1984).

[81] Evans, D.J. and Rondoni, L., J. Stat. Phys. **109** 895 (2002).

[82] Lebowitz, J.L., Physica A **263** 516 (1999).

[83] Ruelle, D., Physica A **263** 540 (1999).

[84] Prigogine, I., Physica A **263** 528 (1999).

[85] Hoover, W.G., *Molecular Dynamics*, Lecture Notes in Physics **258** (Springer-Verlag, Heidelberg, 1986).

[86] Ruelle, D., J. Stat. Phys. **95** 393 (1999).

[87] Goldstein, S. and Penrose, O., J. Stat. Phys. **24** 325 (1981).

[88] Eu, B.C., *Kinetic Theory and Irreversible Thermodynamics* (Wiley and Sons, New York, 1992); *Non-Equilibrium Statistical Mechanics: Ensemble Method* (Kluwer, Dordrecht, 1998).

[89] Eu, B.C., *Generalized Thermodynamics (The Thermodynamics of Irreversible Processes and Generalized Hydrodynamics)*, Fundamental Theories of Physics, Vol. 124, (Kluwer Academic, Dordrecht, 2002).

[90] Zurek, W.H., Phys. Today **44** 36 (1991).

[91] Zurek, W.H., Phys. Today **46** 81 (1993).

[92] Zurek, W.H. and Paz, J.P., Phys. Rev. Letters **72** 2508 (1994).

[93] Sheehan, W.F., *Physical Chemistry* 2nd Ed., (Allyn and Bacon, Inc., Boston, 1970).

2

Challenges (1870-1980)

An overview of second law challenges and their resolutions is given for the period 1870-1980, beginning with Maxwell's demon. Classical second law inviolability proofs are critiqued and from these, candidate regimes are inferred for modern challenges.

2.1 Maxwell's Demon and Other Victorian Devils

Challenges to the second law began soon after it was discovered. The first, most enduring, and most edifying of these is James Clark Maxwell's celebrated demon. Here we only sketch the many lives and reported deaths of this clever *gedanken* heat fairy, since an adequate treatment would fill an entire volume by itself. A superb discussion and anthology is presented by Leff and Rex [1].

Maxwell's demon was born with a letter from Maxwell to Peter Guthrie Tait in 1867. Maxwell's intention was "to pick a hole" in the second law by imagining a process whereby molecules could be processed on an individual basis so as to engineer microscopically a temperature gradient. Maxwell writes:

> ... Let him [demon] first observe the molecules in [compartment] A and when he sees one coming the square of whose velocity is less than the mean sq. vel. of the molecules in B let him open the hole and let it

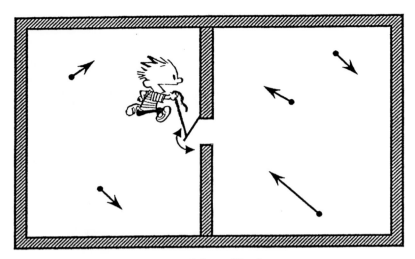

Figure 2.1: Maxwell's demon.

go into B. Next let him watch for a molecule of [compartment] B, the
square of whose velocity is greater than the mean sq. vel. in A, and
when it comes to the hole let him draw the slide and let it go into A,
keeping the slide shut for all other molecules...

See Figure 2.1. Maxwell's original description is both clear and historically important so we quote more extensively from his book [2].

One of the best established facts in thermodynamics is that it is impossible in a system enclosed in an envelope which permits neither change
of volume nor passage of heat, and in which both the temperature and
the pressure are everywhere the same, to produce any inequality of
temperature or of pressure without the expenditure of work. This is
the second law of thermodynamics, and it is undoubtedly true as long
as we can deal with bodies only in mass, and have no power of per-
ceiving or handling the separate molecules of which they are made up.
But if we conceive a being whose faculties are so sharpened that he
can follow every molecule in its course, such a being whose attributes
are still as essentially finite as our own, would be able to do what is
at present impossible to us. For we have seen that the molecules in
a vesselful of air at uniform temperature are moving with velocities
by no means uniform, though the mean velocity of any great number
of them, arbitrarily selected, is almost exactly uniform. Now let us
suppose that such a vessel is divided into two portions, A and B, by a
division in which there is a small hole, and that a being, who can see
the individual molecules, opens and closes this hole, so as to allow only
the swifter molecules to pass from A to B, and only the slower ones to
pass from B to A. He will thus, without expenditure of work, raise the

temperature of B and lower that of A, in contradiction to the second law of thermodynamics.

This is only one of the instances in which conclusions which we have drawn from our experience of bodies consisting of an immense number of molecules may be found not to be applicable to the more delicate observations and experiments which we may suppose made by one who can perceive and handle the individual molecules which we deal with only in large masses.

In dealing with masses of matter, while we do not perceive the individual molecules, we are compelled to adopt what I have described as the statistical method of calculation, and to abandon the strict dynamical method, in which we follow every molecule by the calculus...

This is one of the most cited passages in all of the physical literature. Several points should be made regarding it.

- It is implicit that not only is a temperature gradient possible, but so also a pressure gradient. The latter does not even require measurement of velocity on the demon's part; he merely lets molecules pass one way through his gate, effectively acting as a gas check valve. This pressure (or temperature) difference can be used to perform work and, presumably, can be regenerated at will; thus, it constitutes a *perpetuum mobile* of the second kind.

- In principle, Maxwell's challenge can be extended into the quantum realm.

- It is not required that the demon be alive or sentient; inanimate equipment endowed with faculties of logic will suffice. One can even imagine biomolecular machines — capable of recognizing, binding, and transporting target molecules — performing all the essential actions of the demon [3]. In a letter to Lord Rayleigh, Maxwell remarks, "I do not see why even intelligence might not be dispensed with and the thing made self-acting" [1, 4][1]

- One should not infer from the text that Maxwell was himself seriously intent on breaking the second law; in fact, he was strongly convinced of its statistical truth [4]:

  > The second law of thermodynamics has the same degree of truth as the statement that if you throw a tumblerful of water into the sea, you cannot get the same tumblerful of water out again. (Maxwell's letter to J. W. Strutt in December 1870 [4])

Nevertheless, Maxwell was farsighted enough to realize the limitations of nineteenth century physics so as to entertain the possibility that the second law might one day be violated. This is evident in his writing: "... we conceive of a being whose faculties ... would be able to do what is at present impossible

---

[1]Eventually, Maxwell preferred the term *valve* over *demon*, the latter of which was coined by Kelvin [5].

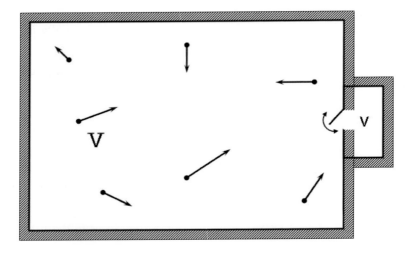

Figure 2.2: Loschmidt's demon.

to us." It is doubtful he could have imagined our present day dexterity in
being able to manipulate individual atoms and elementary particles.

In 1869 Loschmidt proposed a variation of Maxwell's demon (Figure 2.2). Consider a volume V with gas molecules moving with greater and lesser velocities than
the mean velocity. Consider now a small compartment v adjoining V, and a gate
separating it from V. There is a nonzero, though vanishingly small, probability that
v is initially empty. If we can determine the initial conditions of all molecules in V
and the order and direction of their scattering, then the gate could be instructed
to open and close in such a way that would allow only the faster molecules into v.
Thus, one can create a pressure or temperature difference without an expenditure
of work. Clearly, this requires detailed knowledge of the initial conditions, the
dynamical scattering for all the molecules, as well as intricate gate timing. The
physical and computational infeasibility of this scheme was immediately appreci-
ated by Loschmidt and his contemporaries.

Maxwell demons can be catagorized as either sentient or non-sentient. Sentient
ones perceive (make physical measurements) and think (store, manipulate, and
erase information) to guide their molecular separations. Non-sentient demons, on
the other hand, would not necessarily perceive or think, but merely respond to
and manipulate molecules, as by natural processes. Operationally, Maxwell's and
Loschmidt's are sentient demons.

In a very brief article, H. Whiting (1885) [6] proposed what could be considered
a non-sentient demon in the sense that its molecular sorting is replaced by a
natural, automatic process:

> When the motion of a molecule in the surface of a body happens to
> exceed a certain limit, it may be thrown off completely from that sur-
> face, as in ordinary evaporation. Hence in the case of astronomical

bodies, particularly masses of gas, the molecules of greatest velocity may gradually be separated from the remainder as effectually as by the operation of Maxwell's small beings.

Although Whiting proposes a sorting mechanism, he does not couple it to a work extraction process; more than a century later such a mechanism was proposed [7].

2.2 Exorcising Demons

For 130 years, exorcising Maxwell's demon has been a favorite act of devotion among confirmed thermodynamicists. It was understood from its conception that the demon's molecular sorting was beyond the means of 19th century technology, but since this might not always be the case, the demon posed a latent threat to the second law which needed stamping out.

2.2.1 Smoluchowski and Brillouin

In 1912, Marian von Smoluchowski [8] demonstrated that violations of the second law were, in principle, possible in sufficiently small systems since the usual postulates and assumptions of thermodynamics break down as statistical fluctuations become sizable. He found, however, that fluctuations are random and that a *perpetuum mobile* cannot make use of them. The reason is simple: in order to capitalize on the fluctuations, the device's machinery must be comparable in size to the fluctuations themselves, in which case it is subject to the same type of statistical fluctuations it is trying to harness. (Similar arguments are made by Feynman with regard to his celebrated pawl and ratchet device [9].) Modern computer simulations of classical particles and gates strongly support Smoluchowski's exorcism [10].

Importantly, Smoluchowski's analysis omitted quantum effects like quantum correlations and spontaneous processes that figure prominently in several modern second law challenges (§3.6, §4.6). For example, the state of the demon — represented possibly by a two-level quantum system in two topological conformations, corresponding to open- and closed-gate configurations — can be highly correlated with the state of the molecules that it is manipulating. Consequently, the entropy of the (molecules + demon) together is *not* equal to the sum of entropies of the demon and molecules considered separately. (This violates the standard thermodynamic property of additivity.) In modern parlance, the states of the demon and molecules are strongly entangled. Thus, the traditional view that the demon is crippled by its stochastic environment is not justified.

Since it was discovered, each facet of the demon's behavior has been deeply scrutinized for physical reasonableness. Even the most mundane activity of seeing is not as easy for the demon as one might think. In an ideal, isothermal blackbody cavity, the radiation spectrum is uniform and spatially isotropic; that is, in principle everything looks the same in all directions. Individual objects can be

discriminated only if their temperatures are different from the background, or if
they are illuminted by a superthermal radiation source (*e.g.*, a flashlight).

If we imagine a demon at the same temperature as the cavity[2], then, without
a flashlight, he is unable to see the molecules he intends to sort. He is blind. The
flashlight, however, generates entropy. As described originally by Brillouin [11],

> The torch is a source of radiation not in equilibrium. It pours negative
> entropy into the system. From this negative entropy the demon obtains
> "informations". With these informations he may operate the trap door
> and rebuild negative entropy, hence, completing a cycle:
>
> negentropy → information → negentropy.
>
> We coined the abbreviation "negentropy" to characterize entropy with
> the opposite sign. ... Entropy must always increase, and negentropy
> always decreases.

Brillouin's analysis, though compelling, suffers from multiple shortcomings. First,
thermodynamic entropy is conditioned by the validity of the second law. Thus,
his analysis assumes the validity of the law it attempts to rescue. Second, the
statistical entropy increases only after a proper coarse graining procedure, which
is not unique. Hence, the statistical entropy increase is not unique and, thus,
can hardly be directly identified with the thermodynamic entropy. Third, this
defense of the second law is somewhat of a straw man since there are numerous
non-sentient demons that do not require active perception for their successful
operation. Controversy regarding the demon's eyesight continued into the 1980's
[12, 13, 14].

2.2.2 Szilard Engine

In 1929, the demon was launched into the information age by Leo Szilard. He
envisioned a heat engine consisting of a cylinder divided by a partition/piston
with a one-molecule working fluid (Figure 2.3). The piston is driven by the single molecule; however, in order to achieve a thermodynamic cycle the molecule's
position must be coordinated with the placement of the piston. If successful, this
cycle would allow cyclic extraction and conversion of heat from a surrounding heat
bath into work, this in seeming violation of the second law.

Szilard correctly identifies three important and previously unappreciated aspects of the demon — measurement, information, and memory. Serendipitously,
the demon helped lay down the foundation of modern information theory. Szilard
finds that, in order to comply with the second law, the measurement process must
entail a compensatory entropy production of $S = k \ln(2)$. This 'bit' of entropy and
information creates a new link between these two seemingly disparate concepts,
later established more solidly by Shannon [15] and others. It is often mistakenly
assumed that Szilard demonstrated that the second law prevails over the demon;
rather, his principal result was derived *assuming* the second law was absolute.

---

[2]If he is not at the same temperature he plays right into the hands of the second law.

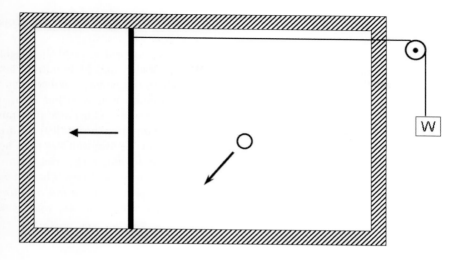

Figure 2.3: Szilard's engine.

Over the last 75 years a tremendous amount has been written on Szilard's engine and the informational aspects of Maxwell's demon [1]; notable researchers include Landauer, Bennet, Penrose, Zurek, Popper, and Feyerabend. It is now the consensus that memory erasure is the tragic flaw of the *sentient* demon. A finite device, operating in an indefinitely long cycle, which must record information about individual molecules it processes, must eventually erase its memory. Erasure creates entropy which at least compensates for any entropy reduction accomplished by the rest of the thermodynamic cycle. In other words, the second law is sustained. One might say the sentient demon is *just too damn smart for his own damn good*. Still, legions of his dumber, non-sentient comrades remain alive, well, and on the loose.

2.2.3 Self-Rectifying Diodes

It is well known that electrical diodes can be used to rectify ac voltage fluctuations, rendering dc voltages. It is also well established experimentally and theoretically that electrical systems exhibit thermally-induced voltage fluctuations. For instance, in the frequency interval $(f \rightarrow f + \Delta f)$ the white noise voltage fluctuations in a resistor are given by the Nyquist theorem as $\langle \Delta V^2 \rangle \simeq 4RkT\Delta f$, where R is resistance and kT is the thermal energy. Given these observations, it is natural to wonder whether diodes can self-rectify their own thermally-induced currents and voltages, thus behaving as micro-batteries. If so, then if arranged suitably they should be able to generate usable power — were it not for the second law [16]. The possibility of self-rectifying diodes stimulated significant interest over the twenty year interval 1950-1970 [16-20]. The consensus is that diodic self-rectification is impossible, but not all loopholes appear to have been closed [20]. The subject has lain dormant for 30 years and to our knowledge no experiments along these lines have been conducted. Recently, diodes have been invoked in other second law challenges [21].

In summary, Maxwell's demon and its many spawn engaged some of the brightest minds of the 19th and 20th centuries. Maxwell probably had little inkling how much moil, toil, and trouble his sentient little devil would cause, nor how abundant its intellectual fruits would be. The sentient demon is now either dead or comatose; however, starting 25 years ago with the proposals of Gordon and Denur, a new wave of non-sentient demons has arrived, sidestepping the critical failures of the sentient variety. These are often phrased in terms of Maxwell's demon not necessarily because of any special similarity to it, but to honor this rich vein of scientific history and scholarship. Moreover, an alignment with Maxwell helps fend off the negative connotations associated with the term *perpetual motion machine*. (After all, Maxwell's demon is a perpetual motion machine, but one with pedigree.)

2.3 Inviolability Arguments

The second law has no general theoretical proof. Except perhaps for a few idealized cases like the dilute ideal gas, its absolute status rests squarely on empirical evidence. As remarked by Fermi [22] and echoed by others, "support for this law consists mainly in the failure of all efforts that have been made to construct a *perpetuum mobile* of the second kind." Certainly its empirical support is vast and presently uncracked; however, one must be careful not to fall prey to the primary fallacy of induction pointed out by Hume: That which was true in the past will be true in the future[3]. Induction is especially dicey in science when new physical regimes and new phenomena are incorporated into existing paradigms. One would be hard-pressed to name *any* physical theory, concept, law or principle that has not undergone major revision either in content or interpretation over the last 100 years. For example, the list of energy types has been updated several times (*e.g.*, with dark energy, vacuum energy, mass energy); angular momentum was quantized ($\hbar$); classical mechanics became a limiting case of quantum mechanics (Bohr's correspondence principle); Galilean kinematics and Newtonian inertial mechanics became the low-velocity limit of special relativity; Newtonian gravity was interpreted as the weak-field limit in general relativity, which itself is now considered the classical limit of a much-anticipated theory of quantum gravity. If one even casually reviews the history of science, particularly over the last century, one should not be surprised that the second law is now in jeopardy. Quite the opposite, one should be more surprised that it has not been put in jeopardy earlier since all the basic science for the modern challenges was in place 40-50 years ago. The damning question is this: Why has it taken so long for its absolute status to be seriously questioned?

---

[3] At the very least, progress and novelty in the world depend on this fallacy being false.

2.3.1 Early Classical Arguments

Over the last 150 years, many attempts have been made to derive a general proof for the second law. All have failed. Among early attempts, Boltzmann's H-theorem (1872) is deservedly the most celebrated [23]. In it, the quantitiy H is defined $H \equiv \sum_i \mathcal{P}_i \ln(\mathcal{P}_i) \equiv \langle \ln \mathcal{P}_i \rangle$, where $\mathcal{P}_i$ is the probability of finding a system in an accessible state i. Boltzmann showed that $\frac{dH}{dt} \leq 0$, and only at equilibrium is $\frac{dH}{dt} = 0$. Clearly, H is related to the standard statistical (not thermodynamic) entropy as $S = -kH$. Though suggestive, the H-theorem does not constitute a second law proof, but it does serve as a fundamental bridge between microscopic (statistical mechanical) and macroscopic (thermodynamic) formalisms.

Early objections to the H-theorem were raised by Loschmidt (1876-77) and Zermelo (1896). Loschmidt pointed out that since the laws of physics upon which the H-theorem is based are time symmetric (reversible), it is not clear that the H-function should model nature in displaying time-asymmetry[4] [24]. Zermello argued, based on Poincaré's recurrence cycles, that systems should display quasi-periodicity; that is, a system's trajectory in 6N-dimensional phase space should eventually pass arbitrarily close to previously visited points, which can represent lower entropy states, in which case the system will have spontaneously evolved toward lower entropy, in violation of the second law [25, 26]. These objections have been largely overcome through proper appreciation of the truly large number of complexions available to thermodynamic systems. For macroscopic systems, the number of complexions at equilibrium is so much greater than the number even slightly away from equilibrium that nonequilibrium states are effectively never seen once equilibrium is achieved. Their relative phase space volume is essentially zero[5]. A commanding discussion of these issues is presented by Lebowitz [27]. Ultimately, the H-theorem is too limited in purview to serve as a general second law proof; however, it is a rallying point for the faithful.

The second law is emergent in nature. This means it does not bend other physical laws to its will, but rather, arises statistically from the interplay of multiple particles that themselves are governed by more basic laws like conservation of energy, linear and angular momentum. The second law must accommodate these more fundamental laws, not vice versa. Perhaps it should be considered not a *law*, but rather a *principle* or *meta-law*. In fact, if the second law were to be *derived* from more basic principles, properly, it should cease to be law. After all, laws are axioms of science. They are not proved, they are observed; they are recognized and assented to. For the present, the second law is absolute, not because it has been proven, but because it is observed to be so.

---

[4]By simply reversing the direction of all particles ($\mathbf{p}_i \rightarrow -\mathbf{p}_i$) — certainly a physically acceptable prescription — the system spontaneously reverts to lower entropy states.

[5]Let's say the average number of air molecules in a faculty member's office is roughly $N_{fo} = 3.3 \times 10^{26}$. For a fixed total system energy, consider two spatial configurations of the air: (i) smoothly distributed through the entire volume; and (ii) smoothly distributed in one half this volume. Assuming equal *a priori* probability, the ratio of the number of complexions in configuration-(i) to configuration-(ii) is $2^{N_{fo}} = 2^{3.3 \times 10^{26}} \simeq 10^{10^{26}}$. Clearly, the number of complexions for configuration-(i) so far outnumber those for configuration-(ii) that (ii) cannot be expected to spontaneously arise. Even small perturbations away from equilibrium are extremely unlikely.

2.3.2 Modern Classical Arguments

Perhaps the best known modern arguments in theoretical support of the second law are those layed out by Martynov [28]. In an isolated classical systems there are at least two integrals of motion: the total energy of the particles and the "phase volume occupied by the system" $\Delta\Gamma$. It follows that $\ln(\Delta\Gamma)$ is also an integral of motion[6].

It is reasonably assumed that the total energy $E(\mathbf{r}_1, \ldots, \mathbf{r}_N, \mathbf{p}_1, \ldots, \mathbf{p}_N)$ and $\ln(\Delta\Gamma)$ are the *only* two independent integrals of motion. (At this point $\ln(\Delta\Gamma)$ has not been identified with entropy.) However, the thermodynamic entropy — whose existence follows from the *presumed* validity of the Second law — provides quasi-equilibrium isolated systems with another integral of motion. As an integral of motion, entropy should be additive[7], in which case it must be a linear combination of E and $\ln(\Delta\Gamma)$. Total energy is excluded on grounds of physical intuition, thus the thermodynamic entropy is identified with $k\ln(\Delta\Gamma)$, by assumption. (The Boltzmann constant k follows from a properly chosen temperature scale.) For macroscopic systems, the phase space volume associated with equilibrium is far greater than the volume for the system out of equilibrium. Hence, up to an additive constant (which, presumably, can be determined via the third law), the entropy can be written $S = k\ln(\Delta\Gamma) = k\ln(\Omega)$, the Boltzmann entropy, where Ω is the *total* number of microstates available to the system.

There are at least two major weaknesses in this development of entropy on the way to the second law. First, entropy is arrived at under statistical assumptions, namely ergodicity, applied to ensembles, whereas it is understood that thermodynamics speaks to individual systems. Furthermore, the ergodic hypothesis is not generally valid beyond equilibrium. As an illustration, consider an ensemble consisting of a single quantum system isolated from its surroundings. This, and only this, would render this first concern moot. On the other hand, if the system is initially out of equilibrium, its thermodynamic entropy should increase with time in accord with the second law. However, if the state of the system is initially pure, it remains so forever, as discussed previously (§1.3). Moreover, in this form of $\rho(t)$, the statisical entropy S_{stat} remains constant for all times, in contrast with the expected behavior of the thermodynamic entropy S_{td}. One might argue — as many authors implicitly or explicitly do — that a coarse graining procedure should be invoked to guarantee S_{stat} increases with time. However, there appears no compelling physical reason to coarse grain a single system in a pure state. Additionally, this procedure would not be unique. Subjective coarse graining is physically dubious.

A second weakness in this development is that *additivity* as a general as-

---

[6]In phase space, an individual system is represented by a single point whose volume is zero. Hence, $\Delta\Gamma$ should be understood as the phase space volume of an ensemble of points. This, of course, invites criticism for equating thermodynamic with statistical entropies.

[7]Total energy should be additive if all sub-systems are statistically independent (macroscopic, separated well in space, and all interactions short-ranged). Since $\Delta\Gamma_Q = \Delta\Gamma_B \cdot \Delta\Gamma_C$, one expects $\ln(\Delta\Gamma)$ also to be additive. Regarding thermodynamic entropy, its additivity can be deduced from the assumed equality of sub-system temperatures and the additivity of heat increments δQ.

sumption is suspect. Additivity for the statistical entropy is acceptable provided that any two subsystems of the system in question are statistically independent. Equally well for thermodynamic entropy, additivity can be deduced from the additivity of heat increments at equilibrium. However, additivity is violated when subsystems are correlated, as in case of the Maxwell gate and the state of the gas with Maxwell's demon. As a result, systems with high degrees of correlation (entanglement) are prime candidates for second law challenges.

A different approach toward entropy and the second law has advantages over this last method and avoids some of its pitfalls [28]. It involves no formal coarse graining; it provides a systematic way how to get from the Liouville equation to a chain of the BBGKY-hierarchy[8] of simpler kinetic equations for particle distribution functions; and it illustrates how, in the thermodynamic limit, the final member of the BBGKY family of equations — the Liouville equation — becomes irrelevant from the point of view of the lowest equations of the hierarchy[9]. This method could, in principle, be adapted the quantum case, however, the necessary formalism has not yet been worked out.

Consider a classical system of N particles. For simplicity, we ignore problems associated with indistinguishability and ascribe to all particles the same microscopic parameters (e.g., mass, charge). One first defines the distribution function in the phase space of N particles, $\mathcal{G}_{(N)}$, by

$$f(q, p, t) \equiv \frac{1}{V^N \mathcal{P}^{3N}} \mathcal{G}_N(\mathbf{r}_1, \ldots, \mathbf{r}_N, \mathbf{p}_1, \ldots, \mathbf{p}_N, t) = \langle \prod_{j=1}^{N} [\delta(\mathbf{r}_j - \mathbf{r}_j(t))\delta(\mathbf{p}_j - \mathbf{p}_j(t))] \rangle,$$

(2.1)

where $\mathbf{r}_j(t)$ and $\mathbf{p}_j(t)$ designate spatial coordinates and momenta of the j-th particle, while the average $\langle \ldots \rangle$ denotes ensemble averaging over, for instance, their initial values. V is the total volume of the system and $\mathcal{P} = \sqrt{2\pi mkT}$. Since the system might be out of equilibrium, T can be a fictious temperature not necessarily connected with the initial state; it establishes proper units for momenta. All quantities — entropy included — are calculated from the ensemble, not the individual system.

There are, in general, at least two conserved quantities for the N-particle system if it is isolated from the surroundings, and if the forces between its constituent particles are conservative. These are the global (total) energy (as determined by the system's Hamiltonian H_N) and entropy. The global energy is:

$$E_{(N)} \equiv \langle H_{(N)}(\mathbf{r}_1, \ldots, \mathbf{r}_N, \mathbf{p}_1, \ldots, \mathbf{p}_N) \rangle$$

$$= \frac{1}{V^N \mathcal{P}^{3N}} \int \mathbf{r}_1 \ldots d\mathbf{r}_N \int_{-\infty}^{+\infty} d\mathbf{p}_1 \ldots d\mathbf{p}_N H_{(N)} \times \mathcal{G}_{(N)}$$

$$= \int_V d\mathbf{r} \left\{ n(\mathbf{r}, t) \frac{mc(\mathbf{r}, t)}{2} + \frac{3}{2} n(\mathbf{r}, t) \Theta(\mathbf{r}, t) + \frac{1}{2} \langle \Phi \rangle \right\}. \qquad (2.2)$$

---

[8] usually Bogolyubov, Born, Green, Kirkwood, and Yvon

[9] The BBGKY hierarchy of equations, on the other hand, determines the basic macroscopic, phenomenological laws of continuous media, including those of thermodynamics.

Here $\langle\Phi\rangle$ is the averaged potential energy for the mutual interaction of the particles while $n(\mathbf{r}, t)$, $\mathbf{c}(\mathbf{r}, t)$, and $\Theta(\mathbf{r}, t)$ are the ensemble-mean local particle concentration, macroscopic velocity, and temperature[10]. The five hydrodynamic parameters $n(\mathbf{r}, t)$, $\mathbf{c}(\mathbf{r}, t)$, and $\Theta(\mathbf{r}, t)$ are basic parameters through which all time-dependent macroscopic parameters can be expressed during slow relaxation of the system to equilibrium.

The global (total) entropy is given by

$$S_{(N)} = -\frac{k}{V^N \mathcal{P}^{3N}} \int d\mathbf{r}_1 \dots d\mathbf{r}_N \int_{-\infty}^{+\infty} d\mathbf{p}_1 \dots d\mathbf{p}_N \ln(\mathcal{G}_{(N)}) \times \mathcal{G}_{(N)}. \tag{2.3}$$

That both $E_{(N)}$ and $S_{(N)}$ are integrals of the motion can be verified through the Liouville equation:

$$\frac{\partial}{\partial t} \mathcal{G}_{(N)} = \sum_{j=1}^{N} [\frac{\partial U_{(N)}}{\partial \mathbf{r}_j} \cdot \frac{\partial \mathcal{G}_{(N)}}{\partial \mathbf{p}_j} - \frac{\mathbf{p}_j}{m} \cdot \frac{\partial \mathcal{G}_{(N)}}{\partial \mathbf{r}_j}]. \tag{2.4}$$

Here $U_{(N)}$ is the total interaction energy of the N particles.

At this point, a key concept is the correlation sphere. This is a sphere around any particle beyond which no correlations exist between the particle and those lying outside the sphere. Clearly, this is a well defined notion for systems like gases and liquids, but it becomes problematic for systems demonstrating long-range order like solids and self-gravitating systems. For simplicity we omit these latter cases. The notion of the correlation sphere allows introduction of the l-particle distribution functions

$$\mathcal{G}_{1,\dots,l} = \frac{1}{V^{N-l}\mathcal{P}^{3(N-l)}} \int \mathcal{G}_{1,\dots,N} d(l+1) \dots dN,$$

$$\mathcal{G}_{1,\dots,N} \equiv \mathcal{G}_{(N)}. \tag{2.5}$$

For all particles $1, \dots l$ inside the correlation sphere, one has $\mathcal{G}_{1,\dots,l} = \mathcal{G}_{1,\dots,l/l+1,\dots N}$, where $\mathcal{G}_{1,\dots,l/l+1,\dots N}$ designates the distribution function of the first l particles *provided* that the $(l+1)$-th up to the N-th particles have their coordinates and momenta correspondingly fixed and lying outside the correlation sphere. Using the definition (2.5), one obtains

$$\frac{\partial \mathcal{G}_{1,\dots,l}}{\partial t} = \sum_{j=1}^{l} \left\{ -\frac{\mathbf{p}_j}{m} \cdot \frac{\partial \mathcal{G}_{1,\dots,l}}{\mathbf{r}_j} + \frac{\partial U_{1,\dots,l}}{\partial \mathbf{r}_j} \cdot \frac{\partial \mathcal{G}_{1,\dots,l}}{\partial \mathbf{p}_j} \right\} +$$

$$\sum_{j=1}^{l} \left\{ \frac{n_0}{\mathcal{P}^3} \int \frac{\partial \Phi_{j,l+1}}{\partial \mathbf{r}_j} \cdot \frac{\partial \mathcal{G}_{1,\dots,l+1}}{\partial \mathbf{p}_j} d(l+1) \right\} ; \qquad l = 1, 2, \dots N. \tag{2.6}$$

Here we have used the notation

$$U_{1,\dots,l} = \sum_{1 \le i < j \le l} \Phi_{ij} \tag{2.7}$$

---

[10]Energy units given by kinetic energy of the thermal motion with velocities measured with respect to $\mathbf{c}(\mathbf{r}, t)$.

for the sum of the pair interactions of particles $1, \ldots, l$. The set (2.6) is the desired BBGKY hierarchy of the coupled particle distribution functions. The last equation of the hierarchy (for $l = N$) reproduces the Liouville equation (2.4).

The thermodynamic limit is now taken by increasing the particle number without bound ($N \to \infty$). This limit must precede any reasoning connected limits of physical time since otherwise, we encounter problems connected with reversibility of the theory. In the thermodynamic limit, the last equation of the hierarchy (the Liouville equation) disappears, thus becoming irrelevant for local dynamics of the system. Likewise, the total energy $E_{(N)}$ and total (global) entropy $S_{(N)}$ become irrelevant for short-range forces; they are proportional to N, as one expects for extensive thermodynamic variables, diverging to infinity in the thermodynamic limit[11]. Now, in order to investigate the validity of standard equations determining, for instance, entropy, one must introduce corresponding *local* quantities anew and determine the relevant equations for their time-development with the BBGKY set (2.6). Using a proper definition of the entropy density $s(\mathbf{r}, t)$ and entropy flow density $\mathbf{I}^{(s)}(\mathbf{r}, t)$, one obtains [28]

$$n(\mathbf{r}, t)\frac{\partial s}{\partial t} + \mathrm{div}\mathcal{I}^{(s)} = q^{(s)}, \tag{2.8}$$

where the right hand side is the entropy production term. This equation is the entropy (non)conservation law. Note, however, that the local entropy has no direct connection to a global entropy that can be compared with the thermodynamic entropy. Thus, in the thermodynamic limit we face the loss of meaning of such global and indispensable notions as the total energy, entropy, and the Liouville equation.

The definition of individual terms in (2.8) is complicated and results from an interplay of physical intuition and formal proof. The result is that in (2.8),

$$s = \sum_{l=1}^{+\infty} s^{(l)}, \quad \mathcal{I} = \sum_{l=1}^{+\infty} \mathcal{I}^{(s^{(l)})} \tag{2.9}$$

and similarly for the source term $q^{(s)}$. Here, for instance, $s^{(l)}$ is the contribution of the l-th order correlations. Specifically, the local correlation entropy is

$$n(\mathbf{r}, t)s^{(l)}(\mathbf{r}, t) = -kn_0^l \frac{1}{l!} \int_{-\infty}^{+\infty} \frac{d\mathbf{p}_1}{\mathcal{P}^{3l}} \int \omega_{1,2\ldots l}\mathcal{G}_{1,2\ldots l}d\mathbf{r}_2 d\mathbf{p}_2 \ldots d\mathbf{r}_l d\mathbf{p}_l \tag{2.10}$$

where

$$\mathcal{G}_1 = \exp(\omega_1), \quad \mathcal{G}_{12} = \mathcal{G}_1\mathcal{G}_2\exp(\omega_{12}), \quad \mathcal{G}_{123} = \mathcal{G}_1\mathcal{G}_2\mathcal{G}_3\exp(\omega_{12}+\omega_{13}+\omega_{23}+\omega_{123}) \tag{2.11}$$

and similarly for other terms in (2.8). One can argue physically that the sums in (2.9) converge rapidly once l exceeds number of particles in the correlation sphere. Insofar as one wishes to reformulate the second law in terms of local entropy

---

[11]One could, in principle, work with total energy or entropy per particle, but, in general, such quantities have limited explanatory utility for nonequilibrium systems.

increase, one should prove that the local entropy production term $q^{(s)} \geq 0$. This condition, however, cannot be established generally since local decreases in entropy are possible, such as in crystallization. Thus, this non-negativity condition should apply at least 'in most places', so that global entropy increase can be preserved. Unfortunately, no general proof of this condition is known at this time. The strengths of this sophisticated formalism notwithstanding, the non-decrease in entropy and the drive toward equilibrium — the validity of the second law — remains unresolved.

2.4 Candidate Second Law Challenges

Thermodynamics, as it applies to the second law, pertains to macroscopic systems consisting of a large *but finite* number of particles N. The microscopic physics, which presumably is time reversible, is hidden by thermodynamic postulates and caveats. In contrast, statistical mechanics grapples with the underlying microscopic physics and becomes meaningful only in the thermodynamic limit, requiring that ($N \rightarrow \infty$, $V \rightarrow \infty$, and $\frac{N}{V} = C$, finite) in such a way that local properties (*e.g.*, particle and mass densities) remain constant. It is remarkable — or distrubing — that such different formalisms purport to be universally valid for the same general phenomena when, for instance, their underlying definitions, like entropy, canot be consistently and unambiguously reconciled. The standard arguments [28, 29] face problems with the subjectivity of coarse graining, as well as those connected with suppression of global conservation laws in the thermodynamic limit. The use of correlation spheres is formally important, but problematic when long-range order is present, particularly when applied beyond simple liquids and gases, into solid state systems, and ones with nonextensive fields (gravitational, electric fields). Dovetailing local properties with global ones remains a theoretical challenge. Boundary conditions present further complications.

The foregoing critique of contemporaty thermodynamics and statistical mechanics offers inspirations for candidate systems for second law challenges. The following are our best guesses where 'the action' might be.

- Classical systems that display long-range order under equilibrium conditions; systems that undermine the notion of correlation spheres; non-extensive systems; systems with strong boundary effects. Natural examples of these might include solid state, self-gravitating, and plasma systems. The several USD challenges fall under this rubric (Chapters 6-9).

- Quantum systems that display long-range order in the form of quantum entanglement[12] As an illustration, consider a light beam split in two by a half-silvered mirror [30]. Formally, the entropy of the two split beams should exceed that of the single beam, if beam-splitting is an irreversible operation.

---

[12]In general, quantum systems have not been vetted as extensively or thoroughly for second law compliance as have classical systems.

However, the two beams can be optically recombined into a beam indistinguishible from the original single beam. This reunion would seem to violate the second law by reducing entropy, but it does not. The split beams are in fact not independent since they are connected by mutual phase relations. Hence, the total entropy of the two beams together is not equal to the sum of their entropies calculated as if the beams were mutually independent. The notion of relative phase does not arise in classical physics, but it is an integral aspect of quantum mechanics. This type of correlation between and among subsystems is known as quantum entanglement. Quantum challenges by Čápek, et al. (§3.6), and Allahverdyan, et al. (§4.6) bank on these. In principle, mutual correlations can persist even in the classical realm[13].

- Systems operating in extreme classical or quantum regimes in which novel collective behaviors arise[14]. For example, challenges by Keefe (§4.3) and Nikulov (§4.4) rely on the differences between classical and quantum statistical behaviors of electrons at the superconducting transition in mesoscopic structures.

- Systems whose statistical entropy (ensembles) is different from their thermodynamic entropy (individual systems); situations where the time development of individual systems in an ensemble differ appreciably from the ensemble average behavior, e.g., particles in the tail of a classical distribution.

- Systems not in equilibrium Gibbsian states. Formally, the classical regime is the high-temperature limit of the quantum one, so we can treat both simultaneously. The Allahverdyan and Nieuwenhuizen theorem (§1.2) states that for closed quantum systems at equilibrium, no work can be extraced by any periodic variation of external parameters. By equilibrium state it is meant the state with the canonical (Gibbsian) form of the density matrix. Thus, one might look for candidates among systems that are definitely *not* in the equilibrium Gibbsian state. This provides two options: (i) open systems where external agitation keeps the system out of equilibrium[15]; or (ii) open systems where sufficiently strong interactions with the bath drives the density matrix away from the Gibbsian form. (Combinations of these mechanisms is also admissible.)

The Gibbsian form of the density matrix

$$\rho_S = \frac{\exp(-\beta H_S)}{\mathrm{Tr}_S(\exp(-\beta H_S))} \tag{2.12}$$

---

[13]Imagine two newlyweds in their first apartment. Although, in principle, each might be found with equal and independent probability in any room, there will be strong correlations between their movements and both are likely to be found in the same room simultaneously more often than by pure chance. They are highly entangled — and perhaps more so in certain rooms than in others ... [Vincent, M., priv. commun., (1998)].

[14]Almost by definition, *extreme* regimes are ones that have not been carefully vetted for second law compliance.

[15]Preferentially, this agitation should be periodic in time in order to obtain a time-periodic gain in useful work, in accord with the Thomson form of the second law.

is currently known to describe well the equilibrium state of quantum systems in situations when the system-bath coupling is negligible (H_S being the corresponding system Hamiltonian). Thus, candidate systems would be those for which coupling to the bath is non-zero (finite). Presumably, strong coupling could accentuate such effects. Still, arbitrarily small but non-zero system-bath couplings should be enough to test second law status.

- Systems reminiscent of the Maxwell demon, with gates opening and closing via quantum processes; self-sustained correlations (entanglement) between the states of the system and gate that do *not* require external measurement, either by external agents or agitation. In other words, the extended system, consisting of the working medium and the gates can continuously perform its internal 'measurements' without the external influences. The issues of entropy connected with measurement, memory, and erasure need not arise[16]. Classical systems reminiscent of Maxwell's demon have been advanced by Gordon (§5.2), Denur (§5.3), and Crosignani (§5.4).

There seem to be a number of avenues open "to pick a hole" in the second law. The last 20 years represent a renaissance in this endeavor begun by Maxwell over 135 years ago.

In summary, we have cursorily reviewed prominent second law challenges from Maxwell (1867) up to roughly 1980, thus setting the stage for the modern challenges. We have briefly critiqued standard arguments in support of the second law and, from their shortcomings, identified several physical regimes where the second law might be tested for violability.

---

[16]Systems with Maxwellian gates are not the only quantum candidates for second law challenges; more typical quantum processes could also suffice (§3.6).

References

[1] Leff, H.S. and Rex, A.F., *Maxwell's Demon 2: Entropy, Classical and Quantum Information, Computing*, (Institute of Physics, Bristol, 2003); *Maxwell's Demon. Entropy, Information, Computing* (Hilger & IOP Publishing, Bristol, 1990).

[2] Maxwell, J.C., *Theory of Heat* (Longmans, Green, and Co., London, 1871).

[3] Gordon, L.G.M., Found. Phys. **13** 989 (1983).

[4] Daub, E.E., Stud. Hist. Phil. Sci. **1** 213 (1970).

[5] Thomson, W., R. Soc. Proc. **9** 113 (1879).

[6] Whiting, H., Science **6** 83 (1885).

[7] Sheehan, D.P., Glick, J., and Means, J.D., Found. Phys. **30** 1227 (2002).

[8] Smoluchowsky, M., Phys. Zeit. **12** 1069 (1912).

[9] Feynman, R.P., Leighton, R.B., and Sands, M., *The Feynman Lectures on Physics*, Vol. 1., Ch. 46, (Addison-Wesley, Reading, MA, 1963).

[10] Skordos, P.A. and Zurek, W.H., Am. J. Phys. **60** 876 (1992).

[11] Brillouin, L., J. Appl. Phys. **22** 334, 338 (1951).

[12] Denur, J., Am. J. Phys. **49** 352 (1981).

[13] Motz, H., Am. J. Phys. **51** 72 (1983).

[14] Chardin, G., Am. J. Phys. **52** 252 (1984).

[15] Shannon, C.E., Bell System Tech. J. **27** 379 (1948); Shannon, C.E. and Weaver, W., *The Mathematical Theory of Communication* (University of Illinois, Urbana, 1949).

[16] Brillouin, L., Phys. Rev. **78** 627 (1950).

[17] Marek, A. Physica **25** 1358 (1959).

[18] Van Kampen, N.G., Physica **26** 585 (1960).

[19] Gunn, J.B., J. Appl. Phys. **39** 5357 (1968).

[20] McFee, R., Amer. J. Phys. **39** 814 (1971).

[21] Sheehan, D.P., Putnam, A.R., and Wright, J.H., Found. Phys. **32** 1557 (2002).

[22] Fermi, E., *Thermodynamics* (Dover Publications, New York, 1936).

[23] Boltzmann, L., Wiener Ber. **66** 275 (1872).

[24] Loschmidt, J., Wiener Ber. **73** 128 (1876).

[25] Zermelo, E., Ann. Phys. **57** 485 (1896).

[26] Poincaré, H., Acta Math. **13** 1 (1890).

[27] Lebowitz, J.L., Physica A **263** 516 (1999).

[28] Martynov, G.A., *Classical Statistical Mechanics* (Kluwer Academic Publishers, Dordrecht, 1997).

[29] Martynov, G.A., Usp. Fiz. Nauk **166** 1105 (1996).

[30] von Laue, M., *Geschichte der Physik* (Ullstein, Frankfurt/Main, 1958).

3

Modern Quantum Challenges: Theory

Quantum formalism is developed and applied to a series of quantum theoretic challenges advanced by Čápek, et al. A primary requirement of these systems is that their coupling to the thermal bath not be weak.

3.1 Prolegomenon

Experiments on real systems are always carried out under conditions that cannot be characterized as limiting cases. For instance, experimental temperatures may be arbitrarily low, but never zero. Or, investigating the quantum mechanical ground state of macroscopic systems may be useful, but it might not lead to relevant results applicable at finite T. In a similar way, experimental temperatures may be high in the standard sense, but this does not necessarily mean full applicability of classical physics in the sense of the disappearance of quantum effects. Recall that, according Bohr correspondence principle, classical physics becomes fully legitimate only in the infinite temperature limit, while at room or even much higher temperatures many macroscopic quantum effects can survive, like ferro- or diamagnetism.

Similar problems appear with characteristics like external fields or phenomeno-

logical coupling constants like deformation potentials in solids. Such constants may
be manipulated by, say, external pressure. However, real first-principle coupling
constants have values that are not subject to variation. Hence, perturbational
treatments in powers of such coupling constants that are based on assumptions
of possible power-expansions in terms of such constants cannot be well checked
by changing their values in any interval. What is, however, more relevant is that
values of such constants are never infinite or zero. In fact, in real theoretical modeling, we usually and implicitly assume that in any given model, we include only
mechanisms that seem relevant or decisive. Finally, finite (non-zero) values of the
constants complicate even the determination of whether we can apply obtained
models or theories in given situations. In other words, we say that a given theory
is, for instance, the weak-coupling one. It is stressed that this characterization preserves its good mathematical meaning only in the sense that a physical prediction
gives a good quantitative character in the limiting sense that the relevant coupling
constant is mathematically limited to zero. To be specific, ensemble theory in statistical physics, which is connected with the Gibbs canonical or grand canonical
distributions permitting statistical introduction of temperature, is mathematically
meaningful only in the limiting sense of weak coupling, *i.e.*, in the mathematical
limit of the coupling constant between the system and the reservoir (bath) being
exactly zero.

Experiment never corresponds to such an ideal mathematical description since
the system-bath coupling constant must always be considered non-zero (though,
perhaps, quite small). Recall that in this opposite case — *i.e.*, no system-bath
coupling — the system and bath would never fully establish mutual thermal equilibrium. Hence, in practice, we are forced to introduce regimes of applicability
of such mathematically well-formulated limiting statements, introducing sharp inequalities (or even exact order of limiting processes). In this section, we list several
requirements for theories aspiring to direct comparison with experiment. This exercise is often considered trivial and, thus, is often ignored. However, because of
the potentially important conclusions, these requirements should be made clear.

In our opinion, theoretical treatments wishing to be taken seriously in potentially important areas like this should fulfill the following minimum criteria:

- They should make clear statements about the effect observed in the model
  considered, with detailed specification of the area of physical parameters
  involved. Proper definition of the physical regime is very important since
  the usual weak-coupling regime is usually too restrictive.

- They should clearly indicate the role of the initial conditions, which should
  — with the exception of certain basic parameters, like initial temperature
  — be minor in stationary situations or, more generally, after the transient
  initial period of time when details of initial conditions might play a role.
  (This requirement is related to the Bogolyubov principle of decay of initial
  correlations.)

- The statements should be related to the macroworld in which thermodynamics governs as a universal theory. For that, careful taking of the thermody-

namic limit of the bath is critically important but, on the other hand, may help to turn microscopic statements to macroscopic ones.

- The considerations should include discussion of stability of the results with respect to other mechanisms or possible higher order effects.

3.2 Thermodynamic Limit and Weak Coupling

There is perhaps one universal feature distinguishing all modern second law challenges. It is that the interaction of the system with its surrounding — *i.e.*, the thermodynamic bath — cannot be *weak*. In order to explain what weak coupling means, let us introduce some basic concepts. Asserting that classical physics is formally a specific (high-temperature) limiting case of the quantum one, we can limit our discussion to the quantum case.

The first step in considering any problem in statistical physics is to define the *system* and the rest of the 'universe', *i.e.*, the *bath* or *reservoir*. Imagine that we are interested in some electronic property of a crystal. Realizing their indistinguishability, the many electrons may be taken as the system of interest, while the bath would include atoms of the lattice (forming a basis for the static crystal potential, phonons, etc.), contacts, crystal holder, the experimental apparatus, laboratory, etc. Two things should be stressed:

- There is always an arbitrariness in the choice of what is to be considered the system. For instance, part of the bath could be joined with the system, thus forming an extended system with which to proceed. This might have some advantages connected with the fact that the system Hamiltonian H_S is usually exactly diagonalized. H_S might thus incorporate a part of the surroundings with important correlations with the (more restricted) system, which might be important for the problem considered. On the other hand, the increased size of the system correspondingly increases the technical problems connected with the diagonalization of H_S. The extreme choice might be to choose the whole (system + bath) complex as an extended system. In this case, all the correlations would be properly considered upon diagonalization of H_S. On the other hand, the technical problems associated with the diagonalization of H_S are the very reason why we split off part of the complex as our system in the restricted case, understanding the rest of the complex as the bath. Hence, it is a joint matter of personal taste, skill, and optimization of the endeavor that determines what is taken to be the system and what is taken to be the bath.

- For any choice of the system (and bath), the (system + bath) complex must be temporarily isolated from the rest of the universe. If not, there would be no possibility of writing down the Hamiltonian for it. In this case, there would be no possibility of treating its time development via the Schrödinger equation or the Liouville equation via its density matrix. Of course, this assumption of isolation from the rest of the universe is not realistic. That is why one must choose the bath to be sufficiently extended. Moreover, when the time intervals considered increase, one must also increase the size of the bath. This is one of two reasons why the thermodynamic limit of the bath must precede any long-time limit considered. The second reason is connected with the necessity to mimic naturally occurring finite widths of overlapping energy levels by a continuum of energy levels.

There is an important consequence of the infinite (thermodynamic) limit of the bath, specifically, we get rid of the so-called Poincaré cycles [1]. To unpack this notion, realize that for any finite (system + bath) complex, the starting Schrödinger or Liouville equations are reversible, while the behavior of realistic macroscopic systems is clearly irreversible. (See von Neumann entropy (§1.3).) The irreversibility of the time-development appears in our formalism[1] owing to the fact that Poincaré cycles — the quasi-periodic return of system observables to their initial values during the course of free time evolution of the (system + bath) complex — develop very quickly to infinity with increasing size of the complex.

This procedure of increasing bath size does not necessarily appear explicitly. However, since the theory itself is not reversible, it must contain sources of irreversibility. It is not important, for instance, whether there is thermalization of particles at an isothermal surface, or relaxation of parts of the system (*e.g.*, molecular groups) playing a role of the Maxwellian gates to more energetically advantageous configurations. Irreversibility could not appear without increasing the size of the (system + bath) complex to infinity. Clearly, we want to treat the system that we have initially specified, which is finite; thus, it is the bath whose size is increased. So, what specifically is the detailed definition of the thermodynamic limit of the bath? The procedure is as follows:

- Choose the system as a finite number of particles and specify the type (not the size) of the bath and its coupling to the system. This includes, *inter alia*, specification of type of forces among all particles.

- For any finite bath size, specify the initial state of the (system + bath) complex in such a way that with the ensuing increase in the size of the bath to infinity, its macroscopic intensive characteristics remain fixed (*e.g.*, mass density, particle density, energy-density, or temperature[2], macroscopic motion with respect to the system).

- Then express the quantity of interest. It must be finite. Often it is simply the density matrix.

- Now increase the size of the bath to infinity for any time t of the time-development investigated (*e.g.*, the Liouville equation for the density matrix of the (system + bath) complex). Insofar as all the interactions are short-ranged, there is no reason for any divergences (infinities) in this limit.

---

[1] In fact, this is the general formalism of statistical physics.
[2] as far as this notion may be, in the given situation, well introduced

- The resulting value of the system quantity of interest is then the quantity required for direct comparison with experiment as a function of time. Or, as in the case of the density matrix of the system, it is used to investigate the time development of the system for a broader class of system observables.

This procedure is generally accepted. To have a solid basis for further discussion, let us perform the above reasoning with application of the thermodynamic limit of the bath on a level of the density matrix of the system $\rho_S(t)$. Though we shall later return to the formalism, here we will derive two basic equations replacing, on the level of ρ_S as an operator in the Hilbert space of the system, the Liouville equation for the density matrix of the (system + bath) complex $\rho_{S+B}(t)$ in the Hilbert space of the (system + bath) complex together. Such equations bear the name of the Generalized Master Equations (GME). Because of a degree of arbitrariness, they are infinitely many of them (both convolution and convolutionless types) according to the specific choice of projection and information we are interested in. We start from the Liouville equation for the complex $\rho_{S+B}(t)$

$$ i\frac{d}{dt}\rho_{S+B}(t) = \frac{1}{\hbar}[H, \rho_{S+B}(t)] \equiv \mathcal{L}\rho_{S+B}(t). \qquad (3.1) $$

We always assume a time-independent Hamiltonian. This means that there are no external time-variable fields acting on the system. If there are, the sources of such fields would have to be included in our extended (system + bath) complex which would restore the time-independence in (3.1). Further, we have introduced the so-called Liouville superoperator $\mathcal{L}\ldots = \frac{1}{\hbar}[H,\ldots]$. If we remember that operators are general prescriptions ascribing to (wave) functions of the corresponding Hilbert space other functions in the same space, we see that general superoperators ($\mathcal{L}$ is one such example) may be understood as prescriptions ascribing to any operator A another operator; for the Liouville superoperator, it is the operator $\frac{1}{\hbar}[H, A]$. In fact, the space of all operators (acting on functions in the Hilbert space of wave functions) is formally, from the mathematical point of view, a Hilbert space. Correspondingly, the set of all superoperators, acting on operators in this Hilbert space of wave functions, also form such a Hilbert space. Corresponding scalar product of two operators A and B may be introduced in infinitely many ways; the simplest one (though not the most advantageous) is $(A, B) = \text{Tr}(A^\dagger B)$. The latter Hilbert space of operators is then often called the Liouville space and superoperators are then nothing but operators in the Liouville space.

Arbitrary superoperator $\mathcal{P}$ is called projector or projection superoperator if it has idempotency property, i.e., that

$$ \mathcal{P}^2 = \mathcal{P}. \qquad (3.2) $$

In general, such projection superoperators are used to reduce information required. Realize that without such a reduction, upon increasing the size of the bath to infinity, one would soon encounter an information catastrophe: an overabundance of information that would hinder the use of it. For example, one could imagine the projection of the Euclidian three dimensional vector space onto the x-y plane. Projecting any position vector twice yields the same result as a single projection only. For illustration in our Hilbert and Liouville space, one can use the Argyres-Kelley projector [2]

$$\mathcal{P}\ldots = \rho^B \mathrm{Tr}_B(\ldots), \quad \mathrm{Tr}_B \rho^B = 1. \tag{3.3}$$

Arbitrariness in the choice of ρ^B in (3.3) is the source of arbitrariness in the form of the GME mentioned above. Also worth realizing is that if in general $\mathcal{P}$ is a projector, $(1 - \mathcal{P})$ is also a projector since $(1 - \mathcal{P})^2 = 1 - \mathcal{P}$. So, if $\mathcal{P}$ is a projection on an interesting part of information contained in $\rho_{S+B}(t)$, $1 - \mathcal{P}$ may be interpreted as a projection on an uninteresting part thereof.

Let us take the form of (3.3) as simply an illustration, but let us proceed with a general time-independent projector $\mathcal{P}$. From (3.1), we get

$$i\frac{d}{dt}\mathcal{P}\rho_{S+B}(t) = \mathcal{P}\mathcal{L}\mathcal{P}\rho_{S+B}(t) + \mathcal{P}\mathcal{L}(1 - \mathcal{P})\rho_{S+B}(t),$$

$$i\frac{d}{dt}(1 - \mathcal{P})\rho_{S+B}(t) = (1 - \mathcal{P})\mathcal{L}\mathcal{P}\rho_{S+B}(t) + (1 - \mathcal{L})\mathcal{P}(1 - \mathcal{P})\rho_{S+B}(t). \tag{3.4}$$

Solving the second equation of (3.4) for the 'uninteresting' part $(1 - \mathcal{P})\rho_{S+B}(t)$ of the density matrix $\rho_{S+B}(t)$ we get

$$(1 - \mathcal{P})\rho_{S+B}(t) = \exp[-i(1 - \mathcal{P})\mathcal{L} \cdot (t - t_0)](1 - \mathcal{P})\rho_{S+B}(t_0)$$

$$-i\int_{t_0}^{t} \exp[-i(1 - \mathcal{P})\mathcal{L} \cdot (t - \tau)](1 - \mathcal{P})\mathcal{L}\mathcal{P}\rho_{S+B}(\tau) \, d\tau. \tag{3.5}$$

Here and everywhere below, $\exp[\mathcal{A}] \equiv e^{\mathcal{A}}$ is as usually understood as $e^{\mathcal{A}} \equiv \sum_{n=0}^{+\infty} \frac{1}{n!}(\mathcal{A})^n$.

Now, we can proceed in two ways. The first consists in the direct introduction of (3.5) into the first equation of (3.4). The result is the so-called Nakajima-Zwanzig identity [3, 4, 5]:

$$\frac{d}{dt}\mathcal{P}\rho_{S+B}(t) = -i\mathcal{P}\mathcal{L}\mathcal{P}\rho_{S+B}(t)$$

$$-\int_{t_0}^{t} \mathcal{P}\mathcal{L}\exp[-i(1 - \mathcal{P})\mathcal{L} \cdot (t - \tau)](1 - \mathcal{P})\mathcal{L}\mathcal{P}\rho_{S+B}(\tau) \, d\tau$$

$$-i\mathcal{P}\mathcal{L}\exp[-i(1 - \mathcal{P})\mathcal{L} \cdot (t - t_0)](1 - \mathcal{P})\rho_{S+B}(t_0). \tag{3.6}$$

The second way is to rewrite $\ldots \rho_{S+B}(\tau)$ at the end of (3.5) as $\ldots [\exp[i\mathcal{L} \cdot (t - \tau)](1 - \mathcal{P})\rho_{S+B}(t) + \exp[i\mathcal{L} \cdot (t - \tau)]\mathcal{P}\rho_{S+B}(t)$ [3] and solve the resulting equation for the 'interesting' part of the density matrix $\mathcal{P}\rho_{S+B}(t)$. The results reads

$$(1 - \mathcal{P})\rho_{S+B}(t) = \left[1 + i\int_0^{t-t_0} e^{-i(1-P)\mathcal{L}\cdot\tau}(1 - \mathcal{P})\mathcal{L}\mathcal{P}e^{i\mathcal{L}\cdot\tau} \, d\tau \right]^{-1}$$

$$\cdot \left[-i\int_0^{t-t_0} e^{-i(1-P)\mathcal{L}\cdot\tau}(1 - \mathcal{P})\mathcal{L}\mathcal{P}e^{i\mathcal{L}\cdot\tau} \, d\tau + e^{-i(1-P)\mathcal{L}\cdot(t-t_0)}(1 - \mathcal{P})\rho_{S+B}(t_0) \right]. \tag{3.7}$$

---

[3] Here we use a formal solution of (3.1) in the form of $\rho_{S+B}(\tau) = \exp(i\mathcal{L}(t - \tau))\rho_{S+B}(t)$

Introducing (3.7) into the first equation of (3.4) yields the Shibata-Hashitsume-Takahashi-Shingu (SHTS) identity[4] [6, 7]

$$\frac{d}{dt}\mathcal{P}\rho_{S+B}(t) = -i\mathcal{P}\mathcal{L}\left[1 + i\int_0^{t-t_0} e^{-i(1-\mathcal{P})\mathcal{L}\cdot\tau}(1-\mathcal{P})\mathcal{L}\mathcal{P}e^{i\mathcal{L}\cdot\tau}\,d\tau\right]^{-1}$$

$$\cdot\left[\mathcal{P}\rho_{S+B}(t) + e^{-i(1-\mathcal{P})\mathcal{L}\cdot(t-t_0)}(1-\mathcal{P})\rho_{S+B}(t_0)\right]. \tag{3.8}$$

With this, we now formally perform the thermodynamic limit of the bath. On the level of (3.6) or (3.8), it still cannot be performed in general. However, with projectors of the type of (3.3), the task is solvable. Introduce first (3.3) into (3.6) and perform the trace over the bath, i.e., Tr_B. For the density matrix of the systems $\rho_S(t) = \mathrm{Tr}_B\rho_{S+B}(t)$, we obtain

$$\frac{d}{dt}\rho_S(t) = -i\mathrm{Tr}_B[\mathcal{L}(\rho^B\rho_S(t))] + \int_0^{t-t_0} \mathcal{M}(\tau)\rho_S(t-\tau)\,d\tau + I(t-t_0), \tag{3.9}$$

where the memory superoperator

$$\mathcal{M}(\tau)\ldots = -\mathrm{Tr}_B[\mathcal{L}e^{-i(1-\mathcal{P})\mathcal{L}\cdot\tau}(1-\mathcal{P})\mathcal{L}\{\rho^B\ldots\}] \tag{3.10}$$

is now a superoperator in the Liouville space of operators in the Hilbert space of the system only, and the initial condition term (operator)

$$I(t-t_0) = -i\mathrm{Tr}_B[\mathcal{L}e^{-i(1-\mathcal{P})\mathcal{L}\cdot(t-t_0)}(1-\mathcal{P})\rho_{S+B}(t_0)] \tag{3.11}$$

depends, for fixed $\rho_{S+B}(t_0)$, only on the time difference $t - t_0$ and expresses influence on time development of $\rho_S(t)$ of the 'uninteresting' part $(1-\mathcal{P})\rho_{S+B}(t_0)$ of the initial condition $\rho_{S+B}(t_0)$.[5] It is worth noting that all the terms on the right hand side of (3.9) depend on the detailed choice of ρ^B. Solution $\rho_S(t)$ is, however, ρ^B-independent. One could give all these formulae a standard matrix form. For that, one must introduce matrix elements of linear superoperators.[6] Let $\mathcal{A}$ be a linear superoperator. Then its matrix elements (with four indices) are determined as

$$\mathcal{A}_{a,b,c,d} = [\mathcal{A}(|c\rangle\langle d|)]_{a,b}. \tag{3.12}$$

Thus,

$$(\mathcal{A}B)_{a,b} = \sum_{c,d} \mathcal{A}_{a,b,c,d}B_{c,d}. \tag{3.13}$$

---

[4]A time-local identity for just the 'interesting' part of the density matrix $\mathcal{P}\rho_{S+B}(t)$ was first derived by Fuliński and Kramarczyk [8, 9]. Full equivalence of their equation to (3.8) was proved by Gzyl [10].

[5]The 'interesting' part of the initial condition $\mathcal{P}\rho_{S+B}(t_0) = \rho^B\rho_S(t_0)$ must be added to the integro-differential operator equation (3.9) as the standard initial condition.

[6]All the superoperators here are linear; i.e., they fulfill the condition $\mathcal{A}(aA + bB) = a\mathcal{A}A + b\mathcal{A}B$, where a, b are general c-numbers and A, B are arbitrary operators.

Hence, defining matrix elements of the density matrix of the system as $\rho_{m,n}(t) \equiv (\rho_S(t))_{m,n}$, (3.9) may be rewritten as

$$\frac{d}{dt}\rho_{m,n}(t) = -i\sum_{p,q}\mathcal{L}^{eff}_{m,n,p,q}\rho_{p,q}(t) + \sum_{p,q}\int_0^{t-t_0} w_{m,n,p,q}(\tau)\rho_{p,q}(t-\tau)\,d\tau + I_{m,n}(t-t_0),$$
(3.14)

where $\mathcal{L}^{eff}_{m,n,p,q} = \sum_{\mu}(\mathcal{L}(\rho^B \otimes |p\rangle\langle q|)_{m\mu,n\mu}$, and memory functions $w_{m,n,p,q}(\tau)$ are defined as matrix elements of the above memory superoperator

$$w_{m,n,p,q}(\tau) \equiv \mathcal{M}_{m,n,p,q}(\tau) = -\sum_{\mu,\pi,\kappa}[\mathcal{P}\mathcal{L}e^{-i(1-\mathcal{P})\mathcal{L}\tau}(1-\mathcal{P})\mathcal{L}\mathcal{P}]_{m\mu,n\mu,p\pi,q\kappa}\rho^B_{\pi,\kappa}$$

$$= -\sum_{\mu,\pi,\kappa}[\mathcal{L}e^{-i(1-\mathcal{P})\mathcal{L}\tau}(1-\mathcal{P})\mathcal{L}]_{m\mu,n\mu,p\pi,q\kappa}\rho^B_{\pi,\kappa},$$
(3.15)

and

$$I_{m,n}(t-t_0) = -i\sum_{\mu}[\mathcal{L}e^{-i(1-\mathcal{P})\mathcal{L}\cdot(t-t_0)}(1-\mathcal{P})\rho_{S+B}(t_0)]_{m\mu,n\mu}.$$
(3.16)

One often speaks of (3.9) as (one particular[7] form of) the Time-Convolution Generalized Master Equation (TC-GME). In a completely analogous way, with the Argyres and Kelley projector (3.3), one can rewrite (3.8) as a set of differential equations with time-dependent coefficients

$$\frac{d}{dt}\rho_{m,n}(t) = \sum_{p,q}W_{m,n,p,q}(t-t_0)\rho_{p,q}(t) + J_{m,n}(t-t_0),$$
(3.17)

with *transfer quasi-rates*

$$W_{m,n,p,q}(t-t_0) =$$

$$-i\sum_{\mu,\pi,\kappa}\left[\mathcal{L}\left[1+i\int_0^{t-t_0}e^{-i(1-\mathcal{P})\mathcal{L}\cdot\tau}(1-\mathcal{P})\mathcal{L}\mathcal{P}e^{i\mathcal{L}\cdot\tau}\,d\tau\right]^{-1}\right]_{m\mu,n\mu,p\pi,q\kappa}\rho^B_{\pi,\kappa}$$
(3.18)

and the initial condition term

$$J_{m,n}(t-t_0) = -i\sum_{\mu}\left[\mathcal{L}\left[1+i\int_0^{t-t_0}e^{-i(1-\mathcal{P})\mathcal{L}\cdot\tau}(1-\mathcal{P})\mathcal{L}\mathcal{P}e^{i\mathcal{L}\cdot\tau}\,d\tau\right]^{-1}\right.$$

$$\left.\cdot e^{-i(1-\mathcal{P})\mathcal{L}\cdot(t-t_0)}(1-\mathcal{P})\rho_{S+B}(t_0)\right]_{m\mu,n\mu}.$$
(3.19)

This (matrix) equation is a prototype of a Time-Convolutionless Generalized Master Equation (TCL-GME). Everywhere here, the Greek indices μ, π,, etc. correspond to an arbitrary basis in the Hilbert space of the bath, Latin indices m, n etc. correspond to a general basis in the Hilbert space of the system, and the

---

[7]We have used a very specific choice of $\mathcal{P}$ here.

double-indices $m\mu$, $p\pi\, q\kappa$ etc. correspond to a basis in the Hilbert space of the (system + bath) complex $|m\mu\rangle = |m\rangle|\mu\rangle$.

Now we come to one of the central problems of this section, namely that of taking the thermodynamic limit of the bath. One could take, as a general representative example, the case of a (quantum or classical) gas of molecules interacting with and thermalizing with the walls of the gas container. It is tempting to prematurely identify terms, *e.g.*, the memory (second) term on the right hand side of (3.9) with thermalization terms in the corresponding kinetic equation for the gas (system) in the container (bath). In principle, this is possible, but *only after the thermodynamic limit of the bath* is taken. The point is that thermalization is an irreversible process while our equations (3.9) and (3.17) are so far still reversible. Thus, the thermodynamic limit of the bath is necessary. In order to identify what is the qualitative change expected in (3.9) and (3.17) that cause such a large change of the behavior of the solution $\rho_S(t)$, one should first notice that, because of keeping the system fixed, in the thermodynamic limit of the bath, the only quantities changing during this limit are the memories $w_{m,n,p,q}(\tau)$, the transfer quasi-rates $W_{m,n,p,q}$, and the initial condition terms $I_{mn}(t-t_0)$ and $J_{m,n}(t-t_0)$. Before the thermodynamic limit, the reversible behavior of the solution (reflected in the existence of Poincaré recurrences) is reflected in similar recurrencies in $w_{m,n,p,q}(\tau)$, $W_{m,n,p,q}$, $I_{mn}(t-t_0)$ and $J_{m,n}(t-t_0)$. Formally, it comes from the fact that all these quantities contain in their definitions summations over states of the bath (summations over Greek indices), and the latter summations are discrete, on account of the finite size of the bath before the thermodynamic limit of the bath is taken. That is why, for instance, memories decay with increasing time, but not directly to zero. Small fluctuations around zero are expected to return again to finite amplitudes of w's with the appearance of the above recurrences. In the thermodynamic limit of the bath, however, the recurrence time becomes infinite and correspondingly the summations become integrals. Then, as a consequence of the Riemann-Lebesgue theorem of the theory of the Lebesgue integral[8], the memories $w_{m,n,p,q}(\tau)$ are expected to decay to zero when $\tau \to +\infty$. One then speaks about a finite memory of systems interacting with the bath. In finite-order perturbation theories for the memory functions, such a behavior is actually observed. As for the transfer quasi-rates $W_{m,n,p,q}(\tau)$, one expects that they should turn to constants (as is observed in finite orders of perturbation theories) when $\tau \to +\infty$. On physical grounds, it is tempting to believe that this remains true even in the infinite orders in perturbations but counter-arguments also exist [11]. As for the initial condition terms $I_{m,n}(\tau)$ and $J_{m,n}(\tau)$, they are usually believed to decay sufficiently fast to zero with increasing time argument. Physical arguments exist, however, that this cannot be generally true except for very specific — though physically the most important — choices of ρ^B fitted to, *e.g.*, the initial density matrices of the bath [12]. If these ideas supported by the lowest-order perturbational studies are really universally valid, then, for example, it is the finite extent of the memory functions that distinguishes the irreversible behavior of the infinite bath (thermodynamic limit) from the reversible behavior of the finite bath.

---

[8]Assume that $\int f(x)\, dx$ converges. Then $\int f(x)\cos(xt)\, dx$ and $\int f(x)\sin(xt)\, dx$ also converge and go to zero when $t \to \pm\infty$.

Now, finally, we are able to specify the weak-coupling regime. For the sake of simplicity, we will work with just TC-GME (3.14) and will assume for now that the initial condition term $I_{m,n}(t - t_0)$ in (3.16) is exactly zero. This means to assume the initial condition that: (i) the bath is initially statistically independent of the bath, i.e.,

$$\rho_{S+B}(t_0) = \rho_B(t_0) \otimes \rho_S(t_0), \qquad (3.20)$$

where $\rho_B(t) = \mathrm{Tr}_S \rho_{S+B}(t)$ is the density matrix of the bath; and (ii) that ρ^B in (3.3) is the initial density matrix of the bath (and, thus, in all the formulae containing $\mathcal{P}$), i.e.,

$$\rho^B = \rho_B(t_0). \qquad (3.21)$$

It is easy to verify that if both these conditions are satisfied, then both $I_{m,n}(t - t_0)$ and $J_{m,n}(t - t_0)$ are identically zero. We will also henceforth implicitly assume that the thermodynamic limit of the bath is already performed so that the memories $w_{m,n,p,q}(\tau)$ decay sufficiently fast to zero, becoming thus integrable.

The idea of the weak-coupling regime goes back to van Hove [13] and consists of several steps:

- The Hamiltonian of the (system + bath) complex is written as a sum of those of the system only H_S, the bath only H_B, and the system-bath coupling H_{S-B}, i.e.,

$$H = H_S + H_B + H_{S-B}. \qquad (3.22)$$

- It is assumed that the system-bath coupling is proportional to λ, where λ is a formal coupling parameter, i.e.,

$$H_{S-B} = \lambda \cdot \bar{H}_{S-B}, \qquad (3.23)$$

where $\bar{H}_{S-B}$ is λ-independent.

- A new time unit $\tau_0 \propto \lambda^{-2}$ is introduced, leading to a new (dimensionless) time $t' = (t - t_0)/\tau_0$.

- The formal limit $\lambda \to 0$ in (3.14) is taken utilizing the fact that then

$$\lim_{\lambda \to 0} \lambda^2 \tau_0 = const \neq 0. \qquad (3.24)$$

The idea is that in this limit, all but the lowest (second) order in expansion (in powers of λ of the memory functions $w_{m,n,p,q}(\tau)$) disappears. In order to verify this, let us first prove that expansion of the memory functions really starts from the second order term $\propto \lambda^2$. In order to see that, let us first introduce superoperators

$$\mathcal{L}_S \ldots = \frac{1}{\hbar}[H_S, \ldots], \quad \mathcal{L}_B \ldots = \frac{1}{\hbar}[H_B, \ldots], \quad \mathcal{L}_{S-B} = \frac{1}{\hbar}[H_{S-B}, \ldots] \propto \lambda. \quad (3.25)$$

With that, one can easily verify that[9]

$$\mathrm{Tr}_B(\mathcal{L}_B \ldots) = 0, \quad \mathcal{P}\mathcal{L}_S = \mathcal{L}_S \mathcal{P}, \quad \mathcal{L}_B \mathcal{P} = 0. \qquad (3.26)$$

---

[9]This is still the Argyres-Kelley projector $\mathcal{P}$ in (3.3)

In the last equality we have used that

$$\rho^B = \frac{\exp(-\beta H_B)}{\mathrm{Tr}_B(\exp(-\beta H_B))}, \tag{3.27}$$

$[H_B, \rho^B] = 0$. The form of (3.27) will be used henceforth since it allows us to introduce, in connection with the above initial condition (3.20-3.21), the initial temperature of the bath $T = 1/(k_B\beta)$ (k_B being the Boltzmann constant). Thus, using the idempotency property of $\mathcal{P}$ and from (3.26) and (3.15), we obtain

$$w_{m,n,p,q}(\tau) = -\sum_{\mu,\pi,\kappa} [\mathcal{L}_{S-B} e^{-i(1-\mathcal{P})\mathcal{L}\tau}(1-\mathcal{P})\mathcal{L}_{S-B}]_{m\mu,n\mu,p\pi,q\kappa} \rho^B_{\pi,\kappa}. \tag{3.28}$$

This is proportional to λ^2. Of course, because of the presence of $\mathcal{L}_{S-B}$ in the exponential in (3.28), higher orders in λ are present in $w_{m,n,p,q}(\tau)$, too. We wish to show that in the weak-coupling limit, they simply disappear. Let us introduce the density matrix of the system in the Dirac representations as

$$\tilde{\rho}_S(t) = e^{i\mathcal{L}^{eff}(t-t_0)}\rho_S(t). \tag{3.29}$$

Then, with the initial condition (3.20) and identity (3.21), (3.9) may be rewritten as

$$\frac{d}{dt}\tilde{\rho}_S(t) = \int_0^{t-t_0} e^{i\mathcal{L}^{eff}(t-t_0)}\mathcal{M}(\tau)e^{-i\mathcal{L}^{eff}(t-t_0-\tau)}\tilde{\rho}_S(t-\tau)\,d\tau. \tag{3.30}$$

Introduce now the same density matrix of the system in the Dirac picture, but with time dependence expressed in terms of the new (rescaled) time t' as

$$\tilde{\varrho}_S(t') = \tilde{\rho}_S(t). \tag{3.31}$$

Then (3.30) reads

$$\frac{d}{dt'}\tilde{\varrho}_S(t') = e^{i\mathcal{L}^{eff}t'\tau_0}\int_0^{t'\tau_0}\tau_0\mathcal{M}(\tau)e^{i\mathcal{L}^{eff}\tau}e^{-i\mathcal{L}^{eff}t'\tau_0}\tilde{\varrho}_S(t'-\frac{\tau}{\tau_0})\,d\tau. \tag{3.32}$$

Now, let us take the limit of $\lambda \to 0$ as assumed in the van Hove limit of the weak system-bath coupling. The presumably slowly varying $\tilde{\varrho}_S(t' - \frac{\tau}{\tau_0})$ may be then well replaced by $\tilde{\varrho}_S(t')$ — and notice that $\tau_0 \propto \lambda^{-2}$. For the same reason, $\int_0^{t'\tau_0} \ldots \to \int_0^{+\infty}$ and also $\lim_{\lambda\to 0}\tau_0\mathcal{M}(\tau)e^{i\mathcal{L}^{eff}\tau} = \lim_{\lambda\to 0}\tau_0\mathcal{M}^{(2)}(\tau)e^{i\mathcal{L}_S\tau}$. Here the superscript $\ldots^{(2)}$ means the second order contribution only. Returning then back to variable $\rho_S(t)$, we thus obtain

$$\frac{d}{dt}\rho_S(t) \approx -\mathcal{L}^{eff}\rho_S(t) + \mathcal{R}\rho_S(t),$$

$$\mathcal{R}\ldots = \int_0^{+\infty}\mathcal{M}^{(2)}(\tau)e^{i\mathcal{L}_0\tau}\,d\tau\ldots \tag{3.33}$$

Here $\mathcal{R}$ is so-called Redfield relaxation superoperator (its matrix elements being sometimes called the Redfield relaxation tensor). Thus, we find not only that just the second-order term in the expansion of the memory superoperator contributes,[10] but we also find that the convolution GME goes to a convolutionless (time-local) form, with the variable $\rho_S(t)$ brought out from below the integral bearing the same time as that one on the left hand side. (This is so-called Markov approximation.) Significantly, both approximations simplify the problem considerably; (3.33) in the matrix form is then nothing but the set of so-called Redfield equations (without Redfield secular approximation) [14, 15, 16].

One should understand the physical meaning of the combined weak-coupling van Hove limit. In fact, one can easily understand why, for weak couplings, higher (than second) order effects described by the memory superoperator $\mathcal{M}$ become negligible as compared with the lowest (second) order ones. However, the latter lowest order terms also become slower and slower. The reason why they do not finally disappear is that we simultaneously increase the time scale (keeping the relative time t' constant). Hence, we provide these lowest order processes ($\propto \lambda^2$) with longer and longer time to happen, unlike the higher order processes. The rate of increasing the time scale ($\tau_0 \propto \lambda^{-2}$) is simply insufficient to preserve the latter processes $\propto \lambda^4$. The question arises as to whether it is formally justified to keep the second order processes (in the second term on the right hand side of (3.33)) simultaneously with the zeroth order ones (in the first term on the right hand side of (3.33)). The answer is yes and the arguments are as above or can be found, in a more precise mathematical form, in a series of papers by Davies [17, 18, 19].

Equations (3.33) have been applied to a number of physical phenomena, including nuclear magnetic resonance and excitation transfer in solids. These equations have been very successful from many points of view, however, several deficiencies have been recognized that are of critical importance to our purposes. Let us consider them.

- The Redfield equations, irrespective of their rigorous derivation, do not in general preserve positive semidefiniteness of the resulting density matrix of the system $\rho_S(t)$. The reason consists in the mathematical limit $\lambda \to 0$ involved in the derivation, but necessary application for finite (though perhaps small) values of λ, *i.e.*, the system-bath coupling as it is expected in Nature. The situation was, from the point of view of the positive semidefiniteness of $\rho_S(t)$, analyzed in detail by Dümcke and Spohn [20]; it was found that another (related) form of the kernel in (3.33) is necessary. According to Davies [18], there are more (in fact, infinitely many) such forms that are all equally justified in the limit of $\lambda \to 0$. For finite system-bath couplings (as required in physical reality), however, they differ appreciably and the form proposed by Spohn and Dümcke does not seem to be physically the best one [21].

- Above, we write '... The presumably slowly varying $\tilde{\varrho}_S(...)...$'. By that, we mean that the density matrix of the system is a much smoother function of time in the Dirac representation than in the Schrödinger one. This assumption underlies the reason why the above derivation of the corresponding master equation for ρ_S has been made, as usual, in the interaction pic-

---

[10]This second-order approximation is usually called the Born approximation

ture. Only after performing the $\lambda \to 0$ limit, have we returned to the standard Schrödinger picture. Unfortunately, the assumption about appreciably smoother time dependence of the system density matrix in the Dirac picture (as compared with the Schrödinger one) is not true in general, in particular, once we approach a stationary state. Actually, from the definition of the stationary state, mean values of all physical operators (that are in the Schrödinger picture independent of time) should be time-independent. Hence, the density matrix $\rho_S(t)$ should also becomes time-independent, in contrast to that in the Dirac picture $\tilde{\rho}(t)$. Consequently, a much better approximation to (3.9) in nearly stationary states than (3.33) should be, for low λ's,

$$\frac{d}{dt}\rho_S(t) \approx -\mathcal{L}^{eff}\rho_S(t) + \int_0^{+\infty} \mathcal{M}^{(2)}(\tau)\, d\tau \rho_S(t). \qquad (3.34)$$

In the initial stage of the relaxation, on the other hand, the opposite may be true. In any case, this type of uncertainty shows physical problems hidden behind the otherwise perfect mathematical form of the van Hove weak-coupling limit, as applied to the time-convolution formalism [17, 18].

Problems connected with transition, upon increasing time, between (3.33) and (3.34) seem to be perhaps only artificial and connected with the Born approximation within the time-convolution formalism. In order to support this, we mention that TC-GME formalism based on (3.6) and the so-called Mori formalism (§3.8.1) are fully equivalent [1]. So, one can directly use results known for the Mori theory leading to the same time-convolution form of equations for $\rho_S(t)$, (3.9) or (3.14). In [22], it is shown that: (i) the Born (second order) approximation of the kernel in (3.9) alone (*i.e.*, without performing the Markov approximation) does not yield relaxation to any stationary situation, though the standard Born-Markov approximation in the interaction picture (leading to (3.33)) indicates the opposite; and (ii) the solution then shows an infinitely long memory of details of initial conditions.

These features are clearly unacceptable; the problems thus arising also have unpleasant implications in other areas[11]. Completely different is the situation in time convolutionless theories of the type of (3.8) and (3.17). First, a proof exists that the set (3.17) as provided by identity (3.8) is exactly the same as provided by the Tokuyama-Mori theory (§3.8.2) [24]. That is why we can directly use results obtained in this approach. In [25, 26], it was shown that in such time-convolutionless formalisms, the Born approximation yields full relaxation to a stationary state, and it properly describes forgetting details of initial conditions during the course of the relaxation.

The question of a proper type of the Markov approximation does not even appear, implying that a preference should be given to such time-convolutionless formalisms. Thus, we shall briefly mention the above weak-coupling time-convolution counterpart for (3.33).

In the same way as above, we will work with the initial condition (3.20-3.21), with the Argyres-Kelley projector (3.3), and with the identity (3.27). Applying the

---

[11]See, for example, the asymptotic time symmetry breaking problem in connection with results discussed in [23].

the usual van Hove weak-coupling procedure to (3.17), we fully reproduce (3.33) as a legitimate equation for the short- and intermediate-time regime (as measured by the bath-assisted relaxation times). In order to obtain a counterpart of (3.34) as a master equation for long times, we first perform the infinite time limit in the coefficients in (3.17) and only then expand as usual to the second order in H_{S-B}. We do *not* obtain (3.34) but, instead, again (3.33). Finally, keeping constant time t in the upper limit of the integrals in coefficients in (3.17), and formally expanding the coefficients to the second order in H_{S-B}, we obtain[12]

$$\frac{d}{dt}\rho_S(t) \approx -\mathcal{L}^{eff}\rho_S(t) + \mathcal{R}(t)\rho_S(t),$$

$$\mathcal{R}(t)\ldots = \int_0^t \mathcal{M}^{(2)}(\tau)e^{i\mathcal{L}_0\tau}\,d\tau\ldots \qquad (3.35)$$

Here $\mathcal{R}(t)$ is a time-dependent generalization of the Redfield superoperator $\mathcal{R}$ defined in (3.33) that turns, with increasing time, again to $\mathcal{R}$. The following facts imply that the time-convolutionless form of the Generalized Master Equations (as resulting from the SHTS identity (3.8)) should be preferred to the time-convolution form of the Generalized Master Equations resulting from the Nakajima and Zwanzig identity (3.6):

- $\mathcal{R}(t)$ goes to zero when $t \to 0$; *i.e.*, (3.35) properly describes so called slippage (initial delay of the relaxation);

- (3.35) continuously goes to (3.33) with increasing time;

- there is no necessity to perform (together with the Born approximation) the Markov approximation as above; and

- the Born (second order) approximation in H_{S-B} creates no formal problems.

The fully rigorous mathematical foundation of the van Hove limit, as given by Davies [19] and which can be implemented in both the time-convolution and the time-convolutionless form of the Generalized Master Equations, had an unintented, undesired effect: Many began to believe that certain features, for instance, the relaxation or stationary state as described by the weak coupling theory, are valid in general, in all possible relaxation regimes. One such belief is that the relaxation of open systems in stationary conditions must inevitably go to a canonical Gibbs state. This is not generally true and such a postulate would distort the physics of the systems investigated here. What is the justification for such a strong statement? One should realize that assumption of the weak system-bath coupling means assuming that *all* bath induced processes in the system are (appreciably) slower than competing processes in the system itself. As an illustration, take for our system the gas particles in the Maxwell double compartment, divided by the wall equipped with a gate (Figure 1.1), and let the walls, connected with the rest of the universe, be the bath. Formally, the gate also belongs to the system. At

---

[12]without assuming whether $\rho_S(t)$ has already relaxed to stationary values or not.

this stage, let us not decide who (or what) operates the gate. The act of closing (opening) the gate should be done on grounds of the interaction with the bath as far as the process is a spontaneous process, going downhill (in energy); otherwise, the act of closing and opening the gate would not be spontaneous. If such a bath-assisted process were slower (or even infinitely slower, as assumed in the van Hove weak-coupling regime) than the dynamics of the gas molecules, the separation of the molecules to slower and faster ones (as in the Maxwell demon system) would not be possible. The gate could not open and close in time to accommodate the instantaneous situation of the gas. So, in order to be able to model such 'demonic' systems, we *must* inevitably go beyond limitations imposed by weak-coupling approaches. This is what we now pursue.

3.3 Beyond Weak Coupling: Quantum Correlations

Going beyond the weak-coupling regime appears advantageous for second law challenges. Two questions immediately arise.

- What specific regimes could potentially allow second law violation?

- Are there mathematical justifications for going beyond the limitations of weak coupling?

The answer to the first question is not as simple as it might seem. It is perhaps physically meaningless to consider regimes other than weak coupling in the literal sense. The strong coupling limit in the sense of our coupling parameter λ going to infinity would be oversimplified for our purposes, owing to lack of competition between different processes forming the foundation of our approach. Moreover, from the technical point of view, any expansion would have to be in powers of, for example, a small parameter λ^{-1} around a $\lambda \to +\infty$ solution. That would be complicated both technically and physically. Instead, we need competition between parallel physical processes. Hence, we require, for instance, that the rates of the bath-assisted processes be comparable (in the intermediate regime) or stronger (in the strong coupling case) than those of internal processes (inside) the system. However, the ratios among the processes should stay finite for physical reasonableness. Taking *this* as a definition of the regimes of commensurable or strong coupling, we can then proceed to the second question.

What is important is the relative strength of at least two competing processes, one of them being associated with the coupling to the bath. This, however, says nothing concerning the value of, for instance, a joint small parameter of such competing couplings. Hence, it appears reasonable to use the formalism of the weak-coupling limit, *i.e.*, corresponding to the regime of weak coupling for both/all the competing mechanisms causing the relaxation (kinetic, transfer) processes.[13]

---

[13]This does *not* exclude the possibility, *e.g.*, that the coupling to the bath causes bath-assisted relaxation processes to be, perhaps, several orders of magnitude faster than rates of internal processes in the system. The relative ratios of such rates must, however, always remain finite.

If we speak of the formalism of weak coupling in this sense, keep in mind that the word 'weak' always means in this connection commensurable with $\hbar/\tau_0$ where $\tau_0 \propto \lambda^{-2}$ is the time unit going to infinity, in the $\lambda \to 0$ limit.

With this hint for proper formal description for an intermediate/strong regime kinetics by means of the weak-coupling formalism, we can now speculate as to what might be the most important features distingushing this regime from the usual weak-coupling one. There should be consequences, but we have particularly those in mind that influence the particle (quasiparticle, excitation, etc.) transfer. Refraining from a full list, we distinguish several cases:

- The role of the 'Maxwell gate' could be played, for instance, by topological changes in the molecular systems on which the transfer takes place. Such changes would appear at the moment the particle (or molecular group) arrives at and is detected at a specified place — known in molecular biology as a *receptor*. This is in analogy to what is presently known about mechanisms of activity of molecular systems at the entrance to membrane pumps. In such situations, the desired activity requires:

    - a specific type of correlation between the particle position and the topological form of the molecular system: that the 'entrance bridge' opens whenever the particle arrives at the entrance, but closes once the particle crosses it; and
    - that the dynamics of opening and closing of the bridge is comparable in speed to that of the particle processed.

For reasons explained above, such correlations seem possible. Detailed theory will be developed below.

- The gate system can be formally developed via standard phonons from molecular or solid state lattice dynamics. Imagine, for example, that the processed particle arrives at a receptor that responds to its presence via a molecular bridge which was triggered by the particle far away. Such molecular processes are standard in molecular biology, explaining activities of many biomolecules [27]. Such static shifts can be expressed as a superposition of phonons with perfectly matching mutual phases. Such phonons, on the other hand, also form the usual polaron clouds around particles. Hence, polaron formation may be the mechanism interrupting the return paths for quantum particles, forcing them to go one direction only, even uphill in energy. Energy conservation is preserved, owing to a continuous virtual phonon absorption and emission due to the coupling. Hence, we do not require an external demon to decide whether the 'gate' will be closed or open. The system (the particle) makes the decision spontaneously, on grounds of a spontaneous fall into a polaron well produced by the particle itself. In order to properly describe the polaron states, sufficiently high orders in the particle-phonon coupling are required. Moreover, for the polaron effects to be sufficiently pronounced, one must go beyond weak coupling theory, in which case the correlations that appear and are decisive for the gate effect are: (i) between the particle and position of the polaron deformation cloud; and (ii) among phonons forming the polaron cloud.

In principle, one needs the Maxwell gate simply to induce particles to move preferentially in one direction. This one-directional motion may also be due to other mechanisms with uni-directional properties. We have in mind, for instance, spontaneous processes or on-energy-shell diffusion between tails of two different exciton levels broadened by the exciton-bath coupling. In the former case, we must go beyond the limitations of weak coupling theory since in the lowest (zeroth) order in the system-bath coupling, one obtains the canonical distribution for the particle (system), which leads to zero current or heat flow. For the latter case, the coupling must be sufficiently strong to obtain a pronounced level broadening. The exciton (and energy) flow are then the result of different exciton site populations arising from different site energies. In such situations, there are no pronounced correlations needed for the unidirectional-transfer effect. These cases also go beyond the weak coupling theory. Other situations abetting the inclusion of higher-order effects in the system-bath coupling will be discussed below.

3.4 Allahverdyan-Nieuwenhuizen Theorem

The approach by Allahverdyan and Nieuwenhuizen (A-N) is somewhat different in philosophy. From a general point of view, it provides one of the most decisive treatments because it allows precise and detailed consideration of all basic quantities from phenomenological thermodynamics. On the other had, it is quite special and idealized such that it might not correspond to realistic thermodynamic systems [28, 29].

The authors consider the case of a Brownian particle in a harmonic confining potential, coupled to a thermal bath consisting of harmonic oscillators. The exceptional value of the A-N model is that it is exactly solvable so that it may serve also as a pedagogical example on which basic postulates and statements of thermodynamics can be discussed and scrutinized. Their theoretical quantum mechanical treatment is not limited to the case of the weak system-bath coupling (Brownian particle-harmonic oscillators) for which standard arguments are well supported by statistical physics. Rather, A-N's methods are able to address strong coupling correlation effects (system-bath entanglements) and lower temperature regimes in which deeply quantum effects are pronounced. Note that decreasing temperature to zero does *not* bring the particle (system) to a pure state because it is coupled to the bath (*e.g.*, strongly, weakly).

In this model, only equidistant bath harmonic oscillators and the Drude-Ullersma of the coupling constants are involved[14]. These specifications allow analytic treatment, including the thermodynamic bath limit. Standard formulae are used for connecting the partition function with the Gibbs free energy of the sta-

---

[14] Questions have been raised about the thermodynamic legitimacy of the Ullersma model [30].

tionary (equilibrium) state of the (particle + bath) complex. The statistical (von Neumann) entropy is appropriate.

The analysis of the stationary state finds that effective coordinate and momentum temperatures (in energy units)[15] become different, both of them differing from the bath temperature. Moreover, these differences survive even in the high temperature limit of the bath. This indicates deviation from the usual equipartition theorem. The difference between the total entropy of the (particle + bath) complex and that of the bath without the particle scales as $S_p \propto \gamma T$ (where γ is a dimensionless system-bath coupling constant, and T is the bath temperature). Since the statistical entropy remains finite at $T \to 0$ owing the the system-bath entanglement, the entropy of the (system + bath) complex does *not* additively consist of the bath and particle entropies separately. Entanglement adds a fundamentally new dimension to the thermodynamics. This has been explored by others as a possible energy source [31, 32, 33].

Next, adiabatic changes of parameters is carried out, in particular, adiabatic changes in the spring constant a and the (effective) mass m. For the situation investigated, with two different temperatures T_x and T_p, the Clausius inequality should read:

$$\delta Q \le T_p\, dS_p + T_x\, dS_x, \tag{3.36}$$

where S_p and S_x are the corresponding 'momentum' and 'coordinate' entropies summing up to the Boltzmann entropy $S_B = S_p + S_x$. For the adiabatic (*i.e.*, reversible) changes of a and m, (3.36) should become an equality, however, it does not. The Clausius inequality (3.36) is violated. This fact survives for even high temperatures ($\gamma \ne 0$). It is also remarkable that, even with proper definitions of all the quantities involved, upon comparing two equilibrium systems at two slightly different temperatures T and $T + dT$,

$$\delta Q - T\, dS_{stat} = \frac{\beta^2 \hbar^2 \gamma \Gamma}{24m}\, dT. \tag{3.37}$$

S_{stat} is the von Neumann entropy, β is the reciprocal temperature of the bath in energy units, and Γ is a parameter related to the above assumed Drude-Ullersma spectrum of the harmonic oscillator coupling of the particle to the bath. The right hand side of (3.37) should disappear *provided* the identification of the statistical (von Neumann) entropy with the thermodynamic entropy is justified. The right hand side of (3.37) disappears, however, only in the weak-coupling limit $\gamma \to 0$, in the classical limit ($\hbar \to 0$, Planck correspondence principle) or in the high temperature limit ($T \to +\infty$). (This supports our previous concerns about identification of the statistical entropy with the thermodynamic one.)

The A-N model is also interesting that it allows full analytical treatment of dynamic properties. Among several conclusions obtained, the most salient is that positive entropy production rate is not guaranteed; rather, below a specific temperature determined by parameters of the model, the entropy production rate can

---

[15]$T_x = \langle ax^2 \rangle$ and $T_p = \langle \frac{p^2}{m} \rangle$. Here $V(x) = \frac{1}{2}ax^2$ is the particle potential energy in the confining potential.

be negative. The authors also analyse possible cycles of changes of external parameters of the system. Leaving aside many other results, we conclude by citing
from [28]. For at least the model investigated,

> ...there are principle problems to define thermodynamics at not very
> large temperatures and, in particular, in the regime of quantum entan-
> glement. There is no resolution to this, and thermodynamically unex-
> pected energy flows appear to be possible... Out of equilibrium cycles
> have been designed where a constant amount of heat ... extracted from
> the bath is fully converted into work, while the energy of the subsys-
> tem is at the end of each cycle back at the value of the beginning. In
> this sense, systems described by our models, with parameters in the
> appropriate regime, present at low temperatures true realizations of
> *perpetuum mobile* of the second kind... Our results can be phrased in
> the statement that Maxwell's demon exists.

In conclusion, Allahverdyan and Nieuwenhuizen have provided a strong case
that quantum entanglement can lead to violations of certain formulations of the
second law. Their system is microscopic, but there seems to be no obvious impediment to scaling it the macroscopic regime. Their conclusions support previous
results from other groups. It remains to be seen whether experiments can be devised to test the salient features of this idealized model.

3.5 Scaling and Beyond

Most quantum mechanical second law challenges preceded the harmonic oscillator model by Allahverdyan and Nieuwenhuizen [28, 29], however, they were
more complicated, making exact or rigorous solutions impossible. Because of the
inherent kinetic character of the theories, it was necessary to go beyond the weak-
coupling regime. The proper kinetic theory developed gradually; its final form
can be found in, e.g., [34]. If we wish to translate quantum entanglement into
challenges, we should first identify candidate effects. In most models considered,
the key is a mutual competition between different — possibly interfering — quantum reaction channels. One type relies on bath-induced processes (relaxation)
within the system; thus, *competing* processes might be provided, perhaps, by kinetic processes in the system itself. (This possibility does not exist in the system
investigated by Allahverdyan and Nieuwenhuizen [28, 29] since the internal system consists of just a single particle having no internal degrees of freedom.) To
describe such a competition properly, one must identify the proper terms in the
total Hamiltonian H and treat all such 'kinetic' contributions to H simultaneously
and, as far as possible, on the same footing. Internal kinetics in the system has,
however, nothing to do with the system's coupling to the bath. Consequently,
such 'kinetic' terms must inevitably include not only the system-bath coupling,
but also some parts from the Hamiltonian H_S of the system itself.

Second, in order to preserve the entanglement (competition), one is not necessarily forced to treat such competing 'kinetic' terms (perturbations) exactly. To
the contrary, technical problems force us to use approximations of the type of the
lowest-order (Born-type). Of course, we must be careful not to lose the effects because of these approximations. One rigorous approach is connected with the above
scaling, but this time the scaling cannot be that of the van Hove weak coupling
type. Fortunately, almost all formulae and general theorems provided by Davies
remain valid [17, 18, 19], so this is the form the theory will take.

As usual, first we write the total Hamiltonian of the open system of interest as
$H = H_S + H_B + H_{S-B}$ and rewrite it as

$$H = H_0 + \lambda H_1, \tag{3.38}$$

where now the perturbation λH_1 includes, in addition to (the decisive part) H_{S-B},
also the internal kinetic terms from H_S describing the internal processes in the
system competing with those caused by H_{S-B}. These terms, however, may be
formally of even higher order in λ. This helps to compare, for instance, coherent
transfer rates proportional to $J/\hbar$ (J being resonance transfer, or so-called hopping integral) with bath assisted transfer rates $\propto H_{S-B}^2 \propto \lambda^2$ as dictated by, for
instance, the Golden Rule of quantum mechanics. Thus, (3.38) allows H_1 to be
λ-dependent. Designating the Liouvillians as

$$\mathcal{L}_0 \ldots = \frac{1}{\hbar}[H_0, \ldots], \ \lambda\mathcal{L}_1 \ldots = \frac{1}{\hbar}[\lambda H_1, \ldots], \tag{3.39}$$

one can verify the identity[16]

$$e^{-i(\mathcal{L}_0+\lambda\mathcal{L}_1)t} = e^{-i(\mathcal{L}_0+\lambda\mathcal{P}\mathcal{L}_1\mathcal{P}+\lambda(1-\mathcal{P})\mathcal{L}_1(1-\mathcal{P}))t}$$

$$+\lambda \int_0^t ds e^{-i(\mathcal{L}_0+\lambda\mathcal{P}\mathcal{L}_1\mathcal{P}+\lambda(1-\mathcal{P})\mathcal{L}_1(1-\mathcal{P}))(t-s)}$$

$$\cdot[\mathcal{P}(-i\mathcal{L}_1)(1-\mathcal{P})+(1-\mathcal{P})(-i\mathcal{L}_1)\mathcal{P}]e^{-i(\mathcal{L}_0+\lambda\mathcal{L}_1)s}. \tag{3.40}$$

Here $\mathcal{P} = \mathcal{P}^2$ is an arbitrary projection superoperator which we shall later understand to be that of Argyres and Kelley [2]. This identity is significant. It derives
from the fact that the exponential on the left hand side generates time-development
of the density matrix of the (system+bath) complex, as described by the Liouville equation. By iterating this identity, changing the order of integrations, and
multiplying it from both the left and the right by $\mathcal{P}$, we obtain

$$\mathcal{P}e^{-i(\mathcal{L}_0+\lambda\mathcal{L}_1)t}\mathcal{P} = \mathcal{P}e^{-i(\mathcal{L}_0+\lambda\mathcal{P}\mathcal{L}_1\mathcal{P})t}$$

$$+\lambda^2 \int_0^t du \int_0^{t-u} dx \mathcal{P}e^{-i(\mathcal{L}_0+\lambda\mathcal{P}\mathcal{L}_1\mathcal{P})(t-u-x)}\mathcal{P}(-i\mathcal{L}_1)(1-\mathcal{P})$$

$$\cdot e^{-i(\mathcal{L}_0+\lambda\mathcal{P}\mathcal{L}_1\mathcal{P}+\lambda(1-\mathcal{P})\mathcal{L}_1(1-\mathcal{P}))x}(1-\mathcal{P})(-i\mathcal{L}_1)\mathcal{P}e^{-i(\mathcal{L}_0+\lambda\mathcal{L}_1)u}\mathcal{P}. \tag{3.41}$$

---

[16]Notice that $\lambda\mathcal{L}_1$ may also contain higher powers of λ.

Here, we have assumed that

$$\mathcal{P}\mathcal{L}_0 = \mathcal{L}_0\mathcal{P}. \tag{3.42}$$

Now we rescale the time arguments by setting, á la van Hove,

$$t = \tau/\lambda^2, \ u = \sigma/\lambda^2. \tag{3.43}$$

In physical terms, $1/\lambda^2$ can be understood as a new time unit that renders τ dimensionless. This yields

$$\mathcal{P}e^{-i(\mathcal{L}_0+\lambda\mathcal{L}_1)\lambda^{-2}\tau}\mathcal{P} = \mathcal{P}e^{-i(\mathcal{L}_0+\lambda\mathcal{P}\mathcal{L}_1\mathcal{P})\lambda^{-2}\tau}$$

$$+ \int_0^{\lambda^{-2}\tau} d\sigma \, e^{-i(\mathcal{L}_0+\lambda\mathcal{P}\mathcal{L}_1\mathcal{P})(\lambda^{-2}\tau-\lambda^{-2}\sigma)}\mathcal{H}(\lambda,\tau-\sigma)\mathcal{P}e^{-i(\mathcal{L}_0+\lambda\mathcal{L}_1)\lambda^{-2}\sigma}\mathcal{P}. \tag{3.44}$$

Here

$$\mathcal{H}(\lambda,\tau) = \int_0^{\lambda^{-2}\tau} dx \, \mathcal{P}e^{i(\mathcal{L}_0+\lambda\mathcal{P}\mathcal{L}_1\mathcal{P})x}(-i\mathcal{L}_1)(1-\mathcal{P})$$

$$\cdot e^{-i(\mathcal{L}_0+\lambda\mathcal{P}\mathcal{L}_1\mathcal{P}+\lambda(1-\mathcal{P})\mathcal{L}_1(1-\mathcal{P}))x}(1-\mathcal{P})(-i\mathcal{L}_1)\mathcal{P}. \tag{3.45}$$

Now, we specify $\mathcal{P}$ as the Argyres-Kelley projector (3.3), cancel the initial density matrix of the bath ρ^B on both sides of (3.44), and take the thermodynamic limit of the bath. Following Davies [18], it is permissible that the (super)operator $\mathcal{H}(\lambda,\tau)$ may, in the limit of small λ, be replaced in (3.44) by its formal limit when $\lambda \to 0$, i.e.,

$$\mathcal{H} = \rho^B \int_0^{+\infty} dx \, Tr_B \left(e^{i\mathcal{L}_0 x}(-i\mathcal{L}_1)e^{-i\mathcal{L}_0 x}(1-\mathcal{P})(-i\mathcal{L}_1)\mathcal{P}\right). \tag{3.46}$$

By the word 'formal', we mean that before the integration, the integrand must be already written in the infinite bath (thermodynamic) limit. This makes the integral convergent. With this replacement, (3.44) then reads

$$\mathcal{P}e^{-i(\mathcal{L}_0+\lambda\mathcal{L}_1)\lambda^{-2}\tau}\mathcal{P} = \mathcal{P}e^{-i(\mathcal{L}_0+\lambda\mathcal{P}\mathcal{L}_1\mathcal{P})\lambda^{-2}\tau}$$

$$+ \int_0^{\lambda^{-2}\tau} d\sigma \, e^{-i(\mathcal{L}_0+\lambda\mathcal{P}\mathcal{L}_1\mathcal{P})(\lambda^{-2}\tau-\lambda^{-2}\sigma)}\mathcal{H}\mathcal{P}e^{-i(\mathcal{L}_0+\lambda\mathcal{L}_1)\lambda^{-2}\sigma}\mathcal{P}. \tag{3.47}$$

Its formal solution is

$$\mathcal{P}e^{-i(\mathcal{L}_0+\lambda\mathcal{L}_1)\lambda^{-2}\tau}\mathcal{P} = \mathcal{P}e^{-i(\mathcal{L}_0+\lambda\mathcal{P}\mathcal{L}_1\mathcal{P}+i\lambda^2\mathcal{H})\lambda^{-2}\tau}. \tag{3.48}$$

This means that for initially separable density matrices $\rho_{S+B}(t)$ of the system and the bath:

$$\rho_{S+B}(0) = \rho_S(0) \otimes \rho^B, \tag{3.49}$$

we have the rigorous statement[17]

$$\lim_{\lambda\to 0} \sup_{0\leq\lambda^2 t\leq a} ||\rho_S(t) - e^{-i(\mathcal{L}_0+\langle\mathcal{L}_1\rangle+i\lambda^2\mathcal{K})t}\rho_S(0)|| = 0,$$

---

[17] Here the finite constant a is arbitrary.

$$\mathcal{K}... = \lim_{\lambda \to 0} \int_0^{+\infty} dx \, Tr_B \, (e^{i\mathcal{L}_0 x}(-i\mathcal{L}_1)e^{-i\mathcal{L}_0 x}(1 - \mathcal{P})(-i\mathcal{L}_1)(... \otimes \rho^B),$$

$$\langle \mathcal{L}_1 \rangle ... = Tr_B(\mathcal{L}_1 ... \otimes \rho^B). \tag{3.50}$$

Specifically, we may write, as already indicated above,

$$H_0 = H_S|_{\lambda=0} + H_B, \quad \lambda H_1 = H_{S-B} + H_S|_{\lambda \neq 0} - H_S|_{\lambda=0}, \tag{3.51}$$

with $H_{S-B} \propto \lambda$ and $H_S|_{\lambda \neq 0} - H_S|_{\lambda=0}$ representing the above inherent 'kinetic'[18] terms in H_S that describes bath-nonassisted processes in the system competing with the bath-induced processes, conditioned by the system-bath coupling H_{S-B}. The term $\mathcal{L}_0 + \langle \mathcal{L}_1 \rangle$ in the exponential in the first line of (3.50) give basically the same result as the standard van Hove weak-coupling, with

$$H_0 = H_S + H_B, \quad \lambda H_1 = H_{S-B}. \tag{3.52}$$

Thus, for example, $Tr_B H_{S-B} \rho^B = 0$ (when the effect of H_{S-B} in $\mathcal{L}_0$ disappears), $\mathcal{L}_0 + \langle \mathcal{L}_1 \rangle$ describes a free (non-relaxational) time-development of $\rho_S(t)$ in *extended* eigenstates of H_S. On the other hand, in contrast to the weak-coupling theory with identity (3.52), the superoperator $\mathcal{K}$ in (3.50) does not reduce to the Redfield superoperator $\mathcal{R}$ in (3.33), as it does when (3.52) is accepted. Thus, it no longer describes the bath-assisted relaxation among the extended eigenstates of H_S. To the contrary, in view of the fact that the effect of $H_S|_{\lambda \neq 0} - H_S|_{\lambda=0}$ in $\mathcal{K}$ in (3.50) disappears whenever $\mathcal{P}[H_S|_{\lambda \neq 0} - H_S|_{\lambda=0}, ...] = [H_S|_{\lambda \neq 0} - H_S|_{\lambda=0}, \mathcal{P} ...]$ and also because of the fact that $\mathcal{L}_0$ then describes free relaxation in terms of only localized eigenstates of $H_S|_{\lambda=0}$, $\mathcal{K}$ describes the bath-assisted relaxation among *localized* eigenstates of $H_S|_{\lambda=0}$ only. This remarkable difference has far-reaching consequences that could be ascribed to the above competition between the bath-assisted relaxation processes and inherent kinetic processes in the system.

From (3.50), we conclude that the true dynamics of the system is undistinguishable from the one determined as $\rho_S(t) = e^{-i(\mathcal{L}_0 + \langle \mathcal{L}_1 \rangle + i\lambda^2 \mathcal{K})t} \rho_S(0)$. Differentiating this with respect to t, we obtain

$$\frac{d}{dt}\rho_S(t) = -i(\mathcal{L}_0 + \langle \mathcal{L}_1 \rangle)\rho_S(t) + \mathcal{K}\rho_S(t). \tag{3.53}$$

Here the relaxation superoperator $\mathcal{K}$ describes relaxation in the localized state; *i.e.*, it differs from the Redfield form. Hence, we have obtained, owing to the different regime investigated, a closed equation for $\rho_S(t)$ that is formally and physically different from the Redfield one (3.33). Below, we will use it in several applications.

Two remarks should be made at this juncture. First, Schreiber, et al. [35] recently started investigations, from essentially (3.33) in a weak-coupling regime and finite coupling constant. They also investigated an approximation consisting of omitting, by hand, particle overlap (transfer or hopping) integrals in $\mathcal{R}$ in (3.33). They found violations, in the long-time asymptotics, of the stationary distribution dictated by statistical thermodynamics. This was properly interpreted as an indication of simply the approximate character of the omission. On the other hand,

---

[18]for instance hopping terms of the type $J[c_1^\dagger c_2 + c_2^\dagger c_1]$

the resulting equation is nothing but (3.53), that is, in the regime of competing in-phasing (owing to the overlap integrals) and dephasing (owing to the bath induced relaxation). Thus, although the authors consistently stuck to investigation of the weak-coupling regime only, they obtained violations of statistical thermodynamics that could not be disregarded as due to unjustified approximations. These violations should be taken quite seriously and as supporting our position concerning limits of thermodynamics in the quantum world.

The second comment is connected with the standard practice for kinetic equations of extending the validity of equations like (3.33) or (3.53) to arbitrarily long times. The Davies statement (3.50) means that (3.53) or (3.33) can be validly applied, for any finite strength of the coupling, only for finite time intervals limited from above ($t < \infty$). Beyond a critical time, nothing can be said using this type of argument. Whether this (rather formal) warning should be taken seriously, for the purposes of stationary states and from the point of view of physics,[19] remains an open question. Regardless, conclusions thus obtained can be supported by other arguments, in particular those beyond the above derivation of (3.33) or (3.35) from (3.8).

3.6 Quantum Kinetic and Non-Kinetic Models

We now unpack a series of eight quantum models that challenge statistical thermodynamics, particularly the second law. Some of the models predict violation in the sense of being able to determine, for instance, the heat conversion rate into work and they are able to work cyclically, thereby explicitly violating the Kelvin formulation of the law. The basic ideas will be presented as simple-cycle or simple-step models. Once it is clear that the outcome is positive — acquired work without compensation — and that the system returns to its original state, the process can cycle. For example, in photosynthesis, after photon absorption and electron transfer, the original state of the photosynthetic unit is recovered and there is no question about the process being repeatable. Generalization of our mechanisms to many particles should be straightforward, absent multiparticle effects.

Models in the next subsections could be called 'demon-like models' since their action is reminiscent of the Maxwell demon: gates opening and closing reaction channels to achieve particle transfer. There are, however, other ways of characterizing the operation of such models. For example, operation of the fish-trap model (§3.6.1) can alternatively be characterized as being due to relaxation of the complex scattering center plus the scattered particle in its intermediate state, before the scattered particle is released again.

---

[19]and, what is even more important, for *arbitrary* kinetic equations.

3.6.1 Fish-Trap Model

The fish-trap is the first truly quantum model leading to behavior challenging the second law. Its first version was published in [36]; for more detailed theory see [37]; a simplified version is discussed below [38]. To start, however, we will proceed with its original form. Although it is reminiscent of the Maxwell demon, the inspiration for its Hamiltonian came in 1996 from the molecular biology of membrane pumps[20]. One can profitably compare this to Gordon's classically-developed biochemical models (§5.2).

The system consists of a single particle on three sites and a two-level system with a specific type of instability dependent on the particle position. The Hamiltonian can be written in the usual form

$$H = H_S + H_B + H_{S-B}, \tag{3.54}$$

with

$$H_S = J(c_{-1}^\dagger c_0 + c_0^\dagger c_{-1}) \otimes |d\rangle\langle d| + I(c_1^\dagger c_0 + c_0^\dagger c_1) \otimes |u\rangle\langle u|$$

$$+ \frac{\epsilon}{2}[|u\rangle\langle u| - |d\rangle\langle d|] \otimes [1 - 2c_0^\dagger c_0]$$

$$H_B = \sum_k \hbar\omega_k b_k^\dagger b_k,$$

$$H_{S-B} = \frac{1}{\sqrt{N}} \sum_k \hbar\omega_k (b_k + b_{-k}^\dagger) \otimes \{G_k[|u\rangle\langle d| + |d\rangle\langle u|] + g_k c_0^\dagger c_0\}. \tag{3.55}$$

The sites (particle positions) are designated as $m = -1$, 0 and $+1$; sites '-1' and '+1' represent the left and right (later to be shown to be the input and output) media for the system, respectively. Operators $c_m^\dagger$ and c_m, $m = -1, 0$ or $+1$ designate the creation and annihilation operators at site m. For the single-particle problem, we do not need their commutational (anticommutational) relations or contingent particle spin; thus, spin is fully suppressed. More extended external media can be attached on both sides. For example, this could be achieved by adding to (3.55) a term $\sum_{m,n=-\infty}^{-1} J_{mn} c_m^\dagger c_n + \sum_{m,n=1}^{+\infty} I_{mn} c_m^\dagger c_n$ and assuming an arbitrary number of particles involved. This would, however, only complicate the treatment technically, yielding essentially no change in the physics of the process. In (3.55), J and I are transfer (hopping or resonance) integrals connecting two sites (media) '-1' and '+1' with the central one represented by site '0'. Corresponding terms, that is, the first and second term on the right hand side of (3.55), describe the particle transfer from the left or right media to the site $m = 0$ and *vice versa*. For the process investigated, there is no asymmetry needed between sites '-1' and '+1' introduced by, e.g., values $J \neq I$. One can put simply $J = I$ and the system described works equally well. For any value of J, the '-1'→'0' and '0'→ '-1' transitions are (owing to hermicity of the first term on the right hand side of (3.55)) equally probable if nothing happens upon the particle's arrival at site '0' — and similarly for any value of I, the '0'→'+1' and '+1'→'0' transitions and the second term on the right hand side of (3.55). Thus, the asymmetry of the flow

---

[20]This model cannot be directly applied to any such pump known at present.

reported below is connected exclusively with what is going on with the rest of the system when the particle appears at site '0'.

Site '0' is understood to be attached to a central system representing, for instance, a specific molecule. This system is assumed to have (in a given range of energies of interest) two levels, with energies $\pm\epsilon/2$ (with corresponding states $|d\rangle$ and $|u\rangle$; we assume $\epsilon > 0$). At the moment when a particle is transferred to site '0' attached to the central system, the relative order (on the energy axis) of the two levels of the central system gets interchanged. (This, of course, causes instability of the central system with respect to the $|d\rangle \rightarrow |u\rangle$ transition, which is the same effect as additional-load-induced instability of a ship in water.) Technically, this imbalancing is due to the third term on the right hand side of (3.55) proportional to $1 - 2c_0^\dagger c_0$ and may be in reality due to correlation effects as the particle transferred may, upon its transition to site 0, change the topology or orientation of the central system in space as its originally stable conformation becomes energetically disadvantageous. This is the case in some models of membrane pumps and this particle-induced instability of the central system is the first distinguishing feature of the model. As a simplifying assumption not influencing the physics of the process and accepted here just for simplicity, we have assumed in (3.55) that this imbalancing goes with preserving both the 'centre-of-gravity' as well as the relative distance of the two levels on the energy axis.

The second important feature of the model is connected with the assumption that the central system (together with the site '0' attached to it) is open (i.e., able to accept from or return the particle) to only the left (site '-1') or only the right (site '+1') as far as its state is respectively $|d\rangle$ or $|u\rangle$. (Remember that, as assumed above, these are the stable states of the central system without and with the attached particle, respectively.) In the Hamiltonian (3.55), this feature is ensured by the multiplicative factors $|d\rangle\langle d|$ and $|u\rangle\langle u|$ in the first and second term in (3.55), respectively. One can imagine that before the particle arrives at site '0', the stable configuration corresponds to state $|d\rangle$ of the two-level system. Upon the arrival, such a configuration becomes unstable. Transition of the two-level system to state $|u\rangle$ might then mean, for instance, moving site '0' (together with the particle) somewhere else in space. This could cause interruption of a molecular chain via which the particle originally arrived, which amounts to blocking the particle return channel. Like the Gordon models (§5.2) and the chemical system in Chapter 7, this challenges common expectations of detailed balance.

The terms discussed so far define the Hamiltonian H_S of the system consisting of the central system and the particle transferred. All the rest in the above model is due to the assumed special form of the bath (here harmonic oscillators with phonon energies $\hbar\omega_k$), and its interaction with the central system (here linear in phonon creation and annihilation operators $b_k^\dagger$ and b_k). Constants G_k and g_k represent, respectively, coupling constants of the central system and the particle at site '0' to the individual bath modes; N is the number of the modes. Index k designates the individual modes and is, for technical simplicity, taken as a wave vector of a running plane-wave phonon mode. As we shall argue mathematically and explain in words below, at finite temperatures and under proper dephasing conditions, *this quantum model transfers particles preferably from the left to the right*. After finishing one act of the transfer, the central system returns to its original configuration, (*i.e.*, to the 'down' state $|d\rangle$), prepared to transfer another particle. This is why (3.55) has the appellation *swing* or *fish-trap* model.

The physical regime is critical to this model; it must go beyond the limitations of the weak-coupling regime to the bath. In fact, to get a commensurable dephasing and in-phasing, we should assume

$$H_{S-B} \propto \lambda, \quad J \propto I \propto \lambda^2, \tag{3.56}$$

and use the regime corresponding to (3.51). Simple physical reasoning reveals that using the usual weak coupling theory with $J \propto I \propto \lambda^0$, *i.e.*, (3.52) would yield bath-induced relaxation (*i.e.*, dephasing) infinitely slower than the in-phasing induced by the hopping transfer terms proportional to J, I. If so, the gates handling the particle transfer, provided by terms $|d\rangle\langle d|$ and $|u\rangle\langle u|$ in H_S, would not be able to close (open) in time to hinder the particle back-transfer.

Owing to the form of the Hamiltonian of the model, the particle transfer is accompanied by sufficiently fast re-relaxation between states $|d\rangle$ and $|u\rangle$ of the two-level system. Hence, correlations between the particle position and state of the two-level system are of utmost importance and should be described as rigorously as possible. That is why it is desirable to work with the entire density matrix of the system, which is represented by a 6×6 matrix $\rho_{\alpha,\gamma}(t) = \langle \alpha | \rho_S(t) | \gamma \rangle$, $|\alpha\rangle = |m\rangle \otimes |d\rangle$ or $|m\rangle \otimes |u\rangle$, $m = 1$, 2 or 3 (with $|m\rangle$ being the state with particle on site m). Equations (3.53) are then, in the above regime, fully equivalent to (3.35) when time t in the upper limit of the integral is set to $+\infty$.[21] The results are the same as with application of the Tokuyama-Mori method [37], or with a simple standard equation-of-motion method for Heisenberg operators [39]. In all cases, we get the same set of 36 linear differential equations. For reasons of technical simplicity, we give here their form for just $g_k = 0$. It reads

$$\frac{d}{dt} \begin{pmatrix} \rho_{uu} \\ \rho_{dd} \\ \rho_{ud} \\ \rho_{du} \end{pmatrix} = \begin{pmatrix} \mathcal{A} & \mathcal{B} & 0 & 0 \\ \mathcal{C} & \mathcal{D} & 0 & 0 \\ 0 & 0 & \cdots & \cdots \\ 0 & 0 & \cdots & \cdots \end{pmatrix} \cdot \begin{pmatrix} \rho_{uu} \\ \rho_{dd} \\ \rho_{ud} \\ \rho_{du} \end{pmatrix}. \tag{3.57}$$

Here, we have used the matrix notation

$$(\rho_{uu})^T =$$

$$\left(\rho_{-1u,-1u}, \; \rho_{0u,0u}, \; \rho_{1u,1u}, \; \rho_{-1u,0u}, \; \rho_{-1u,1u}, \; \rho_{0u,-1u}, \; \rho_{0u,1u}, \; \rho_{1u,-1u}, \; \rho_{1u,0u} \right),$$

$$(\rho_{dd})^T =$$

$$\left(\rho_{-1d,-1d}, \; \rho_{0d,0d}, \; \rho_{1d,1d}, \; \rho_{-1d,0d}, \; \rho_{-1d,1d}, \; \rho_{0d,-1d}, \; \rho_{0d,1d}, \; \rho_{1d,-1d}, \; \rho_{1d,0d} \right),$$

---

[21] This requires that t exceeds the initial time $t_0 = 0$ by more than a dephasing time of the bath. In fact, we are interested in only the long-time behavior of the solution.

$$(\rho_{ud})^T =$$

$$\begin{pmatrix} \rho_{-1u,-1d}, & \rho_{0u,0d}, & \rho_{1u,1d}, & \rho_{-1u,0d}, & \rho_{-1u,1d}, & \rho_{0u,-1d}, & \rho_{0u,1d}, & \rho_{1u,-1d}, & \rho_{1u,0d} \end{pmatrix}$$

(3.58)

and similarly for ρ_{du}. Superscript $\cdots^T$ designates transposition. We have also already applied the initial conditions (3.20) and (3.21), implying that there is no inhomogeneous (initial-condition) term in (3.57). In the square matrix in (3.57), all the elements are actually 9×9 blocks. The whole set splits into two independent sets of 18 equations each; we shall be interested only in that one for ρ_{uu} and ρ_{dd}. This reads, as in (3.57),

$$\mathcal{A} = \begin{pmatrix}
-\Gamma_\downarrow & 0 & 0 & 0 & 0 & 0 & 0 & 0 & 0 \\
0 & -\Gamma_\uparrow & 0 & 0 & 0 & 0 & iI/\hbar & 0 & -iI/\hbar \\
0 & 0 & -\Gamma_\downarrow & 0 & 0 & 0 & -iI/\hbar & 0 & iI/\hbar \\
0 & 0 & 0 & k & iI/\hbar & 0 & 0 & 0 & 0 \\
0 & 0 & 0 & iI/\hbar & -\Gamma_\downarrow & 0 & 0 & 0 & 0 \\
0 & 0 & 0 & 0 & 0 & k^* & 0 & -iI/\hbar & 0 \\
0 & iI/\hbar & -iI/\hbar & 0 & 0 & 0 & k^* & 0 & 0 \\
0 & 0 & 0 & 0 & 0 & -iI/\hbar & 0 & -\Gamma_\downarrow & 0 \\
0 & -iI/\hbar & iI/\hbar & 0 & 0 & 0 & 0 & 0 & k
\end{pmatrix}.$$

(3.59)

Here $k = -i\epsilon/\hbar - 0.5(\Gamma_\uparrow + \Gamma_\downarrow)$, and $\cdots^*$ designates complex conjugation. Further,

$$\Gamma_\uparrow = \frac{2\pi}{\hbar} \frac{1}{N} \sum_k |G_k|^2 (\hbar\omega_k)^2 n_B(\hbar\omega_k) \delta(\epsilon - \hbar\omega_k),$$

$$\Gamma_\downarrow = \frac{2\pi}{\hbar} \frac{1}{N} \sum_k |G_k|^2 (\hbar\omega_k)^2 [1 + n_B(\hbar\omega_k)] \delta(\epsilon - \hbar\omega_k), \quad n_B(z) = \frac{1}{e^{\beta z} - 1}. \quad (3.60)$$

In these formulae, we as usual understand the thermodynamic limit of the bath (converting $1/N \sum_k \cdots$ into integrals) as performed in order to avoid problems associated with Poincaré cycles. Clearly, $n_B(z)$ is the Bose-Einstein distribution of phonons with the reciprocal temperature β entering the problem via the initial density matrix ρ_B of the reservoir. In $\Gamma_\uparrow$ and $\Gamma_\downarrow$, one can recognize the lowest-order Golden Rule up- and down transition rates between states $|d\rangle$ and $|u\rangle$. These are known from the Pauli Master Equations, which describe the kinetics in quantum systems involving a finite number of levels. Still, however, our theory is, via inclusion of the off-diagonal elements of $\rho_S(t)$, more rigorous including many higher-order processes.

In an analogous way, we can identify other blocks in (3.57). Block $\mathcal{B}$ is fully diagonal, with diagonal elements $\mathcal{B}_{11}, \ldots, \mathcal{B}_{99}$ equal to $\Gamma_\uparrow$, $\Gamma_\downarrow$, $\Gamma_\uparrow$, $(\Gamma_\uparrow + \Gamma_\downarrow)/2$, $\Gamma_\uparrow$,

$(\Gamma_\uparrow + \Gamma_\downarrow)/2$, $(\Gamma_\uparrow + \Gamma_\downarrow)/2$, $\Gamma_\uparrow$ and $(\Gamma_\uparrow + \Gamma_\downarrow)/2$, respectively. Next,

$$\mathcal{D} = \begin{pmatrix} -\Gamma_\uparrow & 0 & 0 & iJ/\hbar & 0 & -iJ/\hbar & 0 & 0 & 0 \\ 0 & -\Gamma_\downarrow & 0 & -iJ/\hbar & 0 & iJ/\hbar & 0 & 0 & 0 \\ 0 & 0 & -\Gamma_\uparrow & 0 & 0 & 0 & 0 & 0 & 0 \\ iJ/\hbar & -iJ/\hbar & 0 & k^* & 0 & 0 & 0 & 0 & 0 \\ 0 & 0 & 0 & 0 & -\Gamma_\uparrow & 0 & -iJ/\hbar & 0 & 0 \\ -iJ/\hbar & iJ/\hbar & 0 & 0 & 0 & k & 0 & 0 & 0 \\ 0 & 0 & 0 & 0 & -iJ/\hbar & 0 & k & 0 & 0 \\ 0 & 0 & 0 & 0 & 0 & 0 & 0 & -\Gamma_\uparrow & iJ/\hbar \\ 0 & 0 & 0 & 0 & 0 & 0 & 0 & iJ/\hbar & k^* \end{pmatrix}.$$

$$(3.61)$$

As for the block $\mathcal{C}$, it reads the same as $\mathcal{B}$ except for the interchange $\Gamma_\uparrow \leftrightarrow \Gamma_\downarrow$.

It is interesting to consider what the effect might be of terms in H_{S-B} being omitted by setting $g_k = 0$. Without going into details, the physical effect of such terms consists in dephasing, reflected in the site off-diagonal elements of $\rho_s(t)$. A detailed analysis [36] shows that taking this dephasing into account explicitly would only cause at most additive corrections to the $(\ldots)_{44}$ up to $(\ldots)_{99}$ diagonal elements of blocks $\mathcal{A}$ and $\mathcal{D}$ (i.e., increasing only the rate of decay to zero of the off-diagonal elements), but changing nothing in our primary conclusions.

With that, we can now try to find all the matrix elements in $\rho_{uu}(t)$ and $\rho_{dd}(t)$ from (3.57) — which is a double set of twice 18 linear differential equations — for the desired elements of the system's density matrix. The solution for the diagonal elements (probabilities of finding the particle in the respective state) $P_{nu}(t) = \rho_{nu,nu}(t)$ and $P_{nd}(t) = \rho_{nd,nd}(t)$, $n = -1$, 0 and 1 results in a superposition of a constant term and as many as 17 exponentials decaying to a stationary solution. For each set of parameters and initial conditions separately, solutions can be found numerically. As usual, the asymptotic (long-time) form of the solution is determined by the right-eigenvector of the matrix of the set with zero eigenvalue. Its left-right asymmetry can be seen from the asymmetry of the model as well as the set (3.57). This right-eigenvector, however, i.e., the asymptotic form of the solution at long times, can be found easily in the limiting case of very low temperatures $\epsilon/(k_B T) \to +\infty$ without involving any numerics at all. In this case, $\Gamma_\uparrow$ disappears and one can verify the resulting asymmetry in the asymptotic solution, checking by hand that

$$\rho_{1d,1d}(t \to +\infty) = 1, \qquad (3.62)$$

with all other matrix elements of the density matrix ρ going to zero. From a fully analytic solution, using Kramers rule, the error of this result at finite $\epsilon/(k_B T)$ can be seen to be exponentially small ($\propto \Gamma_\uparrow/\Gamma_\downarrow = \exp(-\epsilon/(k_B T))$). Accordingly, at high temperatures $T \gg \epsilon/k_B$, the effect becomes negligible. This is not because of suppression of the above transfer channel, but because of the thermally-activated (bath-assisted) up-in-energy processes increase in efficiency with increasing temperature. These competing processes fully mask the effect when $T \to +\infty$.

Thus, we find that a particle initially anywhere in the system tends, with increasing time, to one (right hand) side of the system, leaving its central part prepared to accept another particle and, thus, ready to repeat the process. Note that (3.62) is not interpretable as a simple renormalization of particle energies. Such cases are known in connection with the Debye-Waller renormalization of the transfer integrals caused by the coupling to the bath, or as a result of a self-consistent nonlinear formulation of the scenario involving of the carrier in a polarizable medium. This renormalization changes the relative asymptotic population of eigenstates of the Hamiltonian of the system, but does not change their order on the energy axis. In terms of an effective temperature T_{eff}, as deduced from the relative asymptotic population of these states, T_{eff} may become even appreciably greater than the physical temperature T, but never becomes negative. The opposite is true in our situation. One can easily check that nothing is qualitatively changed in the solution when we raise the energy of site '+1' by adding an additional term $\delta H_S \delta \epsilon c_1^\dagger c_1$ to H_S. The excited eigenstate of H_S $|1\rangle \otimes |d\rangle$ (with asymptotic population equal to unity) is then intensively fed, in conflict with laws of statistical thermodynamics. As the particle in such a state becomes asymptotically free, its energy (acquired at the cost of only the thermal energy for the bath) can be used for other purposes, in conflict with the second law. With such a surprising conclusion, one should explain in words how the particle transfers from anywhere in the three-site chain to site '+1'.

Assume we have the particle initially at, for instance, site -1 with the central part (molecule) in the down-state with energy $-\epsilon/2$. This situation is energetically advantageous and, thus, almost stationary unless a quantum hop, caused by the transfer term ($\propto J$ in (3.55)), transfers the particle to site 0 joined with the central system. This transfer is, of course, energetically disadvantageous since such a down-state of the central system with the particle attached to it requires as much energy as $+\epsilon/2$, i.e., the energy difference ϵ. Hence, this process should be understood only as virtual (not energy conserving though the bath is not involved in the $-1 \leftrightarrow 0$ transfer). That is the reason we need the coherent hopping term in the Hamiltonian and why one cannot describe this transfer by the Pauli Master Equations.

As soon as the particle virtually attaches to the central system, the latter becomes energetically unstable and (unless the particle leaves it in between returning to site -1) turns to the up-state with lower energy $-\epsilon/2$. This, however, makes the particle unable to leave the central system to the left, but allows it to leave to the right (site $+1$). Such a process is again not energy conserving (if the particle leaves the central system, the latter has, in its up-state, again a high energy $+\epsilon/2$) so the process must be again be looked at as only virtual and be treated in the complicated manner above. For small $|I|$, it may take a long time before such a hop occurs. As far as it actually occurs (owing to the hopping term in (3.55) $\propto I$), however, the central system again becomes unstable and goes again (with the help of the interaction with the bath) to the down-state. This completes the cycle and the system is ready to accept a new particle from the left.

There are several important points concerning this model and the process just described:

- Under the above conditions, the process proceeds nearly uni-directionally. We stress that this surprising feature is *not* owing to any omission of terms in the Hamiltonian (3.55) which would imply prohibiting the backward transfer. To be explicit,

- the terms in our Hamiltonian which are responsible for the particle transfers ' − 1' ↔ '0' and '0' ↔ ' + 1' are the first and the second terms on the right hand side (3.55), respectively. They not only do allow the backward transitions, but are also (necessarily owing to the hermicity of the Hamiltonian) truly *symmetric* with respect to the back and forth particle transitions.

- The system-bath interaction Hamiltonian H_{S-B}, the last term on the right hand side of (3.55) *does* allow, and is the only term in the Hamiltonian which is responsible for, both the $|u\rangle \rightarrow |d\rangle$ and $|d\rangle \rightarrow |u\rangle$ transitions. From these processes, as well as those mentioned in the previous item, the backward transfer is *not* prohibited.

- The latter processes $|u\rangle \rightarrow |d\rangle$ and $|d\rangle \rightarrow |u\rangle$ are, however, not equally probable. This is the very reason for our uni-directional particle flow ' − 1' → ' + 1'. Again, however, we stress that this asymmetry has not been incorporated in the model 'by hand,' *e.g.*, by omission of anything in the Hamiltonian of the model. These are the spontaneous processes making up-in-energy and down-in-energy processes unequally probable. This asymmetry in Nature — the existence of the spontaneous down-in-energy processes in systems interacting with quantum baths — date back to Einstein [40, 41, 42]. The left→right flow is simply a manifestation of the natural up- and down-in-energy asymmetry.

• The transfer proceeds without any gain or loss in energy for the system. If we assume the particle begins on the left (with the central part of the system in the down-state) and ends on the right, the mean energy of the system remains unchanged insofar as $\delta\epsilon = 0$ (determined as a mean of the Hamiltonian of the system H_S). With $\delta\epsilon \neq 0$, however, the particle energy is increased and the question arises as to the source of the energy increase. The answer is that phonon emission or absorption processes mediate the particle transfers in each elementary act of the combined process.

• It might seem that the system is lowering its entropy. Responding similarly to above,

- there are problems with identification of the thermodynamic entropy and the statistical (von Neumann) entropy of the system; and

- in connection with the second law, it is uncertain what is the corresponding entropy of the full, closed (system + bath) complex since the entropy of the bath cannot be determined. In the thermodynamic limit of the bath, the bath's entropy becomes infinite so it cannot be easily discussed with a correspondingly high level of mathematical rigor.

In conclusion, the operation of the fish-trap can be interpreted as a kind of a spontaneous self-organization. It is a quantum mechanical machine, driven by thermal energy.

3.6.2 Semi-Classical Fish-Trap Model

Assume the model above with the Hamiltonian in (3.55). The model is modified in that the state of the central two-level system is now represented by a classical real variable $z(t)$. Its meaning is the mean value of $|u\rangle\langle u| - |d\rangle\langle d|$. This is legitimate provided that the central system is sufficiently large (e.g., being represented by a macromolecule or large protein). If $z = -1$ or $+1$, the system allows for the particle to be transferred from (or to) the left (site '-1') or the right (site '+1'), respectively. The bath then formally falls out of the discussion; however, it still influences the time-development of our swing variable $z(t)$. For purposes of simple modeling, it will be described as a single-exponential relaxation to a state which depends on whether the particle resides on the central system or not. Thus, replacing $|d\rangle\langle d|$ and $|u\rangle\langle u|$ in H_S in (3.55) by $[1 - z(t)]/2$ and $[1 + z(t)]/2$, respectively, the semi-classical version of the fish-trap model is defined by the Hamiltonian

$$H_{semiclass}(t) = \bar{J}(c_{-1}^\dagger c_0 + c_0^\dagger c_{-1})[1 - z(t)]$$

$$+\bar{I}(c_1^\dagger c_0 + c_0^\dagger c_1)[1 + z(t)] - \epsilon z(t)c_0^\dagger c_0 + \delta\epsilon c_1^\dagger c_1, \quad \bar{J} = J/2, \quad \bar{I} = I/2. \tag{3.63}$$

Here $\delta\epsilon$ is an up-in-energy shift of site '+1'. Equation (3.63) is clearly nothing but a mean value (with respect to the state of the central system) of the first two terms in H_S in (3.55). The third term in H_S is now, on the semi-classical level of description, represented by a special form of the right hand side of the assumed dynamical equation for $z(t)$ (see (3.65)), determining to which value $z(t)$ would relax provided that no particle transitions to and from the central system were allowed (i.e., for $J = I = 0$). The effect of H_B and H_{S-B} in (3.55) is then, in this dynamical equation, represented by the relaxation rate constant (γ in (3.65)), the very form of which gives the relaxation of $z(t)$.

The semi-classical Schrödinger equation for one-particle state $\Psi(t)$ reads in the matrix form as

$$i\hbar\frac{\partial}{\partial t}\begin{pmatrix} \psi_{-1}(t) \\ \psi_0(t) \\ \psi_1(t) \end{pmatrix}$$

$$= \begin{pmatrix} 0 & \bar{J}(1 - z(t)) & 0 \\ \bar{J}(1 - z(t)) & -\epsilon z(t) & \bar{I}(1 + z(t)) \\ 0 & \bar{I}(1 + z(t)) & \delta\epsilon \end{pmatrix} \cdot \begin{pmatrix} \psi_{-1}(t) \\ \psi_0(t) \\ \psi_1(t) \end{pmatrix}. \tag{3.64}$$

Here $\psi_n(t)$ are the usual quantum amplitudes of finding the particle at the respective sites. Their definition is given by $|\Psi(t)\rangle = \sum_{n=-1}^{+1}\psi_n(t)c_n^\dagger|vac\rangle$. For the $z(t)$-variable, we choose the above-mentioned dynamical equation in the form

$$\frac{\partial}{\partial t}z(t) = -\gamma[z(t) + 1 - 2|\psi_0(t)|^2]. \tag{3.65}$$

As for the initial condition, arbitrary values of $z(0)$ between -1 and $+1$ and arbitrary complex values of $\psi_{-1}(0)$, $\psi_0(0)$ and $\psi_1(0)$ fulfilling

$$\sum_{n=-1}^{+1}|\psi_n(0)|^2 = 1 \tag{3.66}$$

are admissible. This model was studied in detail in [43].

The system of four equations (3.64-3.65) is nonlinear. This non-linearity is somewhat unphysical since the standard Schrödinger equation with the quantum Hamiltonian (3.54) is linear. The nonlinearity results from replacing the quantum operator $|u\rangle\langle u| - |d\rangle\langle d|$ by its time-dependent mean value $z(t)$.

From (3.64), however, it follows that

$$\frac{\partial}{\partial t} \sum_{n=-1}^{+1} |\psi_n(t)|^2 = 0. \tag{3.67}$$

Thus, $\sum_{n=-1}^{+1} |\psi_n(t)|^2$ (=1 owing to (3.66)) is a conserved quantity. The Hamiltonian in (3.63) is time-dependent, owing to the time-dependence of $z(t)$. Owing to the nonlinearity, no non-stationary analytical solutions to (3.64-3.65) were found explicitly. The first task in investigating the time-dependence of physical properties is an analysis of stationary states.

For simplicity, unless otherwise stated, we set $\delta\epsilon = 0$. There are then three stationary states of our semi-classical problem. The solutions (up to a phase factor) are

$$E_1 = 0, \ z(t) = -1,$$

$$\begin{pmatrix} \psi_{-1}(t) \\ \psi_0(t) \\ \psi_{+1}(t) \end{pmatrix} = \begin{pmatrix} 0 \\ 0 \\ 1 \end{pmatrix}$$

$$\tag{3.68}$$

and[22]

$$E_{2,3} = \pm\sqrt{\bar{J}^2 + \bar{I}^2}, \ z(t) = 0,$$

$$\begin{pmatrix} \psi_{-1}(t) \\ \psi_0(t) \\ \psi_{+1}(t) \end{pmatrix} = \frac{1}{\sqrt{2}} \begin{pmatrix} \pm\bar{J}/(\sqrt{\bar{J}^2 + \bar{I}^2}) \\ 1 \\ \pm\bar{I}/(\sqrt{\bar{J}^2 + \bar{I}^2}) \end{pmatrix} \cdot e^{\mp i\sqrt{\bar{J}^2+\bar{I}^2}t/\hbar}.$$

$$\tag{3.69}$$

In state (3.68), the particle is already on the right hand side and the system is able to accept another particle from the left. This solution corresponds to the expected asymptotic state corresponding to (3.54), as we show below. This due to the fact that the semi-classical replacement $|u\rangle\langle u| - |d\rangle\langle d| \rightarrow z(t)$ underlying (3.63) becomes exact as long as we are in a state where the operator on the left hand side assumes a sharp value (*i.e.*, -1 in our case). Such an fortuitous situation does not appear in other (including non-stationary) states, in particular (3.69). In the corresponding states, the central system is in equilibrium with the particle, which is partly on both sides. Such an 'equilibrium' would immediately

---

[22]For $8|\bar{I}\bar{J}| < \epsilon^2 \neq 4(\bar{J}^2 + \bar{I}^2)$, two other stationary states exist with $z = (-4\bar{I}^2 + 4\bar{J}^2 \pm \sqrt{\epsilon^4 - 64\bar{I}^2\bar{J}^2})/(\epsilon^2 + 4\bar{I}^2 + 4\bar{J}^2) \neq 0$, $z \in (-1,+1)$. For our purposes, their detailed form as well as their properties are unimportant.

become disturbed once quantum fluctuations are taken into account. We have in mind those fluctuations which result from the fact that the occupation-number operator of the particle at site 0, that is, $c_0^\dagger c_0$, does not commute with the quantum Hamiltonian (3.55). From that, quantum fluctuations appear for not only $c_0^\dagger c_0$ but also $|u\rangle\langle u| - |d\rangle\langle d|$ (the quantum analog to $z(t)$). Classical treatment of the central system does not, however, allow such a discussion. In addition to that, owing to a simple form of (3.63) omitting the above dephasing processes important for the very effect, the approximate semi-classical solution (3.69) cannot be taken seriously from the point of view of the original quantum model (3.55).

Let us now, for the sake of generality as well as for physical implications, admit $\delta\epsilon \neq 0$. Negative values of $\delta\epsilon$ are from the point of view of implications less relevant, so we mainly consider $\delta\epsilon \geq 0$. Preliminary numerical studies based on the system (3.64) and (3.65) show that the above fluctuations, omitted upon accepting the semi-classical description of the central system, are in fact indispensable. That is why numerical studies in [43] are based on a scheme whereby equations (3.64) and (3.65) were solved on open intervals of time (t_{i-1}, t_i), $i = 1, 2, \ldots, t_i = \Delta t \cdot i$, $t_0 = 0$, with continuity conditions imposed on $\psi_{-1}(t)$, $|\psi_0(t)|$ and $\psi_1(t)$, as well as $z(t)$ in each point $t = t_i$. As for the phase of $\psi_0(t_i)$, it was always randomly set upon reaching $t = t_i$. As is well known, noise can be admissibly added to the Schrödinger equation provided we model it by an external stochastic potential; the theory and practice are now well established. Physically, one can imagine that such an instantaneous loss of phase of ψ_0 could be owing to, for instance, molecules from external media impinging on the central system. If their potential (as felt by our particle at site '0') is almost shot-like, the above instantaneous loss of phase of ψ_0 becomes justified. It is not important for the final outcome of the particle transfer process whether the impinging times t_i are regular or not[23].

An independent issue is the averaging over the noise. It can, of course, be performed appropriately only on the level of the density matrix. This will not be done here. However, we do not need it for our purposes because we obtain uni-directional transfer for any realization of the noise, assuming properly chosen intensity of the dephasing. Thus, the averaging cannot obviate the uni-directional transfer as the main result of the present model. Implications are at hand for such an uphill uni-directional transfer to site '+1' with $\delta\epsilon > 0$ for the utilization of this site energy acquired at the sole expense of interaction with the heat bath[24]. The above randomization scheme of the phases of ψ_0 is not connected with any unphysical energy flow to the particle. One can show that the only increase of the particle energy, which is solely due to the randomization, is due to the breaking of covalent bonding of the particle at site '0' to its neighbors [43].

Solution to (3.64) and (3.65) can be represented as a path in a seven-dimensional space of real variables $\Re\mathrm{e}\,\psi_n$, $\Im\mathrm{m}\,\psi_n$ $(n = -1, 0, +1)$, and z. The above stationary

---

[23]One obtains dephasing with the fully quantum and responsive bath, too. This is connected with suppression of the off-diagonal elements of the density matrix.

[24]One could object to the idea of using the acquired site energy instead of that corresponding to some extended particle states. The point is, however, that the above dephasing violates the particle bonding between site '+1' and the other sites. This makes the particle at site '+1' dynamically free to participate in ensuing dynamical processes.

solutions are in fact periodic solutions in this space (limit cycles). For stationary solutions of (3.69), all the $\psi_n(t)$ are nonzero and are determined up to a common phase factor. Thus, the relative phase of, for instance, $\psi_0(t)$ and $\psi_1(t)$, is well determined. The randomization of only the phase of $\psi_0(t)$ at any randomization event means a relatively long jump in the 7d-space. In other words, it brings the path out of the neighborhood of the stationary solution and wandering in the 7d-space begins anew. For the stationary solution of (3.68), however, the situation is completely different. Here $\psi_{-1}(t) = \psi_0(t) = 0$. Hence the randomization of the phase of ψ_0 has no effect; *i.e.*, the path becomes resistive with respect to the randomization. In other words, the solution of (3.68) remains stationary (*i.e.*, as a resistive limit cycle in the 7d-space) even with respect to any randomization procedure of the type used above. This helps to explain what is going on even for such long times that numerical data become unreliable.

Numerical studies indicate the following. The model — a semi-classical version of the previous one whereby the opening and closing of the gate is given by a single c-number variable $z(t)$ — diminishes the tendency of the particle toward final localization at the site with the highest site energy compared with that observed in the previous model. The reason is simple: using only one c-number variable $z(t)$, the possibility of describing strong correlations between the particle position and state of the gate is greatly reduced. Nevertheless, for properly chosen chosen rate of dephasing, the particle almost always ends up on site '+1', irrespective of initial distribution, the randomization process of the phase of $\psi(t)$, and the randomization times.

In order to understand the behavior obtained, several problems remain to be solved. The following questions should be answered: (i) how it is that particle energy is raised at the expense of the bath when the reorientations of the central system (closing and opening) are always due to down-in-energy (spontaneous) transitions of the central system; and (ii) how it is that the opening and closing of the central system occurs 'on average' at a time proper to cause the ' $-1'$ $\rightarrow$ '$+1'$ drain reported here. Note, proper timing is a precondition of the process. For the second question, there is no answer except that we have observed this timing to be automatically ensured by (3.64-3.65) themselves. This suggests perhaps deeper physics of the model, in particular the tendency toward one-way transfer ('$1'$ $\rightarrow$ '$+1'$) contained in the original quantum fish-trap model [36, 37]. The fish-trap was constructed specifically to yield the timing desired. As for the energy conservation arguments that could be raised against the process, note that dephasing implies continuous energy exchange between the system and bath, thus, final energy conservation is always possible. These results contradict the standard formulation of the second law in the sense that particles acquire energy at the expense of the bath and make it available for other applications[25].

---

[25] For example, one could add a leakage between sites '+1' and '-1', thus providing the particle with a means to complete the cycle and release.

3.6.3 Periodic Fish-Trap Model

Although we argued above that single-step processes are sufficient for these models, generalizations showing a continuous cyclic activity would be desirable. Perhaps the simplest one is to allow the particle to jump directly from '+1' to '-1' so as to form a circular dc particle flow '-1'→'0'→'+1' →'-1'. Connecting the particle to a screw whose potential energy increases as it turns, collects the particle energy gained from the bath thermal energy in a form of increasing potential energy of the screw. This model was first reported in [44] and later in [45, 46]. An equivalent but simpler derivation for the single-particle density matrix can be also found in [39].

The Hamiltonian for this model can be written:

$$H_S = \sum_{\iota=-\infty}^{+\infty} \{J(c_{-1,\iota}^\dagger c_{0,\iota} + c_{0,\iota}^\dagger c_{-1,\iota}) \cdot |d\rangle\langle d|$$

$$+I(c_{1,\iota}^\dagger c_{0,\iota} + c_{0,\iota}^\dagger c_{1,\iota}) \cdot |u\rangle\langle u| + K(c_{1,\iota}^\dagger c_{-1,\iota+1} + c_{-1,\iota+1}^\dagger c_{1,\iota})$$

$$+\frac{\Delta}{3} \sum_{m=-1}^{+1} \sum_{\iota=-\infty}^{+\infty} [3\iota + m]c_{m,\iota}^\dagger c_{m,\iota} + \frac{\epsilon}{2}[1 - 2\sum_\iota c_{0,\iota}^\dagger c_{0,\iota}] \cdot [|u\rangle\langle u| - |d\rangle\langle d|]. \quad (3.70)$$

Notation is the same as in the fish-trap model. The differences are:

- We have added a screw (*e.g.*, molecular, solid state) to the particle; its turns on the thread are designated by ι.

- The screw can be arbitrarily long; for formal reasons, we take it as infinite. Position on the thread with the particle located at site m is designated as m, ι.

- Each forward step of the particle ($-1, \iota \to 0, \iota$, $0\iota \to +1, \iota$, or $+1\iota \to -1, \iota+1$) implies an increase of the (particle + screw) complex by $\Delta/3$. Thus Δ is the potential/mechanical energy acquired from the heat bath from a single turn of the screw. In connection with this, the hopping integrals J, I, and K represent not merely particle integrals, but also represent inertial properties of the screw.

In (3.70), the back and forth transfers between any pair of the sites occur with the same amplitude, as a consequence of the hermicity of H_S. The uni-directional character of the process is not due to a contingent difference between these amplitudes, but results exclusively from the existence of spontaneous processes between states $|u\rangle$ and $|d\rangle$ of the central system.

As for the Hamiltonians of the bath H_B, we assume the same form as in (3.55). Finally, one should specify the Hamiltonian of the system-bath coupling H_{S-B}. Keep in mind that there are two roles played by H_{S-B}. First is the particle dephasing among different sites that conditions the breaking of covalent bonds among the sites necessary for the particle flow. Second, H_{S-B} causes transitions between states $|u\rangle$ and $|d\rangle$ of our central system. These transfers ensure the dynamic blocking of the particle hopping integrals mentioned above and, in connection with uncertainty relations, makes localization of the particle at one site energetically disadvantageous — the underlying reason for the asymmetric stationary flow. Such transfers between states $|u\rangle$ and $|d\rangle$ of the central system could also cause particle dephasing, *i.e.*, breaking of the particle covalent bonds. Such a dephasing would become effective only in higher orders of the perturbation theory. That is why we assume H_{S-B} consists of two terms which separately give rise to the $|u\rangle \leftrightarrow |d\rangle$ transitions and the on-site dephasing. We take

$$H_{S-B} = \frac{1}{\sqrt{N}} \sum_k \hbar\omega_k (b_k + b^\dagger_{-k}) \{ G_k[|u\rangle\langle d| + |d\rangle\langle u|] + \sum_{m=-1}^{+1} \sum_{\iota=-\infty}^{+\infty} g_k^m c^\dagger_{m,\iota} c_{m,\iota} \}$$

$$\equiv H'_{S-B} + H''_{S-B}. \qquad (3.71)$$

For the sake of technical simplicity, we assume that H'_{S-B} and H''_{S-B} do not interfere (by assuming, *e.g.*, $g_k G_k^* = 0$ for each k). This condition, however, does not imply the existence of two different baths.

Here we report [39, 46] only on the form of the resulting equations of motion for the single-particle density matrix $\rho_{a\alpha,b\beta} = \sum_{a,\iota,\alpha;b\iota,\beta}(t)$[26], again in the regime where the characteristic times of the particle transfer $\hbar/|J|$, $\hbar/|I|$, and $\hbar/|K|$ become at least commensurable with the reciprocals of the bath-assisted relaxation rates, *i.e.*, $(\Gamma_\uparrow + \Gamma_\downarrow)^{-1}$. Here $\Gamma_\downarrow$ and $\Gamma_\uparrow$ are the down-hill and up-hill Golden Rule transfer rates between states $|u\rangle$ and $|d\rangle$ of our central system caused by H'_{S-B}, defined as

$$\Gamma_\downarrow = \frac{2\pi}{N\hbar} \sum_k (\hbar\omega_k)^2 |G_k|^2 [1 + n_B(\hbar\omega_k)]\delta(\hbar\omega_k - \epsilon),$$

$$\Gamma_\uparrow = \frac{2\pi}{N\hbar} \sum_k (\hbar\omega_k)^2 |G_k|^2 n_B(\hbar\omega_k)\delta(\hbar\omega_k - \epsilon) = \exp(-\beta\hbar\omega_k)\Gamma_\downarrow, \qquad (3.72)$$

where $n_B(z) = [\exp(\beta z)-1]^{-1}$ is the Bose-Einstein-Planck distribution for phonons and $\beta = 1/(k_B T)$ is the reciprocal (initial) temperature of our phonon bath. We further require that $|J|/\hbar, |I|/\hbar, |K|/\hbar \ll 2\Gamma$, where 2Γ is the value of all $2\Gamma_{mn} = \frac{\pi}{N\hbar} \sum_k (\hbar\omega_k)^2 |g_k^m - g_k^n|^2 [1 + 2n_B(\hbar\omega_k)]\delta_\gamma(\hbar\omega_k)$, $m \neq n$. Here $\delta_\gamma(x)$ denotes the δ-function properly broadened by all the dephasing processes on the sites involved[27]. Thus, 2Γ is the pure dephasing rate competing with the formation of the valence bonds that would otherwise stop the particle transfer — the whole heat conversion process in which we are interested. These conditions imply that the relevant physical regime is not one of weak coupling to the bath. The result [45, 46, 39] could be again rewritten in a short-hand notations as follows. Again arrange the 36 matrix elements $\rho_{\alpha\gamma}(t)$ in groups of nine designating

$$(\rho_{uu})^T =$$

---

[26]Here, a, $b = 0, \pm 1$ designate the site, α, $\beta = u$ or d designate the state of the central system, and ι designates the turn of the screw.

[27]Those who are uncomfortable incorporating the broadening could, *e.g.*, replace H''_{S-B} in (3.71) by another form incorporating two-phonon processes. This leads, however, to no qualitative change.

$$\Big(\rho_{-1u,-1u},\ \rho_{0u,0u},\ \rho_{1u,1u},\ \rho_{-1u,0u},\ \rho_{-1u,1u},\ \rho_{0u,-1u},\ \rho_{0u,1u},\ \rho_{1u,-1u},\ \rho_{1u,0u}\Big),$$

$$\left(\rho_{dd}\right)^T =$$

$$\Big(\rho_{-1d,-1d},\ \rho_{0d,0d},\ \rho_{1d,1d},\ \rho_{-1d,0d},\ \rho_{-1d,1d},\ \rho_{0d,-1d},\ \rho_{0d,1d},\ \rho_{1d,-1d},\ \rho_{1d,0d}\Big),$$

$$\left(\rho_{ud}\right)^T =$$

$$\Big(\rho_{-1u,-1d},\ \rho_{0u,0d},\ \rho_{1u,1d},\ \rho_{-1u,0d},\ \rho_{-1u,1d},\ \rho_{0u,-1d},\ \rho_{0u,1d},\ \rho_{1u,-1d},\ \rho_{1u,0d}\Big)$$

$$\tag{3.73}$$

and similarly for ρ_{du}. Again, superscript $\cdots^T$ designates transposition. With that, the resulting set reads

$$\frac{d}{dt}\begin{pmatrix}\rho_{uu}\\ \rho_{dd}\\ \rho_{ud}\\ \rho_{du}\end{pmatrix} = \begin{pmatrix}\mathcal{A} & \mathcal{B} & 0 & 0\\ \mathcal{C} & \mathcal{D} & 0 & 0\\ 0 & 0 & \cdots & \cdots\\ 0 & 0 & \cdots & \cdots\end{pmatrix}\cdot\begin{pmatrix}\rho_{uu}\\ \rho_{dd}\\ \rho_{ud}\\ \rho_{du}\end{pmatrix}.$$

$$\tag{3.74}$$

In the square matrix, all elements are 9×9 blocks. Again, the whole set splits into two independent sets of 18 equations each. We are interested here in only that one for ρ_{uu} and ρ_{dd}. This reads as in (3.74) with typical forms for the 9×9 block matrices $\mathcal{A}, \mathcal{B}, \mathcal{C}$ and $\mathcal{D}$. In order to render the presentation as simply as possible, we first completely neglect the Hamiltonian H''_{S-B} describing a direct coupling of the particle to the bath, responsible for the above additional dephasing. A direct calculation yields: $\mathcal{A} =$

$$\begin{pmatrix}
-\Gamma_{\downarrow} & 0 & 0 & 0 & iK/\hbar & 0 & 0 & -iK/\hbar & 0\\
0 & -\Gamma_{\uparrow} & 0 & 0 & 0 & 0 & iI/\hbar & 0 & -iI/\hbar\\
0 & 0 & -\Gamma_{\downarrow} & 0 & -iK/\hbar & 0 & -iI/\hbar & iK/\hbar & iI/\hbar\\
0 & 0 & 0 & k_+ & iI/\hbar & 0 & 0 & 0 & -iK/\hbar\\
iK/\hbar & 0 & -iK/\hbar & iI/\hbar & -\Gamma_{\downarrow}^+ & 0 & 0 & 0 & 0\\
0 & 0 & 0 & 0 & 0 & k_+^* & iK/\hbar & -iI/\hbar & 0\\
0 & iI/\hbar & -iI/\hbar & 0 & 0 & iK/\hbar & k_-^* & 0 & 0\\
-iK/\hbar & 0 & iK/\hbar & 0 & 0 & -iI/\hbar & 0 & -\Gamma_{\downarrow}^- & 0\\
0 & -iI/\hbar & iI/\hbar & -iK/\hbar & 0 & 0 & 0 & 0 & k_-
\end{pmatrix}.$$

$$\tag{3.75}$$

Here $k_{\pm} = -i\epsilon/\hbar - 0.5(\Gamma_{\uparrow}+\Gamma_{\downarrow})\pm i\Delta/(3\hbar)$ and $\Gamma_{\downarrow}^{\pm} = \Gamma_{\downarrow}\pm i\Delta/(3\hbar)$, with all the rates defined above, and $\cdots^*$ denotes complex conjugation. Via $\Gamma_{\uparrow}$ and $\Gamma_{\downarrow}$, the initial bath temperature T (entering the problem through the initial bath density matrix ρ_B) enters the game. As for block $\mathcal{B}$, it is fully diagonal with diagonal elements

$\mathcal{B}_{11}, \ldots, \mathcal{B}_{99}$ equal to $\Gamma_\uparrow$, $\Gamma_\downarrow$, $\Gamma_\uparrow$, $(\Gamma_\uparrow + \Gamma_\downarrow)/2$, $\Gamma_\uparrow$, $(\Gamma_\uparrow + \Gamma_\downarrow)/2$, $(\Gamma_\uparrow + \Gamma_\downarrow)/2$, $\Gamma_\uparrow$ and $\mathcal{D} =$

$$
\begin{pmatrix}
-\Gamma_\uparrow & 0 & 0 & iJ/\hbar & iK/\hbar & -iJ/\hbar & 0 & -iK/\hbar & 0 \\
0 & -\Gamma_\downarrow & 0 & -iJ/\hbar & 0 & iJ/\hbar & 0 & 0 & 0 \\
0 & 0 & -\Gamma_\uparrow & 0 & -iK/\hbar & 0 & 0 & iK/\hbar & 0 \\
iJ/\hbar & -iJ/\hbar & 0 & k_-^* & 0 & 0 & 0 & 0 & -iK/\hbar \\
iK/\hbar & 0 & -iK/\hbar & 0 & -\Gamma_\uparrow^+ & 0 & -iJ/\hbar & 0 & 0 \\
-iJ/\hbar & iJ/\hbar & 0 & 0 & 0 & k_- & iK/\hbar & 0 & 0 \\
0 & 0 & 0 & 0 & -iJ/\hbar & iK/\hbar & k_+ & 0 & 0 \\
-iK/\hbar & 0 & iK/\hbar & 0 & 0 & 0 & 0 & -\Gamma_\uparrow^- & iJ/\hbar \\
0 & 0 & 0 & -iK/\hbar & 0 & 0 & 0 & iJ/\hbar & k_+^*
\end{pmatrix}.
$$

$$(3.76)$$

In a similar way, here $\Gamma_\uparrow^\pm = \Gamma_\uparrow \pm i\Delta/(3\hbar)$. As for the block $\mathcal{C}$, it reads the same as $\mathcal{B}$ except for the interchange $\Gamma_\uparrow \leftrightarrow \Gamma_\downarrow$. The last issue is how to properly include the second term H''_{S-B} of H_{S-B}. The resulting additive contributions are simple: nothing but an additional -2Γ factor appears in 44, 55, 66, 77, 88, and 99 terms of blocks $\mathcal{A}$ and $\mathcal{D}$. This is assumed here.

The question arises: What is the mechanical output — rate of increase of the potential energy — from heat conversion? We are interested in the stationary case in which the left hand side of (3.74) goes to zero. Because the rank of the set of (3.74) is less by one than the order, the set must be complemented by the normalization condition

$$
\sum_{m=-1}^{+1} [\rho_{mu,mu} + \rho_{md,md}] = 1. \tag{3.77}
$$

Then (3.74) and (3.77) determine the relevant components of the density matrix uniquely. From the physical meaning of the transfer rates, one can take the flow as

$$
\mathcal{J} = \Gamma_\downarrow \rho_{0d,0d} - \Gamma_\uparrow \rho_{0u,0u}. \tag{3.78}
$$

Using (3.74), (3.78) can be given a number of other forms, e.g.,

$$
\mathcal{J} = \frac{2I}{\hbar} \Im m \rho_{0u,1u} = \frac{2J}{\hbar} \Im m \rho_{-1d,0d} = \frac{2K}{\hbar} \Im m[\rho_{1u,-1u} + \rho_{1d,-1d}] \equiv \frac{2K}{\hbar} \Im m \rho_{1,-1}. \tag{3.79}
$$

Numerically, one can now determine $\mathcal{J}$ and, from this, the mechanical output (rate of heat conversion); it is

$$
W = \mathcal{J} \cdot \Delta. \tag{3.80}
$$

Numerical results can be found in [45, 46]. Summarizing,

- For $\Gamma \to 0$, the mechanical output disappears. This confirms the above reasoning that valence bonds (becoming less and less violated by decreasing $\Gamma > 0$) can stop the circulation of the particle, and with it, the whole conversion process. With very high values of $\Gamma > 0$, suppression of the off-diagonal elements of the density matrix can also lead to suppression of the flow.

- For constant $\Gamma > 0$, the mechanical output increases with increasing $\Delta > 0$ up to a critical value. (It is always positive.) This indicates that the system works as a (periodic) *perpetuum mobile* of the second kind, converting heat from the bath into the potential energy of the screw.

- Increasing Δ above the critical value, the rate of conversion decreases, but remains nonzero.

- In the limit of high temperatures $k_B T \gg \epsilon$ (classical limit), the rate of conversion becomes zero, for any $\Delta > 0$. Thus, the systems operates at finite temperatures on purely quantum effects. This supports the discussion above on the role of quantum correlations[28].

Significantly, this mechanism was used in a many-body version to challenge other laws of thermodynamics [47].

3.6.4 Sewing Machine Model

The sewing machine model is also a single-cycle model, this time working with pairs of particles. Two sorts of particles (designated c- and g-particles) are assumed They can form an energetically disadvantageous, but perhaps long-lived bound state. The latter would almost never appear were it not for an 'agent' (central system) in the system acting as an endothermic catalyst binding the particles together, leaving them coupled, while itself preparing to couple another unbound pair. The second law violation arises after many cycles in utilizing the excess concentration of energy-rich pairs, which store chemical energy accummulated at the expense of the heat from the bath. The idea has appeared in simple [48] and complex [49] forms.

The total Hamiltonian can be written as in (3.54). The Hamiltonian H_S of the system is the sum:

$$H_S = H_{cen-sys} + H_{part} + H_{cs-part}. \qquad (3.81)$$

$H_{cen-sys}$ describes a central system (molecule acting as a catalyst), which we assume to have just two eigenstates $|u\rangle$ and $|d\rangle$ with energies $\pm\epsilon/2$. Thus,

$$H_{cen-sys} = \frac{\epsilon}{2}[|u\rangle\langle u| - |d\rangle\langle d|]. \qquad (3.82)$$

As for the particle Hamiltonian H_{part}, describing the above two types of particles, we write it via the corresponding creation (or annihilation) operators $c_m^\dagger$ and $g_m^\dagger$ (c_m and g_m). The c-operators commute with g-operators as usual. As for the commutation relations 'c vs. c' or 'g vs. g' (or anti-commutation), these will be unimportant since, for the sake of simplicity, we assume only one c- and one g-particle. Though generalization to greater number of particles and greater particle

---

[28] Here correlations are between the particle position and occupation of states $|d\rangle$ and $|u\rangle$ of the central system, serving here as a Maxwellian gate.

reservoirs is straightforward, for simplicity we deal with only a particle reservoir consisting of just two sites (labeled 1 and 2). Thus, we have

$$H_{part} = J(g_1^\dagger g_2 + g_2^\dagger g_1 + c_1^\dagger c_2 + c_2^\dagger c_1)$$

$$+V(c_1^\dagger c_1 g_1^\dagger g_1 + c_2^\dagger c_2 g_2^\dagger g_2). \tag{3.83}$$

We always assume for the c-g interaction, $V > 0$. As for the interaction Hamiltonian between the central system and reservoir of particles, we assume it consists of two terms,

$$H_{cs-part} = H'_{cs-part} + H''_{cs-part},$$

$$H'_{cs-part} = \epsilon[|d\rangle\langle d| - |u\rangle\langle u|]c_0^\dagger c_0 g_0^\dagger g_0. \tag{3.84}$$

In connection with this, we assume our model also contains the third site (labeled as 0, with creation operators of particles $c_0^\dagger$ and $g_0^\dagger$) which we do not ascribe to the particle reservoir, but which we assume is tightly connected with the central system. The form $H'_{cs-part}$ is chosen in such a way that whenever the site 0 accepts both one c- and one g-particle, the central system with the Hamiltonian $H_{cen-sys} + H'_{cs-part}$ becomes unstable in the sense that the states $|u\rangle$ and $|d\rangle$, having originally eigenenergies $\pm\epsilon/2$, acquire energies $\mp\epsilon/2$. (We assume that $\epsilon > 0$.) Physically, this can conceivably occur as a change of a stable molecular configuration upon accepting a pair of particles. As for the $H''_{cs-part}$ term in (3.84) (part of the interaction between the central system and particle reservoir transferring the particles between them), it will be specified below.

Before saying anything about the bath and system-bath coupling, we should explain the motivation and physical ideas behind the model. The central system is unstable upon accepting the c-g pair to site 0. Among macromolecules, many are known to change their topology upon accepting ions or molecular species; common examples would be ion channels in cell membranes. In the language of biology, site 0 would be a receptor for the species in question. This idea for a *receptor instability* is doubly incorporated into this Hamiltonian.

First, the change in the topology could physically bring the c-g pair together, forcing the particles to form a bound state by overcoming the contingent potential barrier (or potential step) between the particles. For instance, the central molecule, being originally rod-like, and holding the pair at opposite ends could flex, bringing them together. In this case, the central molecule must be sufficiently stiff in its new topology; in other words, ϵ should exceed the height of the potential barrier.

Second, if a process requires energy, in principle, the lacking energy could come from the bath, making the process bath-assisted. This would be ineffective except at very high temperatures when all the states of the system would become, at long times, roughly equally populated. This is an uninteresting case. However, in the microworld governed by quantum dynamics, processes are allowed which seemingly contradict the energy conservation law. Tunneling is one such example, but this is not a situation of interest here. Rather, we have in mind the principle that localization of a particle can increase its kinetic and total energy owing to the quantum uncertainty relations[29]. Hence, insofar as there are terms in the Hamiltonian H_S allowing delocalization (in our case these will be provided by $H''_{cs-part}$), the eigenstates of H_S will be, in accord with the variational principle, at least partially delocalized. Thus, if we compile the eigenstates into a time-dependent solution of the Schrödinger equation for the particles in the isolated system $i\hbar d|\Psi(t)\rangle/dt = H_S|\Psi(t)\rangle$ and if we initially put our particles outside site 0, we get from the solution that, with a non-zero probability, they will definitely appear later at site 0 (separately as well as simultaneously). This is a purely quantum process which may bring both the c- and the g-particles to site 0, irrespective of how much site energy it costs. Notice that for this process, no energy from the reservoir (bath) is needed since the bath was completely split off in the Schrödinger equation. The bath energy and the interaction with the bath enters the process only at the moment of turning it from the virtual-type to the real-type process, as discussed below. This change of the character of the process *after* bringing both particles to site '0', without requiring the bath energy, will be connected with the above instability of the central system upon accepting both the 'c' and the 'g' particle, and the form of $H''_{cs-part}$ to be introduced below (3.84). We propose dynamic closing of back reaction channels (forbidding the particles to leave the site '0' individually in the same way as they arrived here) once the central system reorganizes on account of the receptor instability. So, it is not the bringing of the particles to site '0', but the ensuing 'closing the gate behind them' and opening of a new reaction channel for the particles to proceed in their bound state that requires the interaction with the bath. As also argued below, however, no activation energy is needed for this closing and opening the reaction channels as these are mostly spontaneous (down-in-energy) processes with respect to the bath.

Our system, working as a real molecular machine, could serve as an active catalyst for reactions which would otherwise be completely impossible. By the word 'active' we mean a catalytic property which is *not* reducible to merely lowering of potential barriers. We mean an active collecting the thermal energy if need be, or at least borrowing it for awhile for virtual processes. Let us construct the Hamiltonian.

For simplicity, we assume the $1 \leftrightarrow 2$ symmetry for our Hamiltonian. That is why we can limit our considerations to just symmetric states. The symmetric eigenstates of our particle Hamiltonian and the corresponding eigenenergies read

$$|\phi_1\rangle = \frac{1}{\sqrt{[V + \sqrt{V^2 + 16J^2}]^2 + 16J^2}} \left[[V + \sqrt{V^2 + 16J^2}] \cdot \frac{1}{\sqrt{2}}(c_1^\dagger g_1^\dagger + c_2^\dagger g_2^\dagger)|vac\rangle \right.$$

$$\left. +4J \cdot \frac{1}{\sqrt{2}}(c_1^\dagger g_2^\dagger + c_2^\dagger g_1^\dagger)|vac\rangle \right],$$

---

[29] As a textbook example, consider the zero-point energy for a quantum oscillator. Its energy lies above the minimum of the potential and the wave function has a Gaussian form around the potential energy minimum. This is so because further localization of the particle at the potential energy minimum would further lower the mean potential energy, but appreciably increase the mean kinetic. The total energy must be minimal in the ground state and the compromise is the zero-point energy.

$$E_1^{part} = \frac{1}{2}[V + \sqrt{V^2 + 16J^2}],$$

$$|\phi_2\rangle = \frac{1}{\sqrt{[V + \sqrt{V^2 + 16J^2}]^2 + 16J^2}} \left[-4J \cdot \frac{1}{\sqrt{2}}(c_1^\dagger g_1^\dagger + c_2^\dagger g_2^\dagger)|vac\rangle \right.$$

$$\left. + [V + \sqrt{V^2 + 16J^2}] \cdot \frac{1}{\sqrt{2}}(c_1^\dagger g_2^\dagger + c_2^\dagger g_1^\dagger)|vac\rangle \right],$$

$$E_2^{part} = -\frac{8J^2}{V + \sqrt{V^2 + 16J^2}},$$

$$|\phi_3\rangle = \frac{1}{\sqrt{2}}[c_1^\dagger + c_2^\dagger]g_0^\dagger|vac\rangle,$$

$$|\phi_4\rangle = \frac{1}{\sqrt{2}}[g_1^\dagger + g_2^\dagger]c_0^\dagger|vac\rangle,$$

$$E_3^{part} = E_4^{part} = J,$$

$$|\phi_5\rangle = c_0^\dagger g_0^\dagger|vac\rangle,$$

$$E_5^{part} = 0. \tag{3.85}$$

Ignoring the antisymmetric states, H_{part} can be rewritten as

$$H_{part} = \sum_{i=1}^{5} |\phi_i\rangle E_i^{part} \langle\phi_i|. \tag{3.86}$$

With this, we can now rewrite $H'_{cs-part}$ as

$$H'_{cs-part} = \epsilon[|d\rangle\langle d| - |u\rangle\langle u|] \otimes |\phi_5\rangle\langle\phi_5| \tag{3.87}$$

and specify $H''_{cs-part}$ as

$$H''_{cs-part} = P\left([|\phi_2\rangle + |\phi_5\rangle][\langle\phi_3| + \langle\phi_4|] + H.C \right) \otimes |d\rangle\langle d|$$

$$+ Q\left(|\phi_5\rangle\langle\phi_1| + H.C. \right) \otimes |u\rangle\langle u|. \tag{3.88}$$

(Here *H.C.* means Hermitian Conjugate.) In words, we assume that the central system allows different types of particle transitions (channels of the particle scattering) in different physical configurations. The detailed form of (3.88) is chosen for our model in the simplest version yielding the desired effect.

As for the thermodynamic bath, its detailed form is unimportant[30]. We only require that, in connection with the system-bath coupling, it yields the desired

---

[30]However, it cannot be dispersionless.

transitions sufficiently fast among the different states of the central system. The simplest version is that of non-interacting bosons (*e.g.*, phonons)

$$H_B = \sum_k \hbar \omega_k b_k^\dagger b_k. \tag{3.89}$$

The same notation as above is used and the same applies to the system-bath interaction. In its simplest form, causing relaxation between states of the central system, it can be written

$$H_{S-B} = \frac{1}{\sqrt{N}} \sum_k \hbar \omega_k (b_k + b_{-k}^\dagger)$$

$$\cdot \{ G_k [|u\rangle\langle d| + |d\rangle\langle u|] + g_k |\phi_5\rangle\langle\phi_5| \}. \tag{3.90}$$

Here N and G_k are the number of bath modes (with index 'k' taken as a wave-vector here) and a set of interaction constants. In the thermodynamic limit of the bath, N tends to infinity and the sums $(1/N) \sum_k \cdots$ go to the usual integrals. Since (3.90) allows $|u\rangle \leftrightarrow |d\rangle$ relaxation, our problem can be also viewed as a slow combined particle scattering on a central system with relaxation between its intermediate states. Special attention should be devoted to the term in (3.90) containing g_k. This term is the simplest one rendering transversal relaxation (dephasing) processes. In order to understand the importance of dephasing, one should realize that with (3.85), one can easily diagonalize the whole Hamiltonian of the system H_S (3.81). If the dephasing (transversal relaxation) were fully omitted, one would get transitions among the corresponding eigenstates of H_S as the only effect of the coupling to the bath. This is the standard way by which the weak-coupling theories (second-order in H_{S-B}) yield transitions to the canonical state. This means relaxation to practically the ground state at low enough temperatures. The ground state of H_S is, however, not the desired asymptotic state here. It contains components with our particles outside site '0' as well as at this site, with definite phase relations among them. These relations are due to $H''_{cs-part}$ in (3.88) which must be, in the weak coupling (to the bath) theories, taken as dominating over H_{S-B} in (3.90). In our model, however, we assume the opposite relation between the roles of $H''_{cs-part}$ and H_{S-B}. This means that higher orders in H_{S-B} also become effective which cause, in addition to transitions, the transversal relaxation. Physically, the meaning of the sufficiently strong transversal relaxation (dephasing) consists in destroying the above tough phase relations among individual components of the eigenstates of H_S, *i.e.*, turning the above transitions to those which are between the two eigenstates $|u\rangle$ and $|d\rangle$ of (3.82), as already suggested by the form of the first term in (3.90) containing G_k. Such a dephasing would be provided already by higher order terms in G_k, in particular when these coupling constants are sufficiently strong. In order to see this effect explicitly, one would, however, need a detailed higher-order theory, while the term proportional to g_k in (3.90) yields such dephasing processes immediately. Technically, the importance of such terms in H_{S-B} (3.90) proportional to g_k becomes clear by realizing that with such a dephasing, the memory functions to be invoked become more strongly decaying functions of time; *i.e.*, their time-integrals converge better. From the point of the energy conservation, the importance of these terms becomes clear from the observation that the asymptotic state of the system lies at an energy well above the ground state of H_S. Thus, in order to make active binding of the particles really effective, an intense energy exchange with the bath is required[31]. This means strong absorption as well as strong emission of boson excitations in our bath. These processes are effectively provided by the second term in H_{S-B} (3.90) $\propto g_k$.

Let us briefly mention the order of energies of the Hamiltonian of the system H_S, which is an important issue for weak-coupling (in H_{S-B}) kinetic theories. In such approaches, system relaxation goes mostly to — and at low temperatures, exclusively to — the ground state of H_S. Our theory, however, is not the weak-coupling variety. Rather than strength of the system-bath coupling, the parameters P and Q in $H''_{cs-part}$ in (3.88) play the role of the small parameters, although no real expansion in powers of P and Q is used. As we will argue below, $|\phi_1\rangle \otimes |d\rangle$ is actually the asymptotic state of the system (*i.e.*, the particle bound state $|\phi_1\rangle$ is practically the asymptotic state of the particle). This state is one of the eigenstates of H_S at $P = Q = 0$ and remains approximately such at low (but still finite) values of P and Q. The corresponding particle energy E_1 is appreciably above all other particle energies E_i, $i = 2, ...5$ in (3.85) whenever $V \gg |J| > 0$ — the regime we have in mind here. This is what makes our model so challenging.

Technical details of the solution can be found in [49]. In short, time-convolution (Nakajima-Zwanzig) Generalized Master Equations (TC-GME) are formulated only for probabilities $P_{im}(t)$ of finding our system (central system plus particles) in states $|im\rangle = |\phi_i\rangle \otimes |m\rangle$, i=1,...5, $m = u$ or d.[32] When the physical time t goes to infinity, the TC-GME becomes a homogeneous set of linear algebraic equations for stationary probabilities $P_{im}(t \to +\infty)$. Complementing it by the normalization condition, the set is solved and the stationary probabilities are expressable via time-integrals of different memory functions entering the problem. Then the critical points arrives: One cannot expand in powers of H_{S-B}; that would again turn the treatment into the weak-coupling one. In accordance with the discussion above concerning competition between bath-induced relaxation channels and those corresponding to internal dynamics of the system, $H_{S-B} + H''_{cs-part}$ (the part of the total Hamiltonian causing transitions) is taken as a perturbation. Expansion in terms of $H_{S-B} + H''_{cs-part}$ then reveals that:

- The usual detailed balance conditions relating equilibrium populations (via the ratio of bath-assisted transfer rates) are not applicable to the present situation. The reason is simple: We now treat stationary populations of states that are *not* eigenstates of Hamiltonian H_S of the system. In accordance with that, not only bath-assisted transfer rates influence populations of individual states considered.

---

[31]The bath is the only source of energy at our disposal for the endothermic process investigated.
[32]Off-diagonal elements of the density matrix of the system are projected off; in this respect, the method slightly differs from those above.

- The solution yields, at low temperatures $k_B T \ll \epsilon$,

$$P_{1d}(+\infty) \approx 1, \qquad (3.91)$$

with all other asymptotic probabilities being practically zero. Thus, owing to action of the central system, the particles become coupled (in a bound state) with the central system, while it prepares to start action on another pair.

A picture of the process now emerges. The unbound c- and g-particles can appear simultaneously at site 0 joined with the central system. This leads to an instability of the central system, *e.g.*, a change of topology, which in turn may bring the particles together, leading to the formation of a bound state. Owing to the new topology, the particles can then leave site 0 only as a bound pair. With the desorption of the bound pair, the central system becomes unstable again, returns to its original topology, and waits for another c-g pair to bind.

This system could play the role of a quantum microscopic *sewing machine*. It is emphasized that the asymptotic state does not differ from the equilibrium state due to some transitions lacking from excited states to the equilibrium state[33], nor due to any type of energy renormalization shifting the asymptotic state sufficiently down in energy. The paradoxical result is solidly due to the active role played by the central system and the proper combination of two properties: (1) instability of its intermediate state (scatterer) during transition (scattering of the pair); and (2) strong dependence of the scattering channels on the state of the central system, owing to the matrix elements. These results contradict standard thermodynamics by means of an up-in-energy transition at the cost of thermal energy of the bath. This is an isothermal Maxwell demon.

3.6.5 Single Phonon Mode Model

This model is based on the observation that the existence of Maxwellian gate states is not necessarily connected with the two-state character of the 'central system' as in the fish-trap family of models above. The closing or opening of the gate can also be related to two different forms of the polaron deformation cloud around the processed particle. Again, we discuss here just a single-cycle process [50, 51], however, a continuously (periodically) operating system should also behave as a bona fide *perpetuum mobile* of the second kind. Recently, a generalization to the case of a gradual excitation on a ladder of states leading to a spontaneously generated population inversion has been found to be possible [52].

As with the fish-trap model, the central part of this system is an oscillator. Again, splitting the total Hamiltonian, as in (3.54), we have

$$H_S = J(c^\dagger_{-1} c_0 + c^\dagger_0 c_{-1}) \otimes [b + b^\dagger + 2\gamma] + I(c^\dagger_0 c_1 + c^\dagger_1 c_0) \otimes [b + b^\dagger]$$

$$+ \delta\epsilon c^\dagger_1 c_1 + \hbar\omega(b^\dagger + \gamma c^\dagger_0 c_0)(b + \gamma c^\dagger_0 c_0). \qquad (3.92)$$

---

[33]This can be verified by complementing the model with any transitions desired.

Thus, we have one particle at three possible sites $m = 0$ or ± 1. Here c_m (or $c_m^\dagger$), are the annihilation (or creation) operators of our particle at site m. Next, b ($b^\dagger$) designates the phonon annihilation (creation) operator of the oscillator, while $\hbar\omega$ is the phonon energy. The site energies of the particle located at $m = -1$, 0 or +1 are 0, 0 or $\delta\epsilon$, respectively. Thus, a particle located initially off site +1 requires $\delta\epsilon > 0$ energy to transfer to site +1. Undergirding the effect is an arbitrary real parameter γ that determines the shift of the oscillator coordinate upon formation of a small polaron around particle at site '0.'

As for H_B and H_{S-B}, we assume that the bath is connected to the system only by its coupling to the system oscillator. Thus, for particle relaxation, the oscillator plays the role of a bottle-neck. We do not require that J and I be very small, but this regime is the easiest to understand. The point is that the oscillator always relaxes to the canonical state in the representation of eigenstates of $H_S|_{J=I\approx 0}$. This means it relaxes to

$$\rho_{osc}^{can1} = [1 - e^{-\beta\hbar\omega}] \sum_{\nu=0}^{\infty} |\nu\rangle e^{-\beta\nu\hbar\omega} \langle\nu| \tag{3.93}$$

when the particle is at sites -1 or 1, or to

$$\rho_{osc}^{can2} = [1 - e^{-\beta\hbar\omega}] \sum_{\nu=0}^{\infty} |\nu'\rangle e^{-\beta\nu\hbar\omega} \langle\nu'| \tag{3.94}$$

when the particle is at site 0. Here

$$|\nu\rangle = \frac{1}{\sqrt{\nu!}}(b^\dagger)^\nu |0\rangle, \quad |\nu'\rangle = \frac{1}{\sqrt{\nu!}}(b^\dagger + \gamma)^\nu |0'\rangle, \tag{3.95}$$

and $|0\rangle$ and $|0'\rangle$ are the corresponding oscillator ground states, defined by $b|0\rangle = 0$ and $(b + \gamma)|0'\rangle = 0$. Clearly, $|0'\rangle = \exp(\gamma(b - b^\dagger))|0\rangle$. Again, $\beta = (k_B T)^{-1}$ is the reciprocal temperature in the energy units. Ignoring details of H_B and H_{S-B}, we assume the oscillator relaxation in a simple Landau and Teller form, known for more than 50 years [53]. This means exponential relaxation in the oscillator bases.

Let the Latin indices $m, n, \ldots = -1$, 0 or +1 designate the sites and let the Greek indices $\mu, \nu, \ldots = 0, 1, 2, \ldots$ be the quantum numbers of the oscillator (phonon occupation numbers). Let $\rho(t)$ be the density matrix of the (particle+oscillator) system; i.e., $\rho_{m\mu,n\nu}(t)$ is its matrix in the representation of states $|m\mu\rangle = |m\rangle \otimes |\mu\rangle$. Then equation (3.53) reads

$$i\frac{d}{dt}\rho_{m\mu,n\nu}(t) = \sum_{p\pi,q\kappa} \mathcal{L}_{m\mu,n\nu,p\pi,q\kappa} \rho_{p\pi,q\kappa}(t). \tag{3.96}$$

Here $\mathcal{L}_{m\mu,n\nu,p\pi,q\kappa}$ is the four-(double)index matrix of the Liouville superoperator $\mathcal{L}$ consisting of two parts as

$$\mathcal{L} = \mathcal{L}_S + \mathcal{L}^{rel}. \tag{3.97}$$

Here $\mathcal{L}_S \ldots = [H_S, \ldots]/\hbar$; that is,

$$(\mathcal{L}_S)_{m\mu,n\nu,p\pi,q\kappa} = \frac{1}{\hbar}\{(H_S)_{m\mu,p\pi}\delta_{q\kappa,n\nu} - (H_S)_{q\kappa,n\nu}\delta_{m\mu,p\pi}\}. \tag{3.98}$$

As for the oscillator relaxation part of the Liouvillian $\mathcal{L}^{rel}$, it has been assumed to be described by the Landau-Teller relaxation,[34] which is different if the particle resides at the site 0 or outside. Thus,

$$\mathcal{L}^{rel}_{m\mu,n\nu,p\pi,q\kappa} = \delta_{mp}\delta_{nq}\{(1-\delta_{m,0})(1-\delta_{n,0})\mathcal{K}_{\mu,\nu,\pi,\kappa}$$

$$+0.5[(1-\delta_{m,0})\delta_{n,0}+\delta_{m,0}(1-\delta_{n,0})][\mathcal{K}_{\mu,\nu,\pi,\kappa}+\mathcal{K}'_{\mu,\nu,\pi,\kappa}]+\delta_{m,0}\delta_{n,0}\mathcal{K}'_{\mu,\nu,\pi,\kappa}\}. \quad (3.99)$$

Describing this relaxation in terms of the Generalized Stochastic Liouville equation model [54, 55] in the Haken-Strobl-Reineker-like parametrization [56, 57], we get $\mathcal{K}_{\mu,\nu,\pi,\kappa}$ as

$$\mathcal{K}_{\mu,\nu,\pi,\kappa} = i(2\delta_{\mu,\nu}\delta_{\pi,\kappa}[\gamma_{\mu\pi} - \delta_{\mu\pi}\sum_{\lambda}\gamma_{\lambda,\mu}] - (1-\delta_{\mu,\nu})\delta_{\mu,\pi}\delta_{\nu,\kappa}\sum_{\lambda}[\gamma_{\lambda,\mu}+\gamma_{\lambda,\nu}]),$$
$$(3.100)$$

with the Landau-Teller [53] formula for the relaxation rate

$$\gamma_{\mu,\nu} = \tilde{k}[(\mu+1)\delta_{\nu,\mu+1} + \mu\exp(-\beta\hbar\omega)\delta_{\nu,\mu-1}]. \quad (3.101)$$

Notice that, in contrast to the original Stochastic Liouville equation model, the transfer rates $2\gamma_{\mu,\nu}$ are, in general, asymmetric here because of inclusion of spontaneous transfer processes $\nu \to \mu$ (with respect to the quantum bath). The constant $\tilde{k}$ is the only one reflecting the strength of the oscillator coupling to the bath. It need not be small, in principle. As for $\mathcal{K}'_{\mu,\nu,\pi,\kappa}$, it describes the same relaxation, but with the particle located at site 0. Thus, this relaxation is no longer to ρ^{can1}_{osc}, but rather to ρ^{can2}_{osc}. Hence, the matrix of the $\mathcal{K}'$ relaxation superoperator should be, in the basis of $|\nu'\rangle$ states, the same as that of the $\mathcal{K}$ in the basis of $|\nu\rangle$ states. Hence,

$$\mathcal{K}'_{\mu,\nu,\pi,\kappa} = \sum_{\zeta,\eta,\iota,\lambda} \langle\mu|\zeta'\rangle\langle\eta'|\nu\rangle\langle\iota'|\pi\rangle\langle\kappa|\lambda'\rangle\mathcal{K}_{\zeta,\eta,\iota,\lambda}. \quad (3.102)$$

In order to see that the long-time values of the particle site-occupation probabilities are not merely quasistationary values, but correspond to a new asymptotic state, one can check by hand that the long-time asymptotic form of the density matrix $\rho(t)$ as obtained from (3.96) and (3.97-3.102) reads, for $\tilde{k}$ dominating over $\delta\epsilon/\hbar$, $|J|/\hbar$ and $|I|/\hbar$ as

$$\rho(+\infty) = \sum_{m=\pm 1} |m\rangle P_m(+\infty)\langle m| \otimes \rho^{can1}_{osc} + |0\rangle P_0(+\infty)\langle 0| \otimes \rho^{can2}_{osc} + \mathcal{O}(\frac{1}{\tilde{k}}). \quad (3.103)$$

Analytically determining the site occupation probabilities $P_m(+\infty)$ in (3.103), even in the above regime, is difficult. Anyway, (3.103) becomes diagonal in a basis other than H_S. Thus, it appreciably differs from the standard canonical density matrix proportional to $\exp(-\beta H_S)$ to which $\rho(+\infty)$ necessarily goes for small enough $\tilde{k}$ [19].

In order to understand the effect, let us start from an initial condition with the particle at site '-1' and the bath in its canonical state (3.93). This means that the

---

[34]Other types of relaxation yield qualitatively the same results.

average value of the first term on the right hand side of (3.92), with respect to the oscillator state, is equal to $H_S = 2\gamma J(c^\dagger_{-1}c_0 + c^\dagger_0 c_{-1})$. In other words, the particle is permitted to come to site '0' and return back to '-1' unless something happens with the oscillator in between. On the other hand, the mean value (with respect to the oscillator) of the second term on the right hand side of (3.92) is zero; *i.e.*, the particle cannot (on average) immediately proceed to site '+1'. Assume, however, that the particle really appeared partially at site '0'. Fast dephasing suppresses the site off-diagonal elements in the particle density matrix $\rho_{mn}(t) = \sum_\mu \rho_{m\mu,n\mu}(t)$. Thus, the particle is deprived of any phase relations between sites (bonding); *i.e.*, it is localized only at '-1' or '0'. If the particle remains or succeeds in returning from site '0' to '-1', the story begins again. In the other case, the oscillator succeeds in re-relaxing to the canonical state (3.94). Then the mean value of $b+b^\dagger$ with respect to the canonical oscillator state (3.94) becomes equal to -2γ. Hence, the same mean values of the first and second terms on the right hand side of (3.92) become 0 and $-2\gamma I(c^\dagger_0 c_1 + c^\dagger_1 c_0)$, respectively; *i.e.*, effectively, the particle cannot return to site '-1', but can freely proceed to site '+1'. Then, however, the oscillator again re-relaxes to the canonical state (3.93). Hence the return channel of the particle to site '0' effectively gets closed and the particle is forced to remain at site '+1'.

Such arguments apply of course only 'on average'. A possible objection is that the gates between sites '-1' and '0', or '0' and '+1' never become fully closed owing to fluctuations of the oscillator coordinate proportional to $b + b^\dagger$ around its mean values. One should expect, in appropriate situations, an appreciable increase of population at site '+1' with respect to standard values provided by the quantum statistics. Two points are stressed here. First, the expected increase in the site occupation probability $\rho_{+1,+1} \equiv P_{+1}(t)$ is only on account of the above dynamic behavior of the oscillator and can be effective even if the transfer to site '+1' is connected with an appreciable increase of the (site-)energy of the particle transferred ($\delta\epsilon$ in (3.92) becomes positive). Second, as there is no site-diagonal coupling of the particle at site '+1' to the oscillator, the increase of $P_{+1}(t)$ *cannot* be explained as a polaron energy shift. (See also numerical data for P_{+1} obtained from the equilibrium canonical distribution below which would have to reflect this polaron shift.) In order to check for the effect expected, a reliable numerical solution to (3.96-3.102) is necessary. This was solved by a special procedure projecting off the oscillator. The result is that:

- Stationary values of particle population appreciably differ from those determined from the canonical density matrix.

- For $\delta\epsilon > 0$, the highest site (site '+1') may easily become populated with probabilities exceeding $80 - 90\%$.

- Owing to the dephasing, bonding of the particle at site '+1' to other sites becomes appreciably violated. Thus, the particle is finally free to pass its energy wherever it wishes; upon a simultaneous return of the particle to site '-1', we obtain a periodically working *perpetuum mobile* of the second kind, similarly as for the fish-trap model above.

- The effect of final transfer of the particle to the highest site is not forbidden by energy conservation. On the contrary, because of the system dephasing (reducing to a continuous exchange of energy of the system with the reservoir), the bath energy becomes the source of the increased site energy for the particle.

- The system phonon mode works as a Maxwellian gate. Like with the fishtrap model above, however, it does not need any external being (demon) determining the proper timing of the gate, nor does it need any other external interventions. The binary decision (close or open) goes on automatically, as if the system alone performed the necessary measurement on the particle position.

3.6.6 Phonon Continuum Model

A continuum of oscillators — not simply a single oscillator — can play the role of the Maxwellian gate [58]. The bath then ceases to behave in a passive manner, but becomes quite active. The cooperative self-organizational effects induced in the bath by the transferred particle can appreciably modify detailed balance. Possible effects include: (i) single-particle rectification (preferred unidirectional transfer); (ii) prevailingly uphill particle transfer at the cost of energy of a single and typically nonequilibrium infinite bath; and (iii) induced particle self-organization. The model provides a basis for treating dynamic self-organization in open quantum systems, as well as other mechanisms that could lead to violations of basic statistical-thermodynamic principles.

3.6.7 Exciton Diffusion Model

This model has been given a detailed mathematical theory [59, 60], but can be easily understood on grounds of well-founded theory and experimentally established facts. These include:

- The equilibrium population of localized exciton levels is standardly negligible and sharply decreases with increasing exciton level. Thus, higher-in-energy exciton levels have lower exciton population ($\epsilon_1 < \epsilon_2$ implies populations $P_1 > P_2$).

- This inequality survives for even moderate local temperatures ascribed to the levels (i.e., from statistical mechanics, for $T_1 < T_2 < T_1\epsilon_2/\epsilon_1$). Also, when exciton (energy) flows are sufficiently small at small deviations from equilibrium, the inequality $P_1 > P_2$ is preserved.

- The leading term connecting two local but spatially neighboring exciton levels is due to resonance interaction J that is elastic[35] ; thus, the transfer

---

[35]Usually, the corresponding transfer is written in the total Hamiltonian as $J(c_1^\dagger c_2 + c_2^\dagger c_1)$, where $c_m^\dagger$ and c_m, $m = 1, 2$ are the local exciton creation and annihilation operators.

between such levels is energy conserving. Energy conservation for such real — not virtual — transitions requires either broadening of the levels or assistance of bath-assisted inelastic processes.

- For the case when the finite life-time effects for excitons (owing to bath-assisted local exciton creation and annihilation processes) are commensurable with the $1 \leftrightarrow 2$ transfers, real (non-virtual) processes with active participation of the bath (that could make such processes inelastic, *i.e.*, non-symmetric) becomes a process of higher order in the perturbation, thus, they become negligible with respect to elastic processes between tails of broadened exciton levels. Such transfer processes are, however, elastic; that is, they might be given by a transfer rate W that is the same for $1 \to 2$ and $2 \to 1$.

- Consequently, the net exciton flow $1 \to 2$ of the diffusive type reads $W(P_1 - P_2)$; it is positive in this situation.

- Detailed reasoning [59, 60] shows that this prevailing $1 \to$ exciton flow also causes energy flow of the same orientation. Thus, if an exciton-transfer channel is the only one connecting two baths at moderately different local temperatures, the energy flow can go against the temperature step — in sharp contradiction with the Clausius form of the second law.

This model can also be used to undercut the zeroth law of thermodynamics. Both types of challenges, however, disappear in the high-temperature limit.

The Hamiltonian for the model (two exciton levels, each with its own bath of harmonic oscillators) can be written in the form (3.54) or in one that makes the two subsystems I and II explicit. That is,

$$H = H_I + H_{II} + J(c_1^\dagger c_2 + c_2^\dagger c_1),$$

$$H_I = \epsilon_1 c_1^\dagger c_1 + \sum_\kappa \hbar\omega_\kappa b_{1\kappa}^\dagger b_{1\kappa} + \frac{1}{\sqrt{N}} \sum_\kappa g_\kappa \hbar\omega_\kappa (c_1 + c_1^\dagger)(b_{1\kappa} + b_{1\kappa}^\dagger),$$

$$H_{II} = \epsilon_2 c_2^\dagger c_2 + \sum_\kappa \hbar\omega_\kappa b_{2\kappa}^\dagger b_{2\kappa} + \frac{1}{\sqrt{N}} \sum_\kappa G_\kappa \hbar\omega_\kappa (c_2 + c_2^\dagger)(b_{2\kappa} + b_{2\kappa}^\dagger). \quad (3.104)$$

For mathematical details see [59, 60]. Regardless of the transparent physical interpretation and the mathematically rigorous treatment, real experimental tests of this mechanism seem limited because the magnitude of the effect is expected to be small.

3.6.8 Plasma Heat Pump Model

This is perhaps the only model of this genre that corresponds directly to laboratory experiments. The system is a low-temperature plasma in a blackbody cavity (See §8.2 and §8.3). The quantum model of this classically conceived system comes in two slightly different forms [61, 62]. The Hamiltonian of the three-site model of [61] reads, as in (3.54),

$$H_S = \sum_{j=2}^{3} \epsilon_j a_j^\dagger a_j + J(a_1^\dagger a_2 + a_2^\dagger a_1) + K(a_2^\dagger a_3 + a_3^\dagger a_2), \qquad (3.105)$$

where the zero of energy is taken at site 0, representing the walls of the container. Site 2 represents the plasma (ϵ_2 means electron energy inside bulk of the plasma) while site 3 represents a physical probe (See Figure 8.1). For the plasma one has $\epsilon_2 > \epsilon_3 > 0$. In this model the wall-plasma thermionic and plasma-probe transitions are elastic; this is physically reasonable and corresponds to the standard model of electron emission from solids. The load is the location at which the electron can inelastically scatter during $3 \leftrightarrow 1$ transitions. (This corresponds to dissipative electron flow from the probe to the walls through an electrical load.) In the present model, the load is not conservative (e.g., a lossless motor) and, thus, not a challenge to the Kelvin-Planck form of the second law; rather it is purely dissipative (e.g., a resistor) and, therefore, a challenge to the Clausius form of the second law. This requires that the resistor becomes slightly warmer than the surrounding heat bath.

Relatively high temperatures are required for this model to support thermionic emission. This suggests that inside the load, inelastic phonon-assisted scattering processes prevail. This is why we assume, in (3.105), that the $3 \leftrightarrow 1$ transitions are phonon-assisted. The spontaneous phonon-emission processes in the load and the $3 \leftrightarrow 1$ transfer rate imbalance are quantum mechanical effects; in fact, as will be shown, they disappear in the classical limit (infinite temperature). For simplicity, we assume only one electron in the system. The spontaneous phonon-emission transitions $3 \to 1$ are the only source of imbalance that can cause the electron flow $1 \to 2 \to 3 \to 1$, provided that the latter is not blocked by the electron bonding among the sites. This means that we must definitely go beyond the weak coupling theories, where the equilibrium canonical Gibbs distribution entails fully developed bonding. Going beyond this limit corresponds to the modeled plasma experiment, where any such bonding is unphysical.

In the quantum model and experimental system, the cyclic electron flow acts as a heat pump. This requires two heat baths (I and II). Bath I heats the electron at the wall before it enters the plasma ($1 \leftrightarrow 2$ transitions), while Bath II coincides with the load and is coupled to the electron during $3 \leftrightarrow 1$ transitions. Both baths are represented by harmonic phonons; hence,

$$H_B = H_B^I + H_B^{II},$$

$$H_B^I = \sum_{\kappa} \hbar \omega_\kappa b_\kappa^\dagger b_\kappa, \qquad H_B^{II} = \sum_{\kappa} \hbar \omega_\kappa B_\kappa^\dagger B_\kappa. \qquad (3.106)$$

The electron-bath coupling, H_{S-B}, is given by

$$H_{S-B} = H_{S-B}^I + H_{S-B}^{II},$$

$$H_{S-B}^I = \frac{1}{N} \sum_{\kappa_1, \kappa_2} \hbar \sqrt{\omega_{\kappa_1} \omega_{\kappa_1}} g_{\kappa_1, \kappa_2} a_2^\dagger a_2 (b_{\kappa_1} + b_{\kappa_1}^\dagger)(b_{\kappa_2} + b_{\kappa_2}^\dagger),$$

$$H^{II}_{S-B} = \frac{1}{\sqrt{N}} \sum_{\kappa} \hbar\omega_\kappa G_\kappa (a^\dagger_3 a_1 + a^\dagger_1 a_3). \tag{3.107}$$

Identifying the perturbation with the sum of H_{S-B} with terms $J(a^\dagger_1 a_2 + a^\dagger_2 a_1) + K(a^\dagger_2 a_3 + a^\dagger_3 a_2)$ from H_S and proceeding as above, from (3.53) we get

$$i\hbar\frac{d\rho}{dt} = \left(\begin{array}{ccccc} & & \cdot & & \\ & & \cdot & & \\ & \mathcal{A} & \cdot & \mathcal{B} & \\ & & \cdot & & \\ \cdot & \cdot & \cdot & \cdot & \cdot \\ & & \cdot & & \\ & \mathcal{B}^T & \cdot & \mathcal{C} & \\ & & \cdot & & \end{array} \right) \left(\begin{array}{c} \rho_{11} \\ \rho_{22} \\ \rho_{33} \\ \rho_{12} \\ \rho_{21} \\ \rho_{13} \\ \rho_{31} \\ \rho_{23} \\ \rho_{32} \end{array} \right) \cdot$$

$$\tag{3.108}$$

The blocks $\mathcal{A}$, $\mathcal{B}$, $\mathcal{C}$ are given as

$$\mathcal{A} = \left(\begin{array}{ccccc} -i\hbar\Gamma_\uparrow & 0 & i\hbar\Gamma_\downarrow & -J & J \\ 0 & 0 & 0 & J & -J \\ i\hbar\Gamma_\uparrow & 0 & -i\hbar\Gamma_\downarrow & 0 & 0 \\ -J & J & 0 & -2i\hbar\Gamma - \frac{i\hbar}{2}\Gamma_\uparrow - \epsilon_2 & 0 \\ J & -J & 0 & 0 & -2i\hbar\Gamma - \frac{i\hbar}{2}\Gamma_\uparrow + \epsilon_2 \end{array} \right),$$

$$\mathcal{B} = \left(\begin{array}{cccc} 0 & 0 & 0 & 0 \\ 0 & 0 & -K & K \\ 0 & 0 & K & -K \\ -K & 0 & 0 & 0 \\ 0 & K & 0 & 0 \end{array} \right),$$

$$\mathcal{C} = \left(\begin{array}{cccc} -\frac{i\hbar}{2}\Gamma_\downarrow - \epsilon_3 & \frac{i\hbar}{2}\Gamma_\uparrow & J & 0 \\ \frac{i\hbar}{2}\Gamma_\downarrow & -\frac{i\hbar}{2}\Gamma_\uparrow + \epsilon_3 & 0 & -J \\ J & 0 & -2i\hbar\Gamma - \frac{i\hbar}{2}\Gamma_\downarrow + \epsilon_{23} & 0 \\ 0 & -J & 0 & -2i\hbar\Gamma - \frac{i\hbar}{2}\Gamma_\downarrow - \epsilon_{23} \end{array} \right).$$

$$\tag{3.109}$$

Here, $\epsilon_{23} = \epsilon_2 - \epsilon_3$, $\Gamma_\updownarrow = \Gamma_\uparrow + \Gamma_\downarrow$, and we have also used the notation

$$\Gamma_\uparrow = \frac{2\pi}{\hbar}\frac{1}{N}\sum_\kappa |\hbar\omega_\kappa|^2 |g_\kappa|^2 n_B(\beta_{II}, \hbar\omega_\kappa)\delta(\hbar\omega_\kappa - \epsilon_3),$$

$$\Gamma_\downarrow \equiv \frac{2\pi}{\hbar}\frac{1}{N}\sum_\kappa |\hbar\omega_\kappa|^2 |g_\kappa|^2 [1 + n_B(\beta_{II}, \hbar\omega_\kappa)]\delta(\hbar\omega_\kappa - \epsilon_3) = \Gamma_\uparrow \cdot e^{\beta_{II}\epsilon_3},$$

$$2\Gamma = \frac{2\pi}{\hbar}\frac{1}{N^2}\sum_{\kappa_1,\kappa_2} |g_{\kappa_1,\kappa_2}|^2 (\hbar^2\omega_{\kappa_1}\omega_{\kappa_2})^2 n_B(\beta_I, \hbar\omega_{\kappa_1})[1 + n_B(\beta_I, \hbar\omega_{\kappa_2})]\delta(\hbar\omega_{\kappa_1} - \hbar\omega_{\kappa_2}),$$

$$n_B(\beta, z) = \frac{1}{e^{\beta z} - 1}, \qquad (3.110)$$

where $T_{I(II)} = 1/(k_B\beta_{I(II)})$ are the initial temperatures of Baths I and II; and $n_B(\beta, z)$ is the Bose-Einstein phonon distribution function. $\Gamma_\uparrow$ and $\Gamma_\downarrow$ are the Golden Rule formulae for transfer rates $1 \to 3$ and $3 \to 1$. Note that $\Gamma_\uparrow$ and $\Gamma_\downarrow$ are different solely in that the latter has a $(1 + n_B)$ term, whereas the former has only n_B. Physically, this corresponds to $\Gamma_\uparrow$ involving only bath-assisted induced transitions, whereas $\Gamma_\downarrow$ involves both bath-assisted induced and bath-assisted spontaneous transitions. Finally, 2Γ determines the rate of dephasing arising from local electron-energy fluctuations from Bath I and, also, the rate of electron heating in the plasma. As usual, we disregard the inhomogeneous initial-condition term on the right hand side of (3.108) by assuming a factorizable form of the initial density matrix of the system and bath.

The system (3.108) can be tested numerically to get (in the stationary situation with the complementing normalizing condition $\sum_{m=1}^{3} \rho_{mm} = 1$) the heat transfer rate from bath I to bath II. We assume that there is no other heat transfer channel possible than that connected with the circular electron flow. Because $3 \to 1$ transitions should dominate over $3 \leftarrow 3$ ones, and because the former (latter) always lead to the phonon emission to (absorption from) bath II, the heat flow I→II

$$Q = \epsilon_3[\Gamma_\downarrow\rho_{33} - \Gamma_\uparrow\rho_{11}] \qquad (3.111)$$

should be nonzero. This can be proved analytically [61]. Analytically, it can also be proved that Q is always positive; *i.e.*, for arbitrary initial temperatures T_I and T_{II} of the baths, the flow is always from bath I to bath II. Hence, for $T_{II} > T_I$, we have a spontaneous heat flow (not aided from outside) against a temperature step from a colder bath to the warmer one. This corresponds to the plasma experiment in which, presumably, heat spontaneously flows from the colder bath to the warmer resistor. This sharply contradicts the second law in the Clausius formulation. The paradoxical behavior is lifted both in the high-temperature and the low-temperature limits. In the high temperature limit ($k_B T_{II} \gg \epsilon_3$) the underlying spontaneous $3 \to 1$ processes becomes negligible. It also can be numerically checked that the effect disappears when $T_1 \to 0$ since there is no heat in bath I available to be transferred [62].

A classical explanation of this plasma system is given elsewhere (§8.2). It is hoped that these two descriptions will set a precedent for understanding the many seemingly disparate second law challenges under a common rubric.

3.7 Disputed Quantum Models

The following are notable quantum models which are less sanguine about the prospects for second law violation.

3.7.1 Porto Model

Recently, Porto [63] suggested a model consisting of a linear track and a rotor, both consisting of charges interacting mutually by Coulombic forces only. When system parameters are adjusted appropriately, the rotor moves uni-directionally, collecting the energy of thermal fluctuations. The system is describable in terms of Langevin equations in the zero-temperature limit. Porto stresses that the system does not violate the second law. However, if one extends the system of Langevin equations describing the rotor dynamics, a challenge to the second law does in fact emerge. The interested reader is referred to [64], which presents results obtained with a high numerical precision. Evidently, even classical counterparts of the Mori or Tokuyama-Mori equations indicate second law violability.

3.7.2 Novotný Model

Although most systems described thus far are tightly connected with the Davies formalism, other methods can obtain these results. Another approach mentioned in the Appendix of [64] is based on the time-convolutionless GME. In this connection, we mention criticism that appeared in [66, 67]. The paper presents a model that allows an exact solution by the non-equilibrium Greens function method. Its conclusions differ markedly from those obtained by the Davies method. Significantly, however, this conflicting model [66, 67] does not describe diffusion upon which the prior results of [66, 67] rely. This makes the criticism problematic [65].

3.8 Kinetics in the DC Limit

The greatest uncertainty associated with kinetic theories is whether they can be justified in the kinetic regime. The kinetic regime is limited at short time scales by the decay of initial conditions, and on long time scales by the hydrodynamic regime. In only a few academic cases can the validity of the kinetic equations be extended to infinitely long times (zero-frequency, dc limit). One such set of cases involves the Generalized Master Equations (GME) (time-convolution or time-convolutionless), provided one takes the limit of infinite time before performing any approximations. Once we use an approximate treatment, however, the infinite-time limit becomes open to discussion — much to the disappointment, for instance, of those who apply the Boltzmann equation to the dc conductivity problem. Fortunately, there is justification for their use in the models we have treated thusfar. Some of the models are based on approaches where the dc limit is performed on the level of the still exact GME, preceding any expansion in small parameters of the problem. Examine, for instance, the model with active binding of particles above [49], or the Appendix in [60].

With perturbations that correspond to the intermediate or strong coupling regimes, one may start from formulations based on the time-convolutionless GME; see, for example, (3.35). Then $\mathcal{R}(t)$ gives the usual Redfield relaxation superoperator (independent of time) when t exceeds the appropriate bath relaxation. There is no limitation on the validity of (3.35) from the side of long times.

This does not remove all doubts concerning kinetic theories.[36] Some of the reasons are as follows:

- All contemporary kinetic theories are either a consequence of the Liouville equation for the density matrix of the (system + bath) complex, or are at least compatible with it. The Liouville equation, as a consequence of the Schrödinger equation, is part of the standard arsenal of quantum theory. As such, there are still discussions about its applicability to arbitrarily long times — at the very least, with respect to irreversibility in the macroworld [68].

- The universally accepted procedure for theoretical modeling is that one considers only those mechanisms that can be argued to be decisive. The basic argument for choices of such mechanisms is the relative weight of the corresponding terms in the total Hamiltonian as measured by, for example, dimensionless values of coupling constants. Unfortunately, arguments exist that even negligible mechanisms can, in some cases, completely alter the asymptotic state of a system [69]. Therefore, enormous care must be exercised in deciding which results are stable in the long-time limit.

These considerations appreciably increase the role of experiment as an arbiter. Below, we review both optimistic and pessimistic views concerning the predictability and competence of theory to provide conclusions concerning the second law.

3.8.1 TC-GME and Mori

The above identity (3.6), valid for arbitrary projector $\mathcal{P}$, may be understood as a full equivalent of the Liouville equation (3.1) for the density $\rho_{S+B}(t)$ of the isolated (system + bath) complex. It is obtained from the Liouville equation and reduces to it for the special choice $\mathcal{P} = 1$. Because of its connections with the Liouville and Schrödinger equation — the former is a consequence of the latter — (3.6) belongs to the basic tools for calculating time-dependence of expectation values for quantum operators in the Schrödinger picture[37]. Expectation values can also be expressed in the Heisenberg picture in which the density matrix becomes stationary and the time development is transferred to the operators

$$\langle A \rangle(t) \equiv \mathrm{Tr}(\rho(t)A) = \mathrm{Tr}(e^{-iHt/\hbar}\rho(0)e^{iHt/\hbar}A)$$

$$= \mathrm{Tr}(\rho(0)A(t)), \tag{3.112}$$

where the Heisenberg operator is

$$A(t) \equiv e^{iHt/\hbar}Ae^{-iHt/\hbar} \equiv e^{\mathcal{L}t}A. \tag{3.113}$$

---

[36]These concerns about kinetic approaches are not connected, *per se*, with the status of the second law.

[37]These operators are, in the Schrödinger picture, always time-independent insofar as there is no external time-dependent field acting on the system, bath, or both. We assume this situation here.

Hence, the question arises whether a counterpart of (3.6) exists that would be equally general, but which would be equally advantageous whenever A acts only on system variables[38]. This counterpart is provided by the Mori identity [70, 1]

$$\frac{d}{dt}G(t) = e^{i\mathcal{L}t}\mathcal{D}i\mathcal{L}G(0)-$$

$$\int_0^t e^{\mathcal{L}(t-\tau)}\mathcal{D}\mathcal{L}e^{i(1-\mathcal{D})\mathcal{L}\tau}(1-\mathcal{D})\mathcal{L}G(0)\,d\tau + e^{(1-\mathcal{D})\mathcal{L}t}(1-\mathcal{D})i\mathcal{L}G(0). \qquad (3.114)$$

Here $\mathcal{D} = \mathcal{D}^2$ is an arbitrary projector and $G(0)$ is an arbitrary operator in the Schrödinger picture. $G(t)$ then designates the corresponding operator counterpart in the Heisenberg picture. For $\mathcal{D} = 1$, (3.114) reduces to a direct consequence of definition (3.113).

Correspondence between (3.6) and (3.114) may be best seen upon: (i) choosing $G(0)$ in the form of set of operators $|m\rangle\langle n|$ (where $|m\rangle$, $|n\rangle$, etc. denote arbitrary states in the Hilbert space of the system only); (ii) choosing $\mathcal{D}\ldots = \sum_{p,q}|p\rangle\langle q|\mathrm{Tr}_S(|q\rangle\rho^B\langle p|\ldots)$, $\mathrm{Tr}_B\rho^B = 1$; (iii) multiplying (3.114) from the left by the density matrix of the (system + bath) complex $\rho_{S+B}(0)$; and (iv) taking the trace (Tr) of the result. What we obtain is nothing but (3.9), however, identity (3.114) provides additional possibilities. For example, the problem of Brownian particle (system) in a liquid (bath), one may choose G as a set of two (vector) operators of the Brownian particle (coordinate and momentum). Then (3.114) provides a unique and rigorous path to the quantum time-convolution counterpart of standard classical Langevin equations for the Brownian particle [70, 1].

Let us return to the problem connected with the kinetic description provided by (3.9). For purposes of evaluating pro and con arguments associated with the violability of the second law, consider the long-time regime. In particular, we would like to add another objection concerning the reliability of the long-time asymptotics of the density matrix $\rho_S(t)$, as deduced from convolution equations of the type (3.9). This type of objection applies not only to the Nakajima-Zwanzig method for deriving asymptotics of $\rho_S(t)|_{t\to+\infty}$ via (3.6), but also against that based on the Mori identity (3.114). The basic observation is that the initial condition term in (3.6); $I(t) \equiv -i\mathcal{P}\mathcal{L}\exp[-i(1-\mathcal{P})\mathcal{L}\cdot(t-t_0)](1-\mathcal{P})\rho_{S+B}(t_0)$ can be rewritten as [71, 72]

$$I(t) = [\frac{d}{dt} + i\mathcal{P}\mathcal{L}]\mathcal{P}e^{-i\mathcal{L}(t-t_0)}\rho_{S+B}(t_0)$$

$$+ \int_{t_0}^t \mathcal{P}\mathcal{L}e^{-i(1-\mathcal{P})\mathcal{L}(t-\tau)}(1-\mathcal{P})\mathcal{L}\mathcal{P}e^{-i\mathcal{L}(\tau-t_0)}\rho_{S+B}(t_0)\,d\tau. \qquad (3.115)$$

Thus, the Nakajima-Zwanzig identity (3.6) becomes

$$[\frac{d}{dt} + i\mathcal{P}\mathcal{L}]\mathcal{P}\rho_{S+B}(t) + \int_{t_0}^t \mathcal{P}\mathcal{L}e^{-i(1-\mathcal{P})\mathcal{L}(t-\tau)}(1-\mathcal{P})\mathcal{L}\mathcal{P}\rho_{S+B}(\tau)\,d\tau$$

$$= [\frac{d}{dt} + i\mathcal{P}\mathcal{L}]\mathcal{P}e^{-i\mathcal{L}(t-t_0)}\rho_{S+B}(t_0)+$$

---

[38]*i.e.*, $A = A_S \otimes 1_B$, where A_S acts in the Hilbert space of the system only.

$$\int_{t_0}^{t} \mathcal{P}\mathcal{L}e^{-i(1-\mathcal{P})\mathcal{L}(t-\tau)}(1-\mathcal{P})\mathcal{L}\mathcal{P}e^{-i\mathcal{L}(\tau-t_0)}\rho_{S+B}(t_0)\, d\tau. \tag{3.116}$$

From that, two immediate conclusions follow. First, the solution to (3.6) for the 'relevant' part of the density matrix $\mathcal{P}\rho_{S+B}(t)$ reads as

$$\mathcal{P}\rho_{S+B}(t) = \mathcal{P}[e^{-i\mathcal{L}(t-t_0)}\rho_{S+B}(t_0)]. \tag{3.117}$$

This is trivial since it follows from comparison with (3.1). It shows, however, that we have not deviated from what is given by standard formalism of quantum theory. The second and more relevant conclusion stemming from (3.116) is connected with the observation that it is the term

$$(\mathcal{C}\ldots)(t) = \int_{t_0}^{t} \mathcal{P}\mathcal{L}e^{-i(1-\mathcal{P})\mathcal{L}(t-\tau)}(1-\mathcal{P})\mathcal{L}\mathcal{P}\cdot\ldots(\tau)\, d\tau \tag{3.118}$$

that gives rise, in (3.33-3.35), to the Redfield relaxation superoperator. The latter is responsible for the reported challenge to the second law. On the other hand, the term $\mathcal{C}(t)$ from (3.118) appears in (3.116) on both sides, *i.e.*, also in the initial condition term $I(t)$ on the right hand side of (3.116). Hence it cancels, having no influence on the solution *provided* we make the same approximations in this term on both sides of the Nakajima-Zwanzig identity (3.116), *i.e.*, in the memory as well as the initial condition term in the Time-Convolution Generalized Master Equations for the density matrix $\rho_S(t)$. However, this is what contemporary kinetic theories do not do. On the way to, *e.g.*, (3.33), one first applies the initial condition (3.20-3.21) implying that $(1-\mathcal{P})\rho_{S+B}(t_0) = 0$. This leads to the full omission of the initial condition term. Only then one performs, for instance, the second-order approximation on the memory kernel. In a large number of numerical studies, the long-time solution strongly depends on the type and form of approximations in the relaxation (Redfield) term. This is formally incompatible with the independence of the solution (3.117) to distortions of the relaxation term. One can understand this contradiction on grounds of the following: The initial condition term is, in the exact formulation, proportional to $(1-\mathcal{P})\rho_{S+B}(t_0)$. That is why it disappears once (3.20-3.21) is accepted. This proportionality may, however, disappear once we perform the same distortion (approximation) in $\mathcal{C}$ (as on the left hand side of (3.116)) in the initial condition term $I(t)$ on the right hand side of (3.116)). Keeping this, after approximating $\mathcal{C}$ on the left hand side, still $I(t) = 0$ as a consequence of (3.20-3.21) could be inconsistent. No studies exist thusfar of the influence on initial condition terms $I(t) \neq 0$ induced by approximations in the relaxation terms of the kinetic equations, even for $(1-\mathcal{P})\rho_{S+B}(t_0) = 0$. This only adds uncertainty to kinetic theories that are based on standard kinetic equations — both those that agree with and those that conflict with the second law[39]. The point is that one can always, at least formally, solve the resulting GME in the long-time limit and only then perform the corresponding expansions (approximations). Methods from [49] or Appendix in [60] show this unambiguously.

---

[39]This uncertainty, however, is insufficient to question all the theoretical evidence mounting against the second law that is derived from these formalisms.

3.8.2 TCL-GME and Tokuyama-Mori

Similar comments to those directly above also apply to the Time-Convolutionless Generalized Master Equation (TCL-GME) formalism based on either older works by Fuliński and Kramarczyk [8, 9, 10] or on more modern ones by Shibata, Hashitsume, Takahashi, and Shingu [6, 7] and the operator formalism by Tokuyama and Mori (TM)[73]. The former TCL-GME formalism works, like the TC-GME one, with the density matrix in the Schrödinger picture and is based on the time-local identity (3.8). On the other hand, the TM formalism works from a time-local operator identity derived from the equations of motion for operators in the Heisenberg picture. Specifically, the Tokuyama-Mori identity reads

$$\frac{d}{dt}G(t) = e^{i\mathcal{L}t}\mathcal{D}i\mathcal{L}G(0) + \left[e^{i\mathcal{L}t}\mathcal{D}\int_0^t e^{-i\mathcal{L}\tau}\mathcal{D}i\mathcal{L}e^{i(1-\mathcal{D})\mathcal{L}\tau}\,d\tau + e^{i(1-\mathcal{D})\mathcal{L}t}\right]$$

$$\times \left(1 - (1-\mathcal{D})\int_0^t e^{-i\mathcal{L}\tau}\mathcal{D}i\mathcal{L}e^{i(1-\mathcal{D})\mathcal{L}\tau}\,d\tau\right)^{-1}(1-\mathcal{D})i\mathcal{L}G(0). \qquad (3.119)$$

Correspondence between (3.8) and (3.119) may be best seen upon: (i) choosing the column of operators $G(0)$ in form of set of operators $|m\rangle\langle n|$ (where $|m\rangle$, $|n\rangle$, etc. denote arbitrary states in the Hilbert space of the system only); (ii) choosing $\mathcal{D}\ldots = \sum_{p,q}|p\rangle\langle q|\mathrm{Tr}_S(|q\rangle\rho^B\langle p|\ldots)$, $\mathrm{Tr}_B\rho^B = 1$; (iii) multiplying (3.119) from the left by the density matrix of the (system + bath) complex $\rho_{S+B}(0)$; and (iv) taking the trace (Tr) of the result. What we obtain is nothing but (3.17) that is, however, the result of application of the Argyres-Kelley projector (3.3) to (3.8). Identity (3.119) provides additional possibilities. Like (3.114), for example, in the case of a Brownian particle (system) in a liquid (bath), one may choose G as a set of two (vector) operators of the Brownian particle (coordinate and momentum). Then (3.119) provides a unique and rigorous way to find the quantum time-convolution*less* counterpart of standard classical Langevin equations [73].

Having established a correspondence between the Tokuyama-Mori formalism and TCL-GME, we now return to the latter formalism with two purposes in mind. First, like with the TC-GME formalism, initial condition terms in the TCL-GME theories are fully analogous to those in the TC-GME approaches mentioned in the previous paragraph. Second, we raise new concerns, for the long-time limit, about expansions in kinetic theories performed at finite times. For that, let us first mention that (3.8) may be rewritten [71] as

$$\left[\frac{d}{dt} + i\mathcal{P}\mathcal{L}\left[1 + i\int_0^{t-t_0} e^{-i(1-\mathcal{P})\mathcal{L}\cdot\tau}(1-\mathcal{P})\mathcal{L}\mathcal{P}e^{i\mathcal{L}\cdot\tau}\,d\tau\right]^{-1}\right]\mathcal{P}\rho_{S+B}(t) = J(t),$$

$$(3.120)$$

where the initial condition term is $J(t) =$

$$-i\mathcal{P}\mathcal{L}\left[1 + i\int_0^{t-t_0} e^{-i(1-\mathcal{P})\mathcal{L}\cdot\tau}(1-\mathcal{P})\mathcal{L}\mathcal{P}e^{i\mathcal{L}\cdot\tau}\,d\tau\right]^{-1}e^{-i(1-\mathcal{P})\mathcal{L}\cdot(t-t_0)}(1-\mathcal{P})\rho_{S+B}(t_0)$$

$$= \left[\frac{d}{dt} + i\mathcal{P}\mathcal{L}\left[1 + i\int_0^{t-t_0} e^{-i(1-\mathcal{P})\mathcal{L}\cdot\tau}(1-\mathcal{P})\mathcal{L}\mathcal{P}e^{i\mathcal{L}\cdot\tau}\,d\tau\right]^{-1}\right]\mathcal{P}e^{-i\mathcal{L}(t-t_0)}\rho_{S+B}(t_0).$$

$$(3.121)$$

The first conclusion is that the solution to (3.8) (*i.e.*, to TCL-GME (3.17)) is again, like above, given by (3.117). (By the way, this provides the simplest proof of equivalence of TC-GME with TCL-GME methods.) The second conclusion is analogous to that above concerning the initial condition term: While $J(t)$ is proportional to $(1-\mathcal{P})\rho_{S+B}(t_0)$ in the exact formulation and thus $J(t) = 0$ whenever initial condition (3.20-3.21) is accepted, it is no longer necessarily the case after performing the same (*e.g.*, second-order Born) approximation of the relaxation term $i\mathcal{P}\mathcal{L}\left[1 + i\int_0^{t-t_0} e^{-i(1-\mathcal{P})\mathcal{L}\cdot\tau}(1-\mathcal{P})\mathcal{L}\mathcal{P}e^{i\mathcal{L}\cdot\tau}\,d\tau\right]^{-1}$ on both the left hand side of (3.120) and the right hand of (3.121). The conclusion is the same as with the TC-GME approach.

In connection with all approximate theories, assume as usual that $\mathcal{L} = \mathcal{L}_0 + \lambda\mathcal{L}_1$, where $\lambda\mathcal{L}_1 = \frac{1}{\hbar}[\lambda H_1, \ldots]$ indicates commutation with that part of λH_1 which is considered a perturbation. Further assume that $\mathcal{P}\mathcal{L}_0 = \mathcal{L}_0\mathcal{P}$. The usual expansion of coefficients in any kinetic equation (*e.g.*, Boltzmann, Pauli) that could be connected with (3.8) goes, in (3.120), as

$$-i\mathcal{P}\mathcal{L}\left[1 + i\int_0^{t-t_0} e^{-i(1-\mathcal{P})\mathcal{L}\tau}(1-\mathcal{P})\mathcal{L}\mathcal{P}e^{i\mathcal{L}\tau}\,d\tau\right]^{-1}\mathcal{P}\rho_{S+B}(t)$$

$$\approx -i\mathcal{P}\mathcal{L}\mathcal{P}\rho_{S+B}(t) - \lambda^2\mathcal{P}\mathcal{L}_1\int_0^{t-t_0} e^{-i\mathcal{L}_0\tau}(1-\mathcal{P})\mathcal{L}_1\mathcal{P}e^{i\mathcal{L}_0\tau}\,d\tau\mathcal{P}\rho_{S+B}(t). \quad (3.122)$$

On the other hand, one should realize that the integral involved on the left hand side can be exactly calculated as

$$1 + i\int_0^{t-t_0} e^{-i(1-\mathcal{P})\mathcal{L}\tau}(1-\mathcal{P})\mathcal{L}\mathcal{P}e^{i\mathcal{L}\tau}\,d\tau = \mathcal{P} + e^{-i(1-\mathcal{P})\mathcal{L}(t-t_0)}(1-\mathcal{P})e^{i\mathcal{L}(t-t_0)}.$$

$$(3.123)$$

On the right hand side of (3.123), $\lambda\mathcal{L}_1$ appears only in exponentials. Hence, (3.122), where the corresponding $\lambda\mathcal{L}_1$ appears only between the exponentials, is formally based on expanding phase factors in the exponential. It is well known that *e.g.*, $e^{-i\phi t} \approx 1 - i\phi t$ becomes a worse approximation as time t increases, irrespective of value of the small phase ϕ. This raises suspicions about all contemporary kinetic theories — particularly in their long-time, dc limit — for which small parameter expansions are performed for finite times. The situation in (3.122) may be different in the thermodynamic limit of the bath when destructive interference of individual terms under the integral in (3.122) could change the situation at long times. In summary, the structure of all terms, particularly higher order ones, and their long-time behavior have not been adequately examined.

3.9 Theoretical Summary

In this chapter, several different theoretical methods have been applied to various individual models, raising formal theoretical challenges to the second law.

Many of the methods deserve further scrutiny, but the general conclusion is that the second law should be considered to be in theoretical jeopardy. Some of the methods can be well justified when the proper order of steps is taken: first the thermodynamic limit of the bath, followed by the long-time limit, and last, by expansions in parameters.

Still, problems remain that are endemic to theoretical modeling and its philosophy, for instance, whether the smallness of parameters in a mechanism justifies its omission, especially when it is attached to a large parameter, like infinite time. Strong coupling has not yet been satisfactorily formulated within GME — and not only within GME. Many technical problems remain. For instance, the use of Born-type approximations are forced upon us in situations where it is known that they are probably inadequate. There is also a problem with proper inclusion of renormalization in GME; e.g., polaron states have been implemented in this chapter, but it is not settled how best to handle them. This problem persists even in the most advanced models. Also, it is supposed that the Davies approach is correct for kinetic theory, but this is doubted by some. Nonetheless, the present bulk and body of theoretical evidence cast serious doubt on the absolute status of the second law in the quantum realm. Clearly, more theoretical work is warranted. This situation also underscores the importance of experiment.

The challenges in this chapter are predominantly *formal* in the sense that they are disembodied; that is, they are neither couched in terms of concrete physical systems, nor in terms of realizable experiments — with the exception of the plasma system (§3.6.8). One should remember that thermodynamics was founded in the nineteenth century as an axiomatic science based on the generalization of experimental observations. Hopefully, this remains true. If so, the meaning of thermodynamics, and in particular the status of the second law, must ultimately be settled in the laboratory. It is to this enterprise that we now turn.

References

[1] Fick, E. and Sauermann, G., *The Quantum Statistics of Dynamic Processes* Springer Series in Solid-State Sciences 86. (Springer-Verlag, Berlin, 1990).

[2] Argyres, P.N. and Kelley, P.L., Phys. Rev. **134** A98 (1964).

[3] Nakajima, S., Progr. Theor. Phys. **20** 948 (1958).

[4] Zwanzig, R., in *Lectures in Theoretical Physics,* Vol. 3, Boulder, Colorado (1960), Brittin, W.E., Downs, B.W., and Downs, J., Editors (Interscience, New York, 1961) pg. 106.

[5] Zwanzig, R., Physica **30**, 1109 (1964).

[6] Hashitsume, N., Shibata, F., and Shingu, M., J. Stat. Phys. **17** 155 (1977).

[7] Shibata, F., Takahashi, Y., and Hashitsume, N., J. Stat. Phys. **17** (1977).

[8] Fuliński, A., Phys. Letters A **25** 13 (1967).

[9] Fuliński, A. and Kramarczyk, W.J., Physica **39** 575 (1968).

[10] Gzyl, H., J. Stat. Phys. **26** 679 (1981).

[11] Čápek, V., Czech. J. Phys. **48** 993 (1998).

[12] Čápek, V., Phys. Stat. Sol. (B) **121** K7 (1984).

[13] Fujita, S., *Introduction to Non-Equilibrium Quantum Statistical Machanics.* (W. B. Saunders, Philadelphia, 1966).

[14] Redfield, A.G., IBM J. Res. Develop. **1** 19 (1957).

[15] Redfield, A.G., in *Advances in Magnetic Resonance.* Vol. 1, Waugh, J.S., Editor (Academic Press, New York, 1965) pg. 1.

[16] Mahler G. and Weberuß, V.A., *Quantum Networks. Dynamics of Open Nanostructures.* (Springer, Berlin, 1995).

[17] Davies, E.B., Commun. Math. Phys. **39** 91 (1974).

[18] Davies, E.B., Math. Annalen **219** 147 (1976).

[19] Davies, E.B., *Quantum Theory of Open Systems.* (Academy Press, London, 1976).

[20] Dümcke, R. and Spohn, H., Z. Physik B **34** 419 (1979).

[21] Čápek, V., Barvík, I. and Heřman, P., Chem. Phys. **270** 141 (2001).

[22] Čápek, V., Physica A **203** 495 (1994).

[23] Čápek, V., Czech. J. Phys. **46** 1001 (1996).

[24] Peřina, J. Jr., Physica A **214** 309 (1995).

[25] Čápek, V., Physica A **203** 520 (1994).

[26] Čápek, V. and Peřina, J., Jr., Physica A **215** 209 (1995).

[27] Goodsell, D.S., *The Machinery of Life*. Springer Verlag, Berlin, 1993).

[28] Allahverdyan, A.E. and Nieuwenhuizen, Th.M.,
     http://xxx.lanl.gov/abs/cond-mat/0011389

[29] Allahverdyan, A.E. and Nieuwenhuizen, Th.M., Phys. Rev. Lett. **85** 1799
     (2000).

[30] Ford, G.W., Lewis, J.T., and O'Connell, R.F., J. Stat. Phys. **53** 439 (1988).

[31] Scully, M.O., Phys. Rev. Lett. **87** 220601 (2001); Scully, M.O. in *Quantum
     Limits to the Second Law*, AIP Conference Proceedings, Vol. 643, Sheehan,
     D.P., Editor (AIP Press, Melville, NY, 2002) pg. 83.

[32] Scully, M.O., Zubairy, M.S., Agarwal, G.S., and Walther, H., Science **299** 862
     (2003).

[33] Jordan, A.N. and Buttiker, M., Phys. Rev. Lett. **92** 247901 (2004).

[34] Čápek, V. and Barvík, I., Physica A **294** 388 (2001).

[35] Kleinekathöfer, U., Kondov, I., and Schreiber, M., Chem. Phys. **268** 121
     (2001).

[36] Čápek, V., Czech. J. Phys. **47** 845 (1997).

[37] Čápek, V., Czech. J. Phys. **48** 879 (1998).

[38] Čápek, V. and Frege, O., Czech. J. Phys. **50** 405 (2000).

[39] Čápek, V. and Frege, O., Czech. J. Phys. (2002) in review.

[40] Einstein, A., Verhandl. Deutsch. Phys. Gesellschaft **18** 318 (1916).

[41] Einstein, A., Mitt. Phys. Gesellschaft (Zürich) **18** 47 (1916).

[42] Einstein, A., Phys. Zeitschrift **18** 121 (1917).

[43] Čápek, V. and Bok, J., J. Phys. A: Math. & General **31** 8745 (1998).

[44] Čápek, V., reported at MECO 23 Conference (Middle European Cooperation
     in Statistical Physics), International Center of Theoretical Physics, Trieste,
     Italy, April 27-29 (1998).

[45] Čápek, V. and Bok, J., http://xxx.lanl.gov/abs/cond-mat/9905232; Czech.
     J. Phys. **49** 1645 (1999).

[46] Čápek, V. and Bok, J., Physica A **290** 379 (2001).

[47] Čápek, V., Molec. Cryst. and Liq. Cryst. **335** 24 (2001).

[48] Čápek, V., J. Phys. A: Mathematical & General **30** 5245 (1997).

[49] Čápek, V., Phys. Rev. E **57** 3846 (1998).

[50] Čápek, V. and Mančal, T., Europhys. Lett. **48** 365 (1999).

[51] Čápek, V. and Mančal, T., J. Phys. A: Math. and Gen. **35** 2111 (2002).

[52] Čápek, V. and Bok, J., Chem. Phys. (2002) in review.

[53] Landau, L.D. and Teller, E., Phys. Z. Sowjetunion **10** 34 (1936).

[54] Čápek, V., Z. Physik B **60** 101 (1985).

[55] Silinsh, E.A. and Čápek, V., *Organic Molecular Crystals. Interaction, Localization, and Transport Phenomena.* (Amer. Inst. of Physics, New York, 1994).

[56] Haken H. and Strobl, G., Proc. Intern. Symp., Amer. Univ. Beirut, Lebanon, 1967. A. B. Zahlan, Ed. (University Press, Cambridge, 1967) pg. 311.

[57] Reineker, P., in *Exciton Dynamics in Molecular Crystals and Aggregates.*, Springer Tracts in Modern Physics **94**, Höhler, G., Editor, (Springer-Verlag, Berlin, 1982) pg. 111.

[58] Čápek, V. and Tributsch, H., J. Phys. Chem. B **103** 3711 (1999).

[59] Čápek, V., Europ. Phys. Jour. B, in review (2002).

[60] Čápek, V. http://xxx.lanl.gov/abs/cond-mat/0012056

[61] Čápek, V. and Sheehan, D.P., Physica A **304** 461 (2002).

[62] Čápek, V. and Bok, J., http://arxiv.org/abs/physics/0110018

[63] Porto, M., Phys. Rev. E **63** 030102 (2001).

[64] Bok, J. and Čápek, V., http://arxiv.org.abs/physics/0110018.

[65] Bok, J. and Čápek, V., Entropy **6** 57 (2004).

[66] Novotný, T., http://arxiv.org/abs/cond-mat/0204302.

[67] Novotný, T., Europhys. Lett. **59** 648 (2002); in *Quantum Limits to the Second Law*, AIP Conf. Proc., Vol. 643 (AIP Press, Melville, NY, 2002) pg. 104.

[68] Grigolini, G., *Quantum Mechanical Irreversibility and Measurement* (World Scientific, Singapore, 1993).

[69] Šanda, F., http://arxiv.org/abs/physics/0201040

[70] Mori, H., Progr. Theor. Phys. **33** 423 (1965).

[71] Čápek, V., Czech. J. Phys. **42** (1992).

[72] Čápek, V., Phys. Rev. B **38** 12983 (1988).

[73] Tokuyama M. and Mori, H., Progr. Theor. Phys. **55** 411 (1976).

4

Low-Temperature Experiments and Proposals

4.1 Introduction

Several independent groups are currently investigating low-temperature ($T \leq 10K$) second law challenges that exploit uniquely quantum mechanical behaviors. Two of these invoke the phase transition from normal to superconducting states and are currently under active experimental investigation (§4.3, §4.4), while others rely on quantum entanglement and constitute theoretical proposals for experiments (§4.6). We begin with a brief review of superconductivity.

4.2 Superconductivity

4.2.1 Introduction

Superconductivity is a macroscopic quantum phenomenon. The first and most obvious evidence of it is the Meissner effect discovered by Meissner and Ochsenfeld in 1933 when they observed that a superconductor in a weak magnetic field completely expels the field from the superconducting bulk except for a thin layer at the surface [1]. This is a more fundamental aspect of superconductivity than the disappearance of electrical resistance[1], first discovered in mercury at low temperatures by Kamerlingh Onnes in 1911. Superconductivity is a state with long-range

---

[1]Electrical resistance is effectively zero ($\rho < 10^{-26}\Omega$m).

phase correlations, which are a consequence of Bose-Einstein condensation of electron pairs [2, 3, 4]. Pairs of electrons (fermions individually) form Cooper pairs (which are bosons) via electron-phonon-electron interactions, as described by the Bardeen-Cooper-Schrieffer theory [5].

Complete flux expulsion in the simple form of the Meissner effect occurs only in weak magnetic fields. If the applied field is sufficiently strong and if demagnetization becomes appreciable, then magnetic flux penetrates through the superconductor. This penetration differentiates the two types of superconductors, which have differing signs of the wall energy associated with the interface between the normal and superconducting domains. In Type-I superconductors the wall energy is positive and, therefore, the magnetic flux contained in a single normal domain consists of many flux quanta. (One magnetic flux quantum ($\Phi_o = \frac{h}{2q}$) is the smallest unit of magnetic flux.) Type-II superconductors are characterized by negative interface boundary surface energy (wall energy). In this case, magnetic flux can be distributed through the superconductor such as to form either normal regions or a mixed phase of superconducting and normal regions. The particular type of superconductivity is determined by a parameter κ of the Ginzburg-Landau theory [6]: at $\kappa < 1/\sqrt{2}$ the wall energy of a normal-superconducting interface is positive and the superconductor is Type-I, while at $\kappa > 1/\sqrt{2}$ the wall energy is negative and the superconductor is Type-II. The Ginzburg-Landau parameter $\kappa = \lambda(T)/\xi(T)$ relates the two characteristic lengths of superconductor: the penetration depth $\lambda(T)$ and the coherence length $\xi(T)$. These have similar temperature dependences

$$\xi(T) \approx \xi_0 [1 - (\frac{T}{T_c})^4]^{-1/2}$$

$$\lambda(T) \approx \lambda_0 [1 - (\frac{T}{T_c})^4]^{-1/2} \tag{4.1}$$

The penetration depth λ_0, the scale length over which an external magnetic field can penetrate into the superconductor, depends only on the density of superconducting pairs. The coherence length, ξ_0, is the maximum scale length over which Cooper pairs interact, or equivalently, it is the scale length over which superconductivity can be established or destroyed. ξ_0 decreases with the electron mean free path in dirty superconductors. Type-II superconductivity occurs preferentially in alloys or, more generally, in impure systems. Pure metals usually display Type-I superconductivity. The penetration depth has approximately the same value $\lambda_0 \approx 10^{-8} - 10^{-7}$m for most superconductors, whereas the coherence length may run from $\xi(0) \approx 10^{-6}$m in pure aluminium [7] down to $\xi(0) < 10^{-9}$m in high temperature superconductors [8].

The free energy of the superconducting phase $f_s(T)$ is less than that of the normal phase $f_n(T)$ for temperatures lower than the critical one $T < T_c$. The superconductor's exclusion of magnetic flux increases its free energy density by $\frac{\mu_0 H^2}{2}$: the price of diagmagnetism. The magnetic field H_c at which the free energy gain associated with electron condensation into Cooper pairs equals the free energy cost of its diamagnetism, $f_n(T) - f_s(T) = \frac{\mu_0 H_c^2}{2}$ [9], is called the thermodynamic critical field. There is another critical field, $H_{c2} = \sqrt{2}\, \kappa H_c$, called second critical field [9]. According to the Abrikosov theory [10] the transition between the normal and superconducting vortex state of Type-II superconductors takes place at this critical field $H_{c2} = \sqrt{2} \, \kappa H_c > H_c$. However, in reality, this transition is observed below H_{c2} [11, 12, 13]. Samples of Type-I superconductors (for which $\sqrt{2} \, \kappa < 1$) undergo transition between normal and superconducting phases at $H_c > H_{c2} = \sqrt{2} \, \kappa H_c$ when their size or demagnetization coefficient are enough small. In the opposite case of a finite demagnetization coefficient D, the intermediate state is observed at $H_c(1 - D) < H < H_c$. This state is a configuration consisting of a mixture of normal and superconducting domains [9]. The critical field H_c, above which superconductivity of Type-I superconductors disappears, has been found to follow, to good approximation, the empirical relation

$$H(T) = H_c[1 - (\frac{T}{T_c})^2].$$ (4.2)

Type-I superconductors have critical fields $H_c \leq 0.2$T, whereas low-temperature Type-II superconductors with high values of the Ginzburg-Landau parameter κ have $H_{c2} = \sqrt{2} \, \kappa H_c$ up to 50 T. High-temperature ceramic superconductors have the second critical fields up to several hundred Teslas.

4.2.2 Magnetocaloric Effect

In non-zero fields, the normal-superconducting transition of Type-I superconductor is first order and has an associated latent heat. A sample heats (cools) when making the transition *to* the superconducting (normal) state. This is the magnetocaloric effect. (A non-quantum mechanical electrostatic analog, the electrocaloric effect, is employed by Trupp in another second law challenge (§5.5 and [14]).

Although superconductors are perfect diamagnets, excluding magnetic flux from their bulk interiors, surface-parallel fields penetrate shallowly into their outer layers, decaying exponentially in strength with a characteristic *penetration depth* (see (4.1)); that is, $H(z) = H_o e^{-z/\lambda}$. Note that $\lambda \to \infty$ as $T \to T_c$ since $\lambda \propto 1/\sqrt{n_s}$ [9], where $n_s \propto T_c - T$ is the density of superconducting pairs; that is, as the penetration depth becomes large at the transition temperature, the sample becomes normal.

During transition between normal to superconducting phases, a sample usually passes through an *intermediate state* wherein lamellae of normal phase riddle the superconducting bulk. Samples of suitably small size ($\xi \geq d \geq 5\lambda$) can undergo the normal-to-superconducting transition en masse, without passing through an intermediate state. Given the inherently small sizes of ξ and λ, d is narrowly restricted to roughly 10^{-6}m$\geq d \geq 10^{-7}$m. In such a transition, there can be no lamellae and the sample instantaneously can snap from one thermodynamic equilibrium to the other. Type-I elemental superconductors that fit this criterion include Sn $((\xi/\lambda) = 4.5)$, In $((\xi/\lambda) = 6.9)$, and Al $((\xi/\lambda) = 32)$.

Whereas the intermediate state observed in large samples of Type-I superconductors have been investigated in detail [9], the thermodynamics of small samples has not been well studied thus far. Although Pippard raised questions about

the irreversible effects in the magnetization cycle of superconducting colloids as
early as 1952 [15], up to now there has been little experimental work devoted to
the magnetization and transition between normal and superconducting states of
small samples of Type-I superconductors. The resistive measurements of thin tin
whiskers made by Lutes and Maxwell as early as 1955 [16] show that an abrupt
transition from the superconducting to normal state can occur without the intermediate state in samples of suitably small size. But only recently have techniques
been developed [17] that allow quantitative studies of thermodynamic properties of
individual superconducting particles at micron and sub-micron scale lengths. The
results of [18] demonstrate the irreversible effects in the magnetization cycle of Al
disks down to diameter > 0.3 μm [17]. However, it is important to emphasize that
this irreversibility is conditioned by a high value of demagnetization coefficient
typical of thin disks. Reversible behaviour can be expected only in small samples
with geometries like spheres.

The combination of the magnetocaloric effect with reversible transition renders
the *coherent magnetocaloric effect* (CMCE). This is the key new insight underlying
Keefe's second law challenge. Inherently, this is a quantum mechanical process that
relies on the superconductor's long-range order parameter (wavefunction).

4.2.3 Little-Parks (LP) Effect

The Meissner effect is a quantum phenomena arising from the quantization of
momentum circulation of superconducting pairs. The generalized momentum of a
charge q is given by $p = mv + qA$, where A is the magnetic vector potential. For
Cooper pairs $q \longrightarrow 2e$, where e is the charge of the electron. The quantization of
momentum circulation along a closed path is [2]

$$\oint p\, dl = nh = \oint mv\, dl + \oint 2eA\, dl = m\oint v\, dl + 2e\Phi, \qquad (4.3)$$

where n is equal to zero for any closed path inside a simply-connected superconductor without a singularity in its wavefunction. Therefore, the persistent
electrical current $j_p = 2evn_s$ should be maintained in outer layers of a superconductor (where the velocity of superconducting pairs v is determined by the relation
$m\oint v\, dl + 2e\Phi = 0$), while in its interior bulk, where $v = 0$, the magnetic flux
should be absent ($\Phi = 0$).

For a closed path in a multiply-connected superconductor — for example in
a loop — the integer n in (4.3) can be any value and the velocity circulation of
Cooper pairs should be

$$\oint v\, dl = \frac{h}{m}\left[n - \frac{\Phi}{\Phi_o}\right] \qquad (4.4)$$

where $\Phi_0 = h/2e$ is the flux quantum (fluxoid). The magnetic flux inside the loop is
$\Phi = BS + LI_p$, where B is the magnetic induction induced by an external magnet;
S is the area of the loop; L is the inductance of the loop; $I_p = sj_p = s2evn_s$
is the persistent current around the loop. The velocity (4.4) and the persistent

current of the loop with weak screening $(LI_p < \Phi_0)$ is a periodic function of the magnetic flux $\Phi \approx BS$ since velocity circulation (4.4) cannot be zero unless $\Phi = n\Phi_0$ and the thermodynamic average value of the quantum number n is close to an integer number n corresponding to minimum kinetic energy Cooper pairs, $i.e.$, to minimum $\mathcal{E} \propto v^2 \propto (n - \Phi/\Phi_0)^2$. This quantum periodicity leads to experimentally observable effects.

The first such effect was observed by Little and Parks in 1962 [19]. The quantum periodicity in the transition temperature T_c of a superconducting cylinder [19] or a loop [20] from enclosed magnetic flux following Φ was explained as a consequence of the periodic dependence of the free energy [19, 21, 2]: $\Delta T_c \propto -\mathcal{E} \propto -v^2 \propto -(n - \Phi/\Phi_0)^2$. For a cylinder or loop with a radius R, the dependence of critical temperature with flux varies as

$$T_c(\Phi) = T_c \left[1 - (\frac{\xi(0)}{R})^2(n - \frac{\Phi}{\Phi_o})^2 \right], \qquad (4.5)$$

where $\xi(0)$ is its coherence length at $T = 0$. The values of $(n - \Phi/\Phi_0)$ is constrained between -0.5 and 0.5. The relation (4.5) describes well the experimental dependencies $T_c(\Phi)$ obtained from resistive measurements [19, 2, 20].

This explanation of the Little-Parks (LP) effect is not complete, however. It does not explain, for instance, why the persistent current I_p has been observed at non-zero resistances $(R > 0)$ in a number of studies. It is emphasized that the observation of a persistent current I_p — $i.e.$, a direct current observed under thermodynamic equilibrium conditions, at a non-zero resistance $R > 0$ — contradicts standard expectations since it implies power dissipation (RI_p^2) and, by inference, a direct current power source under equilibrium conditions. Nikulov advances this as evidence for the potential violability of the second law.

Nikulov's key insight is to reinterpret and extend the results of the LP experiments to consideration of inhomogeneous superconducting loops immersed in magnetic fields near their transition temperatures. From these he concludes that thermal fluctuations can be used to drive electrical currents in the presence of nonzero resistance, and by this achieve nonzero electrical dissipation at the expense of thermal fluctuations alone. In essence, thermal energy is rectified into macroscopic currents, this in violation of the second law. Nikulov proposes a new force, the *quantum force* — which arises from the exigencies of the quantum-to-classical, superconducting-to-normal transition — to explain these fluctuation-induced currents [22]

4.3 Keefe CMCE Engine

4.3.1 Theory

Keefe proposes a simple thermodynamic process in which a small superconducting sample is cycled through field-temperature (H-T) space and performs net work solely at the expense of heat from a heat bath [23, 24]. (We use Keefe's nomen-

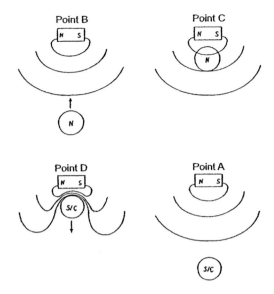

Figure 4.1: Pictorial overview of CMCE cycle.

clature.) It incorporates facets of other standard H-T cycles [25, 26], but also uniquely invokes the coherent magnetocaloric effect (CMCE).

The cycle is given pictorially and graphically in Figures 4.1 and 4.2. Figure 4.1 pictures a small armature of superconductor (meeting CMCE requirements) moving in and out of a magnetic field during a full thermodynamic cycle. Here "N" and "S/C" indicate normal and superconducting states. Figure 4.2 graphs the armature's progress in H-T space and indicates work and heat influxes and effluxes.

The cycle begins with the armature (volume V) in the superconducting state (point A in Figure 4.2) at thermodynamic coordinates (T_1, H_1). Until otherwise noted, the armature is thermally insulated and the process proceeds adiabatically.

The armature is moved slightly closer to the magnet, thus increasing the magnetic field it experiences, so it passes to the normal side of the critical field (Tuyn) curve (point B, Figure 4.2) with coordinates $(T_1, H_1 + \Delta H)$. (The magnetodynamic work to move the armature is assumed to be zero.) The armature coherently transitions to the normal state, evolves latent heat (LH_1) and magnetocalorically cools to T_2, given through

$$LH_1 = T_1(S_{n1} - S_{s1}) = V \cdot \int_{T_1}^{T_2} C_n dT \qquad (4.6)$$

With precisely orchestrated motion, the armature moves inwardly toward the magnet as it cools (Process B, Figure 4.1) so as to skirt the normal side of the Tuyn curve ($B \to C$, Fig 2.). The armature, now fully cooled (point C, Figure 4.2) at coordinates (T_2, H_2), is removed slightly out of the field, thus reducing its

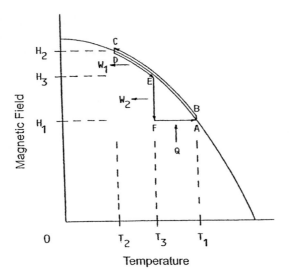

Figure 4.2: Coherent magnetocaloric effect (CMCE) cycle on H-T phase diagram.

field to $H_2 - \Delta H$, and thereby crossing it back to the superconducting side of the Tuyn curve (point D, Figure 4.1) at coordinates $(T_2, H_2 - \Delta H)$. Latent heat is evolved, magnetocalorically heating the armature to T_3, given via

$$LH_2 = T_2(S_{n2} - S_{s2}) = V \cdot \int_{T_2}^{T_3} C_s dT \qquad (4.7)$$

Now on the superconducting side of the Tuyn curve again, the Meissner effect kicks in and forcibly expels the magnetic field from the interior of the armature, whereupon the armature is repelled out of the high field region near the magnet. During its forcible expulsion (path $D \rightarrow E$, Figure 4.2), the armature performs work

$$W_1 = \frac{\mu_o(H_2^2 - H_3^2)}{2} \cdot V \qquad (4.8)$$

Similarly as for path segment $B \rightarrow C$ in Figure 4.2, the armature moves in a precisely timed and coordinated fashion from $D \rightarrow E$ so as to skirt the superconducting side of the Tuyn curve while magnetocalorically heating to T_3 (and also while simultaneously performing work). From point E (Figure 4.2), the superconducting armature is moved further out of the field (Process D, Figure 4.1), performing additional work

$$W_2 = \frac{\mu_o(H_3^2 - H_1^2)}{2} \cdot V \qquad (4.9)$$

and arrives at point F (Figure 4.2) with coordinates (T_3, H_1).

Up to this point, processes have been adiabatic. From $F \to A$ (Fig 2), however, the superconducting armature is thermally coupled to the surrounding heat bath (T_1) and heats $(T_3 \to T_1)$, thus closing the cycle and absorbing heat

$$Q = V \cdot \int_{T_3}^{T_1} C_s dT. \tag{4.10}$$

As described, heat transfer occurs only in the final step of the cycle; here heat is absorbed. Since positive work is performed by the armature elsewhere in the cycle, if the cycle operates in steady-state, the first law implies that the heat absorbed from the heat bath is transformed into work.

Keefe calculated the net work per cycle expected for an exemplary tin armature and cycle [27]. The cycle is specified by the vertex coordinates in Figure 4.2. In terms of tin's critical field (H_c) and the critical temperature (T_c), these are: $(T_1, H_1) = (0.6T_c, 0.64H_c)$, $(T_2, H_2) = (0.186T_c, 0.965H_c)$, $(T_3, H_3) = (0.407T_c, 0.834H_c)$. For this cycle, the latent heat densities are: $LH_1 = 340$ J/m^3, $LH_2 = 50$J/m^3. The work density/cycle is: $W_1 = 88$J/m^3, $W_2 = 107$J/m^3, and the heat density/cycle is: $Q = 195$J/m^3. Satisfying the first law, $W_1 + W_2 = Q$, implies for the second:

$$\Delta S = -\int_{T_3}^{T_1} \frac{dQ(T)}{T} dT < 0 \tag{4.11}$$

In principle, net work can be extracted from the CMCE cycle mechanically (*e.g.*, motor), electronically (*e.g.*, generator), or via a heat pump. Given the theoretical limitation to small armatures, usable power would probably be extracted in large arrays. Since operating frequencies for mechanical devices of this size can be high ($f \sim 10^9 - 10^{12}$Hz), high output power densities might be achieved [28]. For example, assuming an individual tin CMCE motor is 10 times larger (10^3 times greater volume) than its armature ($d \simeq 10^{-7}$m) and operates at $f = 10^{10}$Hz, based on tin's calculated work density/cycle, its power density is estimated to be $\mathcal{P} \simeq f(W_1 + W_2) \simeq 2 \times 10^{12}$W/m^3.

4.3.2 Discussion

The movement of normal phase electrons through the external magnetic field should generate eddy currents, Ohmic heating, and entropy, with magnitude dependent on the rapidity of movement. The armature's coherent transition time could be quite short, perhaps shorter than 10^{-12}s (*i.e.*, 10^{-4} the light travel time across the armature) and the resultant latent heat should manifest itself as a temperature change within a few vibrational periods of the lattice ($\tau_{lattice} \sim 10^{-13}$s), therefore, the armature must cycle quickly to faithfully trace the Tuyn curve, perhaps at THz frequencies. At these frequencies, one expects eddy current heating of the normal electrons (and perhaps even of the superelectrons). Normal electrons are known to interact with ac fields, causing dissipation and entropy production in superconducting samples. Superelectrons can absorb electromagnetic radiation near the necessary projected operating frequency of the armature. Magnetic dipole radiation could also be significant.

The physical magnet giving rise to the armature's external magnetic field should experience a back reaction and possibly internal induced electric fields and dissipation due to the rapid and possibly sizable distortions of field by the action of the armature. Given its small size, account should be taken of thermal fluctuations and whether these might drive it inopportunely across the transition line. Hysteresis should also be considered [29]. Finally, the sophistication in microscopic mechanical engineering required to realize a working CMCE engine is beyond the present state of the art in micro- or nanomanufacturing, but may be on the horizon.

Experiments are currently being pursued in Moscow, Russia to understand better the CMCE effect as it pertains to Keefe's engine. While falling short of an actual engine test, they are laying necessary foundations. Indium spheres ($r \simeq 1.25 \times 10^{-7}$m, $T_c = 3.7$K, $\xi/\lambda = 6.9$) will be analysed with a ballistic Hall micromagnetometer as the sample is cycled through the normal-superconducting transition (2.5K $\leq T \leq 3$K). Keefe, et al. will check predicted values of the transition field, the transition time scale, and investigate hysteresis, which can diminish the efficiency of the thermodynamic cycle. Tight control of the sphere size and purity will be necessary since the CMCE effect is predicted to be robust only within a narrow range of particle sizes.

In summary, the CMCE cycle appears theoretically compelling despite many uncertainties surrounding superconducting and quantum processes in the mesoscopic regime. Experimental considerations are problematic, but are currently being investigated. The technical challenges in fabricating a working mechanical CMCE engine are formidable.

4.4 Nikulov Inhomogeneous Loop

Over the last seven years, Zhilyaev, Dubonov, Nikulov, et al. have conducted laboratory experiments that corroborate the essential features of Nikulov's theory. Recent independent theoretical analysis by Berger [30] also lends support to his position. We introduce this challenge through Nikulov's quantum force.

4.4.1 Quantum Force

Nikulov's proposed quantum force arises from the fundamental differences between classical and quantum states of electrons (or Cooper pairs) in a conducting (superconducting) loop. In their classical state, electrons occupy a continuous energy spectrum. Direct current cannot exist at equilibrium according to classical mechanics because the equilibrium distribution function f_0 for electrons depends on v quadratically through kinetic energy $f_0(v^2)$ such that the average thermodynamic current $j_{av} = q \sum_p v f_0$ for this continuous distribution is an odd integral

$j_{av} = q \int v f_0 dv$, which is equal to zero.

In contrast, in quantum mechanics a persistent current j_{pc} can exist — *i.e.*, direct current observed under equilibrium conditions — since the discrete sum

$$j_{pc} = q \sum_p v f_o(\frac{E(p)}{kT}) = \frac{q}{m} \sum_p (p - qA) f(\frac{E(p)}{kT}) \qquad (4.12)$$

cannot be replaced by a continuous integral as in the classical case. The energy difference between permitted states for a superconducting loop

$$E_p = s \oint 2n_s(\frac{mv_s^2}{2}) dl = \frac{sh}{4lm\langle n_s^{-1}\rangle} \left[n - \frac{\Phi}{\Phi_o} \right]^2 \qquad (4.13)$$

is much higher than the thermal energy $\Delta\mathcal{E} = \mathcal{E}(n+1) - \mathcal{E}(n) \approx sh/4lm\langle n_s^{-1}\rangle \gg kT$ in the closed superconducting state, when $\langle n_s^{-1}\rangle^{-1} = (l^{-1} \oint_l dln_s^{-1})^{-1} \approx n_s$, since the number of Cooper pairs sln_s is very large for any realistic superconducting loop [22]. (Here s and l are the cross-sectional area and length of the loop wire and n_s is the number density of Cooper pairs.)[2] Thus, a transition between the discrete spectrum, with well-spaced energy states $\Delta\mathcal{E} = \mathcal{E}(n+1) - \mathcal{E}(n) \gg kT$, and the continuous spectrum $\Delta\mathcal{E} = 0$ takes place when a loop is switched between superconducting states with different connectivity. The velocity v_s and the momentum p of Cooper pair change at this transition: $\oint_l p_{cl} dl = nh$ and the velocity is defined by (4.4) in the closed superconducting state, whereas in the open superconducting state $v_s = 0$ and $\oint_l p_{un} dl = 2e\Phi$. The momentum circulation changes from $2e\Phi$ to nh at closing of the superconducting state because of the flux quantization: $nh - 2e\Phi = h(n - \Phi/\Phi_0)$. The time rate of change of the momentum due to reiterated switching of the loop between superconducting states (at frequency f) is a force given by

$$\oint F_q dl = \oint (p_{cl} - p_{un}) f dl = h(\langle n\rangle - \frac{\Phi}{\Phi_0}) f. \qquad (4.14)$$

This is coined the *quantum force* [22], $F_q = (p_{cl} - p_{un})f$. Here $\langle n\rangle$ is the thermodynamic average of the quantum number n.

The reiterated switching can be induced by external current [31], by external electrical noise [32], or by equilibrium thermal fluctuations [22]. The quantum force induced by thermal fluctuations is the Langevin force [33, 34]. It maintains the persistent current in the presence of a damping force (dissipation via electrical resistance) just as the classical Langevin force maintains the Nyquist's noise current in a classical normal metal loop. In contrast with the classical Langevin force, however, the average value of the quantum force is not equal to zero at $\Phi = n\Phi_0$ and $\Phi = (n + 0.5)\Phi_0$, when $\langle n\rangle - \Phi/\Phi_0 \neq 0$. Therefore, the persistent current at $R > 0$ is an ordered Brownian motion with non-zero direct component, this in contradistinction to Nyquist's noise, which is completely chaotic. According to

---

[2]The energy difference $\Delta\mathcal{E} = \mathcal{E}(n+1) - \mathcal{E}(n) \approx sh/4lm\langle n_s^{-1}\rangle = 0$ in the open superconducting state since $\langle n_s^{-1}\rangle^{-1} = 0$ when the density of Cooper pairs equals zero $n_s = 0$ in any loop segment. As expected, no current flows in this case.

Nikulov, this phenomenon violates the postulate of randomness under equilibrium conditions, the same that saved the second law of thermodynamics at the beginning of 20th century. Nikulov claims that $I_p \neq 0$ at $R > 0$ is evidence of persistent power generation RI_p^2, the existence of which conflicts with the second law [33, 34]. The first experimental evidence of this phenomenon is, apparently, the original LP experiment itself over 40 years ago [19, 20] — and, therefore, arguably it represents the first experimentally-based second law challenge.

4.4.2 Inhomogeneous Superconducting Loop

By analogy with well-known theoretical and experimental results for normally conductive loops with inhomogeneous (asymmetric) resistivity, voltage oscillations are expected on a segment l_s of an inhomogeneous superconducting loop, satisfying

$$V(\frac{\Phi}{\Phi_0}) = \left[\frac{R_{ls}}{l_s} - \frac{R_l}{l}\right] l_s I_p(\frac{\Phi}{\Phi_0}) \qquad (4.15)$$

Here R_{ls} and l_s are the resistance and length of a loop segment; R_l and l are the resistance and length of the whole of the loop, and I_p is the persistent current. Experimental results corroborate this [31, 32]. The quantum analogy to the classical electrical case appears valid since the quantum force is uniform around the loop [22] just as the Faraday 'voltage' $-d\Phi/dt$ is uniform around a conventional loop.

Segments of a superconducting loop can have different resistances $R_{ls}/l_s \neq R_l/l \neq 0$ at nonzero currents $I_p \neq 0$ if they are in the normal state at different times when the loop is switched between superconducting states with different connectivity. This is possible if loop segments, for example l_a and l_b, have distinct critical temperatures, specifically $T_{ca} > T_{cb}$ (See Figure 4.3.). The limiting case is when one segment (l_b) is switched between superconducting and normal states while the other segment l_a (with $l_a + l_b = l$) remains always in the superconducting state (therefore, with $R_a = 0$ and $R = R_b$). This was considered by Nikulov in [35]. A flat ($h < R$) and narrow ($w \ll R$) loop with $h, w < \lambda$ was analysed. He found that the direct potential difference V_b [35]

$$V_b = R_b I_p \simeq (\frac{l_b \langle n_{sb}\rangle}{l_b n_{sa} + l_a \langle n_{sb}\rangle}) \cdot (\frac{n\Phi_o - \Phi}{\lambda_{La}^2})\rho_b \qquad (4.16)$$

can be observed if the average value of the Cooper pair density $\langle n_{sb}\rangle$ and resistivity ρ_b of the l_b segment do not equal zero; i.e., if the segment is switched between superconducting and normal states. Here $\lambda_{La} = (m/4e^2 n_{sa})^{1/2}$ is the London penetration length for segment l_a. Relation (4.16) is valid for high switching frequency ($f \gg R_b/L$). At a low frequency ($f \ll R_b/L$), the amplitude of the quantum oscillations of the dc voltage with respect to the magnetic field $V_b(\Phi/\Phi_0)$ is proportional to the switching frequency f [35, 22] and is given by

$$V_b = \frac{hf}{2e}(\langle n\rangle - \frac{\Phi}{\Phi_0})\frac{l_b}{l}. \qquad (4.17)$$

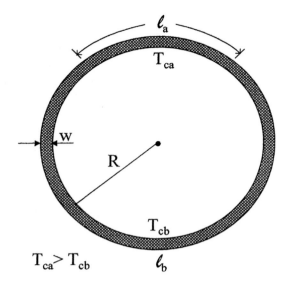

Figure 4.3: Schematic of inhomogeneous mesoscopic superconducting loop with different critical temperatures on segments a and b.

The correlation between the dc voltage and the frequency is similar to that of Josephson [36, 32].

The crux of this second law challenge [35, 22] lies in the switching of the l_b segment by thermal fluctuations at $T \approx T_{cb}$. In this case, V_b is the persistent voltage $V_b \equiv V_p$ and $\mathcal{P}_p = I_p V_p = R_b I_p^2 = V_p^2 / R_b$ is the persistent power [37]. It can be shown in that the persistent power $\mathcal{P}_p$ of a single loop cannot exceed the total power of thermal fluctuations [35, 37, 22]:

$$\mathcal{P}_{thermal} \simeq \frac{(kT)^2}{\hbar}. \tag{4.18}$$

According to (4.18), the persistent power of a single mesoscopic loop made from a high-temperature superconductor (HTSC) with $T_c \simeq 100$K is expected to be quite small; *i.e.*, $\mathcal{P}_p < \mathcal{P}_{thermal} \simeq 10^{-8}$W, while for a low-temperature superconductor ($T_c \simeq 10$)K, one expects even less power: $\mathcal{P}_p < \mathcal{P}_{thermal} \simeq 10^{-10}$W. (Notice that in (4.18), power scales as T^2.) However, since power sources can be stacked, multiple inhomogeneous loops can be arranged in series such that their voltage V_p and power $\mathcal{P}_p$ add. A series of $N = 10^8$ HTSC loops could, in theory, achieve dc power up to $\mathcal{P}_p < N(kT)^2/\hbar \simeq 1$ W [34] in an area ≈ 1cm^2. Power densities of the order of 10^8W/m^3 might be possible [34].

In principle, the persistent voltage V_b can be measured experimentally even on a single loop of low-temperature superconductor; one expects $V_b < R_b((kT)^2/\hbar)^{0.5} \approx 10\mu V$ at $T \simeq 1$ K and $R_b \simeq 10$ Ω. For high-temperature superconductors and loops in series, V_b could be an order of magnitude greater. These voltages are the primary objects of experimental study.

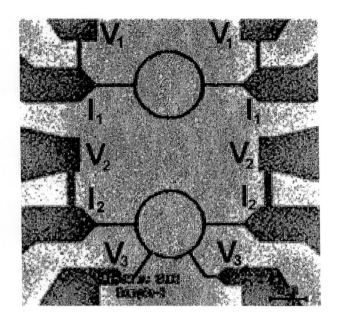

Figure 4.4: Electron micrograph of Series I experiments' symmetric (top) and asymmetric (bottom) mesoscopic aluminum loops. I_j and V_j are current and voltage contacts for each loop. Additional V_3 contacts on lower loop.

4.4.3 Experiments

Experiments investigating Nikulov's paradoxical effect date to 1997 with unpublished observations by I.N. Zhilyaev of dc voltages on segments of mesoscopic superconducting aluminum loops near their transition temperature, in the absence of external current. Since then two mains series of experiments have been conducted. The first (Series I) [33, 32] were to verify Zhilyaev's initial results in light of new theoretical understanding [35], and the second, more detailed series (Series II) examined multi-ring systems and the effects of external ac driving [31]. We will review each.

4.4.3.1 Series I

Series I experiments were conducted on single, symmetric or asymmetric, mesoscopic, high-purity aluminum loops on silicon wafer substrates [33, 32]. Figure 4.4 is an electron micrograph of exemplary symmetric (top) and asymmetric (bottom) loops with current and voltage contacts. Structures were fashioned with electron beam lithography. Loops were 60nm thick and had diameters $2R = 1$, 2, or 4μm and linewidths $w = 0.2\mu$m and 0.4μm. The midpoint of the superconductive resistive transition was roughly $T_c \simeq 1.24$K. Measurements were carried out in a conventional helium-4 cryostat with base temperature of 1.2K. Measurements of voltage oscillations were made in the narrow temperature range

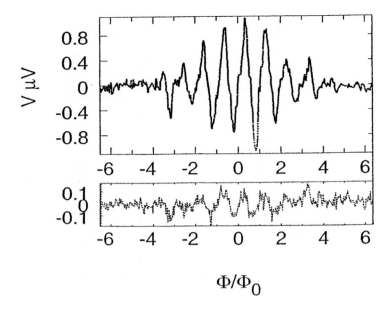

Figure 4.5: Voltage oscillation versus $\frac{\Phi}{\Phi_o}$ on single asymmetric loop measured with V_2 contacts (upper curve) and V_3 contacts (lower curve). Loop parameters: $2R = 4\mu m$, $w = 0.4\mu m$, $I_m = 0$, $T = 1.231K$ at bottom of resistive transition.

$0.988T_c < T < 0.994T_c$. Loop inhomogeneity (asymmetry) was created by reducing ring linewidth.

In principle [35], the dc voltage oscillation *should not* occur in homogeneous (symmetric) loops, but *should* occur in inhomogeneous (asymmetric) loops. Experiments qualitatively confirmed this prediction.

Figure 4.5 displays the dc voltage oscillation $V(\frac{\Phi}{\Phi_o})$ versus $\frac{\Phi}{\Phi_o}$ for a single asymmetric loop ($2R = 4\mu m$, $w = 0.4\mu m$) measured at V_2 contacts (upper curve) and V_3 contacts (lower curve) at $T = 1.231K$. Voltage oscillations are observed both across the whole loop and across the segment that, because of its narrower width, was a normal conductor.

For the symmetric loop, the voltage oscillations, $(V_1 = I_1 R_1)$ followed expectations of standard LP oscillations; that is, when $I_m = I_1 = 0$, the voltage oscillations disappeared ($V_1 = 0$). In contrast, the voltage oscillations on the asymmetric loop did not disappear for $I_2 = 0$. In particular, voltages of magnitude $V_3 \simeq 0.1\mu V$ were observed on the asymmetric segment (Figure 4.6), when $I_2 = 0$. This was observed in the narrow temperature range at the bottom of the resistive transition $\Delta T = T - T_c \simeq 0.1K$. (Larger V_3 were observed at lower temperatures.) This stark difference in behavior between symmetric and asymmetric loops agrees with theoretical predictions [35].

The researchers raise the caution that external noise — rather than purely thermal fluctuations — cannot be ruled out as a cause for the observed dc voltages. This issue partially motivated the next series of experiments.

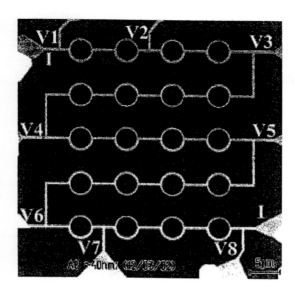

Figure 4.6: Electron micrograph of array of 20 asymmetric aluminum loops for Series II experiments. Current ΔI imposed through $I - I$ contacts. Voltage contacts $V1 - V8$.

4.4.3.2 Series II

Series II experiments were conducted on systems of either 3 or 20 asymmetric mesoscopic aluminum loops, again deposited on Si wafers and fashioned using electron beam lithography [31]. Figure 4.6 shows an array of 20 loops. All loops had diameters $2R = 4\mu$m and thicknesses 40nm. The inhomogenities (asymmetries) consisted of having one half of each loop drawn with linewidth $w = 0.2\mu$m and the other with $w = 0.4\mu$m. Resistance and current oscillations were studied in the range $0.95T_c < T < 0.98T_c$, where T_c is the midpoint of the superconducting resistive transition, $T_c \simeq 1.3$K.

As in Series I, measurements were performed in a helium-4 cryostat and a magnetic field was applied perpendicularly to the rings by a superconducting coil. Unlike Series I, the rings were driven by an external ac current, $I_{ac} = \Delta I sin(2\pi f_{ac}t)$ in the range 10^2Hz$\leq f_{ac} \leq 10^6$Hz, with amplitude 0μA$\leq \Delta I \leq 50\mu$A between contacts $I - I$ in Figure 4.6. I_{ac} was used to understand how noise (thermal or spurious background) induces voltages in the loops. Voltages were measured between contacts labeled $V1 - V8$ in Figure 4.6, thus allowing summation of voltages in series loops to be tested.

DC-voltage oscillations were measured across single and multiple loops at various magnetic field strengths ($B = \Phi/\pi R^2$) as a function of ac-current magnitude ΔI and frequency f_{ac}. The magnitude of voltage oscillations $V(\frac{\Phi}{\Phi_o})$ was found to be independent of f_{ac} over the frequency range explored ($10^2 - 10^6$Hz), but was highly dependent on ΔI. (Independence of dc voltage from f_{ac} is not sur-

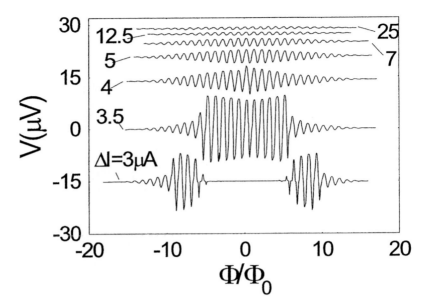

Figure 4.7: Voltage oscillation $V(\frac{\Phi}{\Phi_o})$ on a single asymmetric loop versus $\frac{\Phi}{\Phi_o}$ for different magnitudes of I_{ac} at $f_{ac} = 2.03$kHz and $T = 1.280$K $= 0.97T_c$. All traces except $\Delta I = 3.5\mu$A are displaced vertically.

prising since, compared with the maximum possible switching frequency ($f_{max} \sim 10^{11} - 10^{12}$Hz), the ratio $f_{ac}/f_{max} \sim 0$, such that the driving field is effectively static.)

Figure 4.7 displays plots of $V(\frac{\Phi}{\Phi_o})$ versus $\frac{\Phi}{\Phi_o}$ across a single loop for seven values of ΔI at temperature $T = 1.280$K and frequency $f = 2.03$kHz. For all traces, at large values of Φ/Φ_o (i.e., $\mid \Phi/\Phi_o \mid \geq 10$), one has $V \sim 0$ because of the supression of superconductivity at high imposed field values. The lowest trace ($\Delta I = 3\mu$A) displays no $V(\frac{\Phi}{\Phi_o})$ voltage oscillations below a critical threshold current, ΔI_{cr} for $\mid \Phi/\Phi_o \mid \leq 5$. As the imposed field is increased, the critical current ΔI_{cr} is reduced so that voltage oscillation appear. As noted earlier, however, they disappear again at $\mid \Phi/\Phi_o \mid \geq 10$ as the aluminum superconductivity is supressed by the imposed field.

On the next trace up ($\Delta I = 3.5\mu$A), the voltage oscillations are most robust, just beyond the critical threshold current. For large values of ΔI beyond ΔI_{cr}, the voltage oscillations (higher vertical traces in Figure 4.7) again decrease proportionately because of suppression by ΔI-induced magnetic fields.

Loop oscillation voltages can be summed in series. In Figure 4.8, voltage oscillations $V(\frac{\Phi}{\Phi_0})$ are plotted versus Φ/Φ_0 for two series cases: 3 loops and 20 loops. A comparison is less quantitative than desired since the experimental parameters are distinct for each case (See Figure 4.8 caption for details.), but a trend is evident: The ΔV magnitudes for 20 loops is on the order of 7 times greater than for 3 loops, which in turn is roughly 3 times greater than for 1 loop (Figure 4.4).

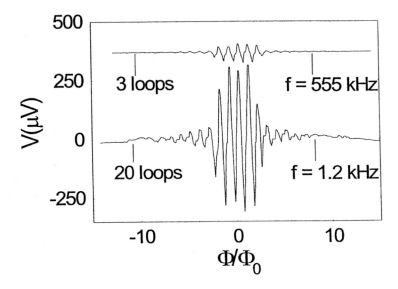

Figure 4.8: Series summation of loop voltage oscillations for 2 loop arrays. 20-loop array [$f_{ac} = 1.2$kHz, $\Delta I = 3.2\mu$A, $T = 1.245$K$=0.97T_c$]; 3-loop array [$f_{ac} = 555$kHz, $\Delta I = 4.5\mu$A, $T = 1.264$K$=0.96T_c$].

Specifically, voltage oscillations were observed up to 10μV for a single loop, up to 40μV for 3 loops in series and up to 300μV for 20 loops in series.

The quantum oscillations in Figures 4.7 and 4.8 can be attributed to loop switchings between superconducting states with different connectivity, as induced by the external current [31], whereas those in Figure 4.5 possibly could be induced by external electrical noise. Neither result directly contradicts the second law because the source of the observed dc power is not equilibrium thermal fluctuations. However, it is significant that the dc voltages observed on Figures 5,7,8 are induced by loop switchings near the critical temperature $T \approx T_c$, thereby corroborating a key aspect of the theory.

Recent theoretical and numerical work by Berger [30] lends strong qualitative support to the experimental work of Dubonos, Nikulov, et al. [31-37]. Berger studied a superconducting loop with two, unequal weak links. The loop was held near T_c and was threaded with magnetic flux. Although it does not match Nikulov's system exactly, it does bear strong physical similarities. Furthermore, it can be modeled by textbook procedures for Josephson junctions and can be compared directly to related work on Josephson rectifying ratchets [38, 39, 40].

Berger found that when loop superconductivity was broken by thermal fluctuations and resistive noise (in the vicinity of T_c), the average dc loop voltage *did not* vanish and showed qualitatively the same $V(\frac{\Phi}{\Phi_0})$ versus Φ/Φ_0 flux dependence and the same frequency independence as was predicted and observed by Nikulov, et al. He also reported the same $V(\frac{\Phi}{\Phi_o})$ dependence in the presence of ac current ΔI. (Compare, for example, Figure 4.5 in [30] with Figure 4.6 [31].)

It is stressed that, while Berger's study shows strong qualitative agreement with the fundamental processes predicted and observed by Dubonos, Nikulov, et al., the quantitative agreement is poor. Some of these differences might be attributable to the differences in the models.

4.4.4 Discussion

The theory and experiments by Nikulov, Dubonos, et al. — and their further independent theoretical corroboration by Berger [30] — represent a cogent challenge to the second law. Conclusive violation, however, cannot be claimed for several reasons. First, only one of the physical variables necessary to establish dissipation was measured experimentally; the other was inferred from theory. Ideally, both should be measured simultaneously by independent means. Second, an unambiguous experimental measurement of dissipation (local heating or radiation emission) should be made and, ideally, some global accounting of energy (work plus heat) should be carried out. Third, in no experiment has it been clearly established that thermal fluctuations were the source of the experimentally measured voltage oscillations and inferred persistent currents. Fourth, the experimental apparatus and experimenter surely generated far more entropy than could be negated by the loops.

The extreme experimental and physical requirements of this system (helium-4 cryostats, vacuum systems, microscale fabrication) probably make it commercially impractical unless perhaps high-temperature superconductors can be employed. On the other hand, as an experimentally-based challenge, it holds much promise.

4.5 Bose-Einstein Condensation and the Second Law

Keefe's and Nikulov's paradoxes share deep similarities. Both capitalize on the normal-to-superconducting transition — the transition from classical mechanical to quantum mechanical behavior — and both operate at the borderline between the two. In many respects, they can be considered quantum thermal ratchets. Keefe's CMCE engine exploits the transition induced by magnetic field strength near the critical field (H_c) and, in its primary incarnation, delivers mechanical work. Nikulov's inhomogeneous mesoscopic loop exploits the transition induced by thermal fluctuations near the critical temperature (T_c) and delivers electrical work. Both exploit the exigency of Bose-Einstein condensation and the persistent current[3]: the spontaneous drive to order as the system falls into a single, macroscopic quantum state. In Keefe's case, this order is found in the diamagnetic persistent current of the Meisner-Oschenfeld effect whereby mechanical work is extracted as the armature is expelled from the high field region. In Nikulov's case,

---

[3]Persistent currents have been claimed in normal metal [41] and semiconductor [42] mesoscopic loops.

again the order is in the form of a supercurrent, this time summoned by thermal fluctuations via the Little-Parks effect.

In light of their simularities, these two challenges suggest a deeper, unifying principle may be connecting them. They also suggest a fundamental limitation to the second law in processes involving the transition from classical mechanical to quantum behavior. Nikulov's *quantum force* may hint at this deeper principle, but it possibly does not go deep enough [22]. At a more fundamental level, the behavior of bosons in Bose-Einstein condensation is antipodal to the behavior of fermions subject to Pauli exclusion. The former intrinsically moves a multiparticle system toward a state of low entropy (single wavefunction), while the latter guarantees a state of relatively high entropy (no two particles in the same state). Neither quantum tendency arises from thermodynamic action; rather, both emerge from the purely quantum mechanical consideration of wavefunction parity.

Most classical systems are dominated by Fermi statistics and Pauli exclusion, rather than by Bose-Einstein statistics or condensation. Given their antipodal thermodynamic tendencies, perhaps it is not surprising that these two second law challenges arise only in systems involving transitions between classical and quantum statistics.

4.6 Quantum Coherence and Entanglement

4.6.1 Introduction

Allahverdyan, Nieuwenhuizen, et al. have written extensively on the limits to various formulations of the second law in the quantum regime, particularly quantum coherence and entanglement. They have been among the most fastidious in recognizing that different formulations can mean different things and that one must be cognizant of the caveats and limitations of each. They have championed the Thomson formulation — *No work can be extracted from a closed equilibrium system during a cyclic variation of a parameter by an external source* — because its basic currency (work) is a well-defined physical quantity, whereas heat and entropy (the more common currencies) are less well-defined and can be context dependent[4]. (These researchers have also shown recently that the quantum mechanical effect of level crossing limits the minimum work principle and that adiabatic processess do not correspond to optimal work if level crossing occurs [43].)

At this time, the hypotheses that quantum coherence or quantum entanglement can lead to violations of various formulations of the second law remain experimentally untested and largely uncorroborated; however, several concepts for experiments have been advanced. We will summarize the two most detailed of these by Pombo, et al. [44] and Allahverdyan and Nieuwenhuizen [45].

---

[4]It is stressed that Allahverdyan, et al. have never claimed violation of the Thomson formulation, but rather, have proved its inviolability for systems starting in equilibrium (§3.4).

4.6.2 Spin-Boson Model

Pombo, et al. [44] propose several general schemes by which two-level systems (modeled as spins) that are quantum mechanically entangled with a bath of harmonic oscillators can extract work from a heat bath. Two-level systems are ubiquitous in nature and technology and are among the most studied quantum systems known. These include nuclear magnetic resonance (NMR), electron spin resonance (ESR) and spintronic systems, two-level Josephson junctions, electrons in quantum dots, and two-level atoms [46, 47].

Pombo, et al. analyse a spin-boson model, which approximates the behavior of several systems listed above, and which also is exactly solvable analytically [46, 47]. The Hamiltonian for a system consisting of a spin 1/2 particle interacting with a bath of harmonic oscillators can be written

$$\mathcal{H} = \mathcal{H}_S + \mathcal{H}_B + \mathcal{H}_I, \tag{4.19}$$

where

$$\mathcal{H}_S = \frac{\epsilon}{2}\hat{\sigma}_z + \frac{\Delta(t)}{2}\hat{\sigma}_x; \quad \mathcal{H}_B = \sum_k \hbar\omega_k \hat{a}_k^\dagger \hat{a}_k; \quad \mathcal{H}_I = \frac{1}{2}\sum_k g_k(\hat{a}_k^\dagger + \hat{a}_k)\hat{\sigma}_z. \tag{4.20}$$

Here $\mathcal{H}_{B,S,I}$ are the Hamiltonians for the bath, the spin-1/2 particle, and their interaction. $\hat{\sigma}$ are the Pauli spin matrices; $\hat{a}_k^\dagger$ and $\hat{a}_k$ are creation and annihilation operators; g_k are the coupling constants between the bath modes and the spin; and $\epsilon = \bar{g}\mu_B B$ is the standard energy of a spin in a magnetic field, where μ_B is the Bohr magneton, $\bar{g}$ is the gyromagnetic constant and B is the magnetic field. $\Delta(t)$ is an interaction potential that is switched on and off quickly from an external source and affects the x-component of spin.

Various modes and protocols of interaction between spin and bath are discussed [44]. We consider an archetypical one in which the spin is subjected to a sudden, brief external pulse, by which the spin is quickly driven about the x-axis. (Here $\Delta \neq 0$ lasts for duration δ_1 and has large magnitude; *i.e.*, $\Delta \sim \frac{1}{\delta_1}$.) In principle, this can be accomplished without changing the energy of the spin ϵ since energy depends on $\hat{\sigma}_z$ and not $\hat{\sigma}_x$. The time-evolution operator associated with the pulse is: $\hat{U}_1 \equiv exp\left[-i\frac{\mathcal{H}t}{\hbar}\right] = exp\left[-i\frac{\mathcal{H}\delta_1}{\hbar}\right]$, which, in this approximation ($\Delta \sim \frac{1}{\delta_1}$), can be written $\hat{U}_1 \simeq exp\left[\frac{i}{2}\Theta\hat{\sigma}_x\right] + \mathcal{O}(\delta_1)$, where $\Theta = -\frac{\delta_1\Delta}{\hbar}$, the x-rotation angle. Note, pulses correspond to a cyclic process of an external work source and they change neither the energy of the spin, nor its statistical (von Neumann) entropy.

Although the energy of the spin need not change during on-off switching of $\Delta(t)$, the total system work involves both the spin *and* bath and it is found that work can be extracted from the heat bath with proper pulsing of $\Delta(t)$. Under the conditions that g is small, $\Theta \equiv -\frac{\delta_1\Delta}{\hbar} = -\frac{\pi}{2}$; and $t \gg \frac{1}{\Gamma}$ = the relaxation time of the bath; and $\epsilon = 0$ (to insure the spin energy does not change), the added work is given by

$$W_1 = \frac{g\hbar\Gamma}{2\pi} + \frac{gkT}{2}\langle\hat{\sigma}_x(0)\rangle exp\left[-\frac{t}{\tau_2}\right], \tag{4.21}$$

where kT is bath thermal energy and τ_2 is the transversal ($\hat{\sigma}_x$) spin decay time. If the spin starts in a coherent state ($\langle\hat{\sigma}_x\rangle = -1$) and if the time is adjusted such that $\frac{kT}{2}exp[-\frac{t}{\tau_2}] > \frac{\hbar\Gamma}{2\pi}$, then (4.21) indicates work can be extracted from the bath ($W_1 < 0$). This demonstrates that the Thomson formulation cannot be applied to the locally equilibrium heat bath. However, it is applicable if the whole system — i.e. the spin and bath together — starts in equilibrium before applying the first pulse. (Other schemes challenge the Clausius inequality, whereby work can be extracted from the bath without changing the entropy of the spin [44].)

Pombo, et al. offer several incentives for pursuing laboratory experiments:
1) Two-level quantum systems and harmonic oscillator heat baths are ubiquitous, within appropriate physical limits [46, 47], e.g., atoms in optical traps, electron spins in semiconductors (injected or photonically excited), excitons in quantum dots, nuclear spins (NMR), or electron spins (ESR) in condensed matter.
2) Experimental detection methods are, in principle, sufficiently well-developed (e.g., ESR, NMR) to make the salient measurements.
3) The main quantum effects survive for completely disordered ensembles of spins.

4.6.3 Mesoscopic LC Circuit Model

Electrical circuits have long been fertile testbeds for thermodynamics and statistical mechanics [48, 49]. Recently, Allahverdyan and Nieuwenhuizen have suggested experiments on mesoscopic or nanoscopic, linear LRC circuits interacting with a low-temperature heat bath, which in principle could test for predicted violations of the Clausius form of the second law in the quantum regime [45].

A classical series LRC circuit can be described in terms of conjugate variables (charge (Q) and magnetic flux (Φ)). These play roles analogous to canonical coordinate and momentum in a mass-spring system. Written side by side the LRC circuit and mass-spring Hamiltonians are written (for zero-damping):

$$H_s = \frac{\Phi^2}{2L} + \frac{Q^2}{2C}; \quad H = \frac{p^2}{2m} + \frac{kx^2}{2} \tag{4.22}$$

From inspection, $p \equiv \Phi$, $m \equiv L$, $k \equiv \frac{1}{C}$; and $x \equiv Q$. Note that the conjugate variables are also related analogously: $\dot{Q} = \frac{\Phi}{L}$ and $\dot{x} = \frac{p}{m}$. For the $R = 0$ case, this classical system can be treated quantum mechanically by allowing Q and Φ to act as operators satisfying the commutation relation: $[Q, \Phi] = i\hbar$.

In either the classical or quantum regimes, a measure of a circuit's disorder (entropy) can be taken to be the volume of phase space (Σ) that it explores. In terms of the LRC circuit variables, this can be written

$$\Sigma = \frac{\Delta\Phi\Delta Q}{\hbar} \equiv \sqrt{\frac{\langle\Phi^2\rangle\langle Q^2\rangle}{\hbar^2}} \tag{4.23}$$

where $\langle\Phi^2\rangle$ and $\langle Q^2\rangle$ are the dispersions (variances) in Φ and Q. In the classical thermodynamic limit, the dispersions take the Gibbsian forms:

$$\langle \Phi^2 \rangle_G = \frac{1}{2} L\hbar\omega_o \tanh(\frac{1}{2}\beta\hbar\omega_o) \tag{4.24}$$

$$\langle Q^2 \rangle_G = \frac{1}{2} C\hbar\omega_o \tanh(\frac{1}{2}\beta\hbar\omega_o)$$

where $\omega_o = \frac{1}{\sqrt{LC}}$ is the undamped LC resonance frequency, and $\beta = \frac{1}{kT}$ is the inverse thermal energy.

Starting from the quantum Langevin equations and assuming quantum Gaussian noise with the Nyquist spectrum having a large cut-off frequency ($\omega_{max} \equiv \Gamma$) [50], the quantum dispersions in Φ and Q can be written

$$\langle \Phi^2 \rangle = \int \frac{d\omega}{2\pi} \frac{\omega^2 k(\omega)}{(1 + \frac{\omega^2}{\Gamma^2}) \left[(\omega^2 - \omega_o^2)^2 + (\frac{\omega R}{L})^2 \right]} \tag{4.25}$$

$$\langle Q^2 \rangle = \int \frac{d\omega}{2\pi} \frac{k(\omega)}{\left[(\omega^2 - \omega_o^2)^2 L^2 + (\omega R)^2 \right]}$$

with $k(\omega) = \hbar R\omega \, coth(\frac{\hbar\omega}{2kT})$.

Comparing (4.24) with (4.25), it is clear that the quantum and classical dispersions are distinct. Notably, the quantum dispersions include damping (R), whereas the classical dispersions do not. In the limit of weak coupling with the heat bath ($R \to 0$) or at high temperatures ($\frac{\hbar\omega}{kT} \to 0$), the quantum cases revert to the classical Gibbsian cases.

The Clausius formulation of the second law can be phrased in the form of the Clausius inequality: $dQ \le TdS$. The heat and entropy changes can also be expressed in terms of changes in phase space volume $d\Sigma$, which in turn can be written in terms of physically measurable variances Q and Φ through (4.23-25). Classically, if the LRC circuit absorbs heat from the heat bath, then its phase space volume will expand; conversely, if heat is lost to the heat bath, its phase space volume will contract. If the dispersions $\langle Q^2 \rangle$ and $\langle \Phi^2 \rangle$ are Gibbsian, then classical thermodynamics applies and the Clausius criterion $dS \ge \frac{dQ}{T}$ is satisfied. However, if the temperature is sufficiently low, then the dispersions follow the quantum prescriptions, (4.25), and as Allahverdyan and Nieuwenhuizen have shown, in this regime the Clausius form of the second law can be violated [51, 52]. In this case, the circuit can absorb heat from the bath while simultaneously contracting in phase space. More precisely, at finite temperatures a cloud of electromagnetic modes forms around the LRC circuit. Its energy should be counted to the bath and may be partly harvested since changing a parameter of the LRC circuit can induce a change in this cloud.

For the LRC circuit, in principle, this second law violation would be realized by varying a system parameter (say, inductance L) via an external agent and, thereby, affecting the heat transferred from the heat bath to the circuit. In the quantum regime, they find that for low quality factor circuits,

$$dQ = \frac{\hbar R}{2\pi L^2} dL > 0 \tag{4.26}$$

and that $\frac{d\Sigma}{dL} < 0$; that is, there is positive heat transfer to the circuit and phase space contraction. These constitute violations of the Clausius form of the second law.

Unlike the classical Clausius constraint $(dQ \leq TdS)$, which requires non-zero temperature for heat transfer, this quantum constraint for heat transfer from bath to system is temperature independent; therefore, in principle, it can occur at zero temperature. This observation spotlights the defining characteristic of this quantum thermodynamic system: the role of entanglement. At first consideration, a heat bath at $T = 0$ should not be able to render up heat since presumably it is in its ground state, possessing only zero-point energy. Equation (4.26) indicates, however, that heat can in fact be rendered from the bath at zero temperature. Appealing to the first law, one concludes that the bath is *not* in its ground state. In fact, because it is always entangled with the circuit — and, therefore, in a mixed state — it is never in its true ground state. This extra *entanglement energy* has no classical analog and lies at the heart of this quantum second law challenge. This entanglement leads to a screening cloud, known elsewhere as a Kondo cloud or polaronic cloud. Not surprisingly, these also arise in some of Čápek's systems (§3.6). Nieuwenhuizen and Allahverdyan speculate that such clouds may well be the sole cause underlying the differences between classical and quantum thermodynamics.

Experimental support for this challenge would consist in measuring heat flow from the heat bath into the LRC circuit while simultaneously measuring decreases in the dispersions $\langle \Phi^2 \rangle$ and $\langle Q^2 \rangle$. Experiments on low-temperature mesoscopic tunnel junctions have reported inferred values of $\langle Q^2 \rangle$ in related sub-Kelvin temperature regimes [53, 54], thus offering hope that full-fledge tests of the Clausius inequality might be possible. Such experiments would be difficult to design, conduct, and interpret, but they appear within the current experimental art.

4.6.4 Experimental Outlook

The above-mentioned experimental concepts and incentives are compelling and should be pursued more fully since they offer the hope of sensitive tests of at least two formulations of the second law (Thomson and Clausius) on many systems for which there is already deep understanding. On the other hand, the level of theoretical analysis and experimental details presented for the experiments thus far [44, 45] are insufficient to determine whether such experiments are truly feasible or even whether their proposed thermodynamic cycles can achieve breakeven in entropy reduction. Let us consider theory first.

The analysis of Pombo's thermodynamic cycle appears incomplete, leaving out key thermodynamic steps. For instance, the work and entropy generation required to prepare the $\langle \hat{\sigma}_x(0) \rangle = -1$ state has not been assessed, although τ_2 is admittedly finite, such that spins must be restored on a regular basis for the cycle to repeat. (If, on the other hand, random spins are used, then the conditions for (4.21) to return strictly negative work $(W_1 < 0)$ appear compromised since $W_1 < 0$ requires the simultaneous tuning of phase-dependent parameters. Or, if random spins are used judiciously, these spin states must be measured, presumably by an energy-consuming, entropy-generating agent.) Additionally, no assessment has been made of the work or entropy generation associated with the nonequilibrium Δ pulse that couples the spin and bath. Similarly, for the quantum LRC circuit no thermodynamic assessment has been made of the dL-varying agent[5]. In short, several key elements of this cycle have not been assessed thermodynamically. On the other hand, if the experiments aim merely to test for the fundamental effects — full second law challenges aside — then these objections may be moot.

There are also many experimental issues that have not been addressed — too many to list here — so let this summarize: No realistic experimental system for either proposal (spin-boson or LRC) has explicitly been shown to meet the criteria for work extraction from a heat bath, nor have explicit experimental designs with realistic experimental parameters been vetted adequately. For neither model has it been well-established that experimental techniques are adequate to make the necessary measurements, nor that the negentropy of the cycle can outweigh the the entropy production of the apparatus, even in principle. In contrast, Nikulov, et al. (§4.4), Keefe (§4.3), and Sheehan (Chapters 7-9) have treated well-defined experimental circumstances, realizing that with experiments, *the devil is in the details*. In summary, within the idealizations of their theoretical development, compelling cases have been made for second law challenges in the quantum regime; however, these experiments are still in the conceptual stage.

The thermodynamic requirements for these and the previous challenges (§4.3, §4.4) are extreme. The superconducting ones require both low-temperatures and micro- or mesoscopic structures, which burden experimental techniques and hamper direct, unambiguous measurements of predicted entropy reductions or heat fluxes — whichever is necessary. It is not enough for the CMCE engine to simply run, or for voltage oscillations to be measured in an inhomogeneous superconducting loop near its transition temperature. To challenge the second law successfully, the involvement of *all* other possible free energy sources must be ruled out and, ideally, direct measurements must be made of sustained heat fluxes or entropy reductions that are causally connected to the forbidden work. These types of energy and entropy determinations are not trivial even for room-temperature, everyday macroscopic devices like flashlights; thus, for a vacuum-packed, microscopic superconductor near absolute zero, it is likely to be more difficult. With regard to the entanglement proposals, here the difficulties associated with microscopic devices are traded for the intricacies and uncertainties associated with entanglement. As a simple example, how does one quantitatively measure system-bath entanglement and demonstrate unambiguously that it has been transformed into work?

Despite these hurdles, these low-temperature challenges are among the most compelling of the modern era. At a deep level they underscore the fundamental differences between classical and quantum thermostatistical behaviors. Of course, quantum statistics were unknown at the time thermodynamics and statistical mechanics were being forged, and since then quantum systems have grown up to be disrespectful of the zeroth and third laws. Perhaps it should not be too surprising that they now show occasional disrespect for the second.

---

[5]In principle, a work source generates negligible entropy, but for real experimental systems this is rarely the case.

References

[1] Meissner, W. and Ochsenfeld, R., Naturwiss. **21** 787 (1933).

[2] Tinkham, M., *Introduction to Superconductivity* (McGraw-Hill, New York, 1975).

[3] Langenberg, D.N. and R.J. Soulen, Jr. in *McGraw-Hill Encyclopedia of Science and Technology*, 7th ed., Vol. 17, (McGraw-Hill, New York, 1992) pg. 641.

[4] Kittel, C., *Introduction to Solid State Physics*, 7th Ed., Ch. 12, (Wiley, New York, 1996).

[5] Bardeen J., Cooper L.N., Schrieffer J.R., Phys. Rev. **108** 1175 (1957).

[6] Ginzburg, V.L. and Landau, L.D., Zh. Eksp. Teor. Fiz. **20** 1064 (1950)

[7] Hauser, J.J., Phys. Rev. B **10** 2792 (1974).

[8] Salamon, M.B., in *Physical Properties of High Temperature Superconductors I*, Ginsberg, D.M., Editor, (World Scientific, Singapore, 1989).

[9] Huebener, R.P., *Magnetic Flux Structures in Superconductors* (Springer-Verlag, Berlin, 1979).

[10] Abrikosov, A.A., Zh. Eksp. Teor. Fiz. **32** 1442 (1957); (*Sov. Phys.-JETP* **5** 1174 (1957)).

[11] Marchenko, V.A. and Nikulov, A.V., Pisma Zh. Eksp. Teor. Fiz. **34** 19 (1981); (JETP Lett. **34** 17 (1981)).

[12] Nikulov, A.V., Supercond. Sci. Technol. **3** 377 (1990).

[13] Safar, H., Gammel, P.L., Huse, D.A., Bishop, D.J., Rice, J.P. and Ginzberg, D.M., Phys. Rev. Lett. **69** 824 (1992).

[14] Trupp, A., in *Quantum Limits to the Second Law*, AIP Conf. Proceed. Vol. 643, Sheehan, D.P., Editor, (AIP Press, Melville, NY, 2002) pg. 201.

[15] Pippard, A.B., Phil. Mag. **43** 273 (1952).

[16] Lutes, O. and Maxwell, E., Phys. Rev. **97** 1718 (1955).

[17] Geim, A.K., Dubonos, S.V., Lok, J.G.S., Grigorieva, I.V., Maan, J.C., Hansen, L.T., and Lindelof, P. E., Appl. Phys. Lett. **71** 2379 (1997).

[18] Geim, A.K., Dubonos, S.V., Palacios, J.J., Grigorieva, I.V., Henini, M., and Schermer J.J., Phys. Rev. Lett. **85** 1528 (2000); Geim, A.K., Dubonos, S.V., Grigorieva, I.V., Novoselov, K.S., Peeters, F.M., and Schweigert V.A., Nature **407** 55 (2000).

[19] Little, W.A. and Parks, R.D., Phys. Rev. Lett. **9** 9 (1962).

[20] Vloeberghs, H., Moshchalkov, V.V., Van Haesendonck, C., Jonckheere, R. and Bruynseraede, Y., Phys. Rev. Lett. **69**, 1268 (1992).

[21] Tinkham, M., Phys. Rev. **129** 2413 (1963).

[22] Nikulov, A.V., Phys. Rev. B **64** 012505 (2001).

[23] Keefe, P., in *Quantum Limits to the Second Law*, AIP Conf. Proceed. Vol. 643, (AIP Press, Melville, NY, 2002) pg. 213.

[24] Keefe, P., J. Appl. Optics **50** 2443 (2003); Entropy **6** 116 (2004).

[25] Chester, M., J. Appl. Phys. **33** 643 (1962).

[26] Keefe, P., Masters Thesis, University of Detroit, MI (1974).

[27] Bryant, C. and Keesom, P., Phys. Rev **123** 491 (1961).

[28] Roukes, M.L., *Tech. Dig. of the 2000 Solid-State Sensor and Actuator Workshop*, Hilton Head Isl. SC, June (2000).

[29] Caswell, H.,J., Appl. Phys. **36** 80 (1965).

[30] Berger, J., http://xxx.lanl.gov/abs/cond-mat/0305130; in press Phys. Rev. B **70** (2004).

[31] Dubonos, S.V., Kuznetsov, V.I., Zhilyaev, I.N., Nikulov, A.V., and Firsov, A.A., JETP Lett. **77** 371 (2003).

[32] Dubonos, S.V., Kuznetsov, V.I., and Nikulov, A.V., in *Proceedings of 10th International Symposium "Nanostructures: Physics and Technology,"* St. Petersburg Ioffe Institute, (2002) pg 350.

[33] Nikulov, A.V., in *Quantum Limits to the Second Law*, AIP Conference Vol. 643, (AIP Press, Melville, NY, 2002) pg. 207.

[34] Aristov, V.V. and Nikulov A.V., in *Quantum Informatics*, Proceedings of SPIE Vol. **148** 5128 (2003); http://xxx.lanl.gov/abs/cond-mat/0310073

[35] Nikulov, A.V. and Zhilyaev, I.N., J. Low Temp. Phys. **112** 227 (1998).

[36] Barone, A. and Paterno, G., *Physics and Applications of the Josephson Effect* (Wiley-Interscience, New York, 1982).

[37] Nikulov, A.V., in *Supermaterials*, Cloots, R., et al., Editors, Proceedings of the NATO ARW, (Kluwer Academic, 2000) pg. 183.

[38] Zapata, I., Bartussek, R., Sols, F., and Hänggi, P., Phys. Rev. Lett **77** 2292 (1996).

[39] Falo, F, Martinez, P.J., Mazo, J.J., and Cilla, S., Europhys. Lett. **45** 700 (1999).

[40] Weiss, S., Koelle, D., Müller, J., Gross, R., and Barthel, K., Europhys. Lett **51** 499 (2000).

[41] Levy, L.P. et al., Phys. Rev. Lett. **64** 2074 (1990); Chandrasekhar, V. et al., Phys. Rev. Lett. **67** 3578 (1991); Jariwala, E.M.Q. et al., Phys. Rev. Lett. **86** 1594 (2001).

[42] Mailly, D., Chapelier, C. and Benoit, A., Phys. Rev. Lett. **70** 2020 (1993); Reulet, B. et al, Phys. Rev. Lett. **75** 124 (1995); Rabaud, W. et al., Phys. Rev. Lett. **86** 3124 (2001).

[43] Allahverdyan, A.E. and Nieuwenhuizen, Th.M, cond-mat/0401548.

[44] Pombo, C., Allahverdyan, A.E., and Nieuwenhuizen, Th.M. in proceedings of *Quantum Limits to the Second Law*, AIP Conference Proceedings, Vol. 643 (AIP Press, Melville, NY, 2002) pg. 254; Allahverdyan, A.E. and Nieuwenhuizen, Th.M., J. Phys. A: Math. Gen. **36** 875 (2004).

[45] Allahverdyan, A.E. and Nieuwenhuizen, Th.M., Phys. Rev. B **66** 115309 (2003).

[46] Leggett, A.J., et al., Rev. Mod. Phys. **59** 1 (1987).

[47] Luczka, J., Physica A **167** 919 (1990).

[48] Nyquist, H., Phys. Rev. **32** 110 (1928).

[49] Brillouin, L., Phys. Rev. **78** 627 (1950).

[50] Weiss, U., *Quantum Dissipative Systems* (World Scientific, Singapore, 1999).

[51] Allahverdyan, A.E. and Nieuwenhuizen, Th.M., Phys. Rev. Lett. **85** (2000); Nieuwenhuizen, Th.M. and Allahverdyan, A.E., Phys. Rev. E **66** 036102 (2002).

[52] Allahverdyan, A.E. and Nieuwenhuizen, Th.M., Phys. Rev. E **64** 056117 (2001).

[53] Cleland, A.N., Schmidt, J.M., and Clarke, J., Phys. Rev. B **45** 2950 (1992).

[54] Hu, G.Y. and O'Connell, R.F., Phys. Rev. B **46** 14219 (1992).

5

Modern Classical Challenges

An assortment of modern classically-based second law challenges are reviewed and critiqued. Most are theoretical, but one has undergone exploratory laboratory tests (§5.5).

5.1 Introduction

Since 1980 over two dozen challenges have been raised against the second law, documented in roughly 50 journal articles. In this chapter we review a half-dozen representative challenges, beginning with the earliest of the modern era by Gordon (§5.2) and Denur (§5.3), up through some of the most recent by Crosignani, et al. (§5.4) and Trupp (§5.5). Aside from their classical natures, this eclectic group appears to have few common threads: Gordon theorizes about the Maxwell demon-like possibilities of chemical membranes; Denur extends the ideas of Feynman's pawl and ratchet toward a truly thermally-driven device; Crosignani, et al. trod the well-worn path gas fluctuations and discover a thermodynamic gem in the mesoscopic regime; Trupp finds surprises involving classical dielectrics; and Liboff considers the mechanical scattering of disks in special geometries. We begin with Gordon, who lobotomized Maxwell's demon once and for all and, thereby, surmounted a 50-year preoccupation with information theory.

5.2 Gordon Membrane Models

5.2.1 Introduction

The challenges of L.G.M. Gordon, beginning in 1981 [1], commence the modern movement to investigate the absolute status of the second law [2, 3, 4, 5]. This series of challenges is based on asymmetries in chemical reactions associated with membranes. Using plausible chemical structures and reaction mechanisms, Gordon posits a breakdown in the principle of detailed balance.

Gordon's seminal contribution was to reduce the Maxwell demon [6] to mere chemicals, specifically to replace the demon's sentience and intelligence with linked chemical reactions that serve as algorithms, and to substitute trapdoors, pawls, and pistons with functional membranes and structured molecules. In doing so, Gordon sidestepped issues of information and measurement theory that had sidetracked second law discussions for half a century — although these digressions were fruitful in other areas [6]. Gordon's general tack of undermining detailed balance was later arrived at independently by other workers, notably Čápek [7] and Sheehan [8], each of whom developed this approach in different directions.

All of Gordon's challenges involve chemical membranes, which can be as simple as monolayers of a pure liquid or as fantastically complex as biomembranes, whose structures and functions still hold fascination after more than 70 years of study [9]. Their properties vary considerably; the following values are representative. Biomembranes have typical thicknesses on the order of 10^{-8}m. Embedded pore proteins have typical masses of 10^6amu, constructed from on the order of 10^5 atoms and having sizes on the order of $5 - 10 \times 10^{-9}$m. Conformational opening and closing times for channels are of the order of $10^{-3} - 10^{-4}$s.

Biomembranes can be likened to molecular machines; they recognize molecules, selectively bind and transport them, modify their own structure in response to their environment. In essence, they can execute all the necessary functions of a Maxwell demon. What is unclear — and the fundamental question that Gordon raises — is whether biomembranes can accomplish these tasks solely using ambient heat, rather than standard free energy sources like ATP.

Ion sorting and transport is one of the primary functions of biomembranes. In doing so, they are able to establish potential gradients between the inside and outside of cells (typically $\Delta V = 60-80$mV) and thereby store significant capacitive electrostatic energy that can be tapped instantaneously to drive other membrane and cell processes. The common cellular Na^+/K^+ gradient will relax away through diffusion unless it is is maintained through active ion transport. This requires energy which is typically supplied by light or ATP.

The proteins that carry out membrane functions are typically embedded in the membrane itself. They act as barriers to the exterior environment (nasty chemicals, viruses, *etc.*), as binding sites, gates, channels and conveyer belts for the influx and efflux of nutrients and wastes. Molecular sorting and transport is accomplished through binding specificities and conformational changes in their structures. While this usually requires energetic free energy sources (ATP or light), some appears to require less. The glucose transporter protein in the cell membrane, for instance, is inactive toward the Na^+ ion alone and toward the glucose molecule alone, but when confronted with a Na^+-glucose complex, it binds it, spontaneously changes conformation, thereby allowing passage of the Na^+-glucose complex into the cell. While this seems to require nothing more than thermal energy to proceed, the net flux is driven by concentration gradients of Na^+-glucose, thus by chemical potential gradients, and therefore does not constitute a second law paradox.

This raises a conundrum. If the opening and closing of the glucose transporter channel is driven by thermal energy and is merely unlocked by the presence of the Na^+-glucose complex (requiring no free energy expenditure), then mechanistically, there appears to be no reason why the necessary conformational changes cannot be triggered from one side of the membrane only; that is, the receptor site to trigger passage could be found on one side only. If this were the case, however, then glucose would pass in one direction, constituting a solute pump, in violation of the second law. Experimentally, however, this is not observed, either for the glucose transporter or any other protein. Experiments find that the maximum glucose transfer rates are the same in both directions across the membrane in the absence of concentration gradients. No examples are known of net particle flux without gradients or free energy driving them, which of course, conforms to the second law, but which is not to say that they cannot exist, in principle.

Gordon's challenges fall into two broad catagories: those that rely primarily on chemical kinetics [1, 3] and those that rely strongly on physical or mechanical properties [2, 4]. Of these, the former type seems robust, while the latter is suspect. We will consider examples of each. Although they invoke characteristics of real chemical entities, Gordon's challenges should properly be considered theoretical. Similarly to Čápek's, they are formal and not known to apply to any specific membrane or chemical system, and no experiments are recommended.

Most of Gordon's work [2, 3, 4, 5] stem from ideas in his first, seminal paper in 1981 [1]. It is motivated by the famous Szilard [10] and Szilard-Popper [11] engines which are also abstract and purely mechanical. For the following discussion, we mostly adopt Gordon's notation.

5.2.2 Membrane Engine

Consider an isothermal, isobaric gas B, partitioned by a membrane having a gated channel, as depicted in Figure 5.1. The gate (vertical dotted line) can exist in two states (left blocked/right open (X1); and left open/right blocked (X2)) and the channel can exist in two states of constriction (wide (Y1) and narrow (Y2)). Gas B can pass through only an open gate or a wide channel. The channel widens and narrows through conformational changes propagating in the direction indicated by the arrows near the channel walls.

Together there are four possible states of the gated channel, as indicated in Figure 5.1. The gas concentration associated with each state $X_i Y_j$ is given by C_{ij}. The rate constant (transition probability) coupling state $X_i Y_p$ to state $X_j Y_p$ is given by k_{ij}, with $(i, j = 1, 2)$. The rate constant (transition probability) coupling state $X_p Y_m$ to state $X_p Y_n$ is given by h_{mn}, with $(m, n = 1, 2)$.

The channel is encased by a cylindrically N-lobed clathrate holding a single molecule M. M links the chemical dynamics of states X and Y, which are other-

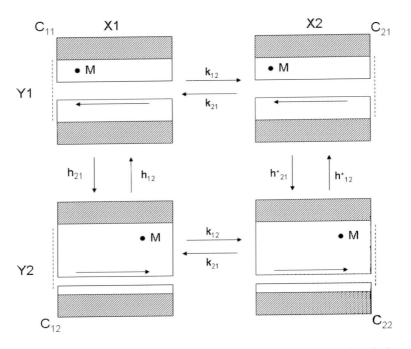

Figure 5.1: Gordon's membrane engine, with four states detailed.

wise independent. By construction, as state X1 transitions to X2, M is trapped in one of the N lobes. This reduces the entropy of the system (hence its free energy, G), increases the local pressure on the trapping lobe, and decreases pressure in the other lobes. This stress favors state Y2 by energy $kT \ln[N]$. Thus, while the rate constants linking the X states (k_{12} and k_{21}) are independent of the channel configuration Y, molecule M renders distinct the rate constants h_{ij} and $h^{\dagger}_{ij}$. Specifically, one has

$$\frac{h_{21}}{h_{12}} = exp[-\frac{\Delta G}{RT}] \tag{5.1}$$

and

$$\frac{h^{\dagger}_{21}}{h^{\dagger}_{12}} = exp[-\frac{\Delta G + RT \ln(N)}{RT}] \tag{5.2}$$

from which one obtains

$$\frac{h^{\dagger}_{21}}{h^{\dagger}_{12}} = \frac{h_{21}}{h_{12}}\frac{1}{N} \tag{5.3}$$

This will be used shortly.

The steady-state concentrations C_{ij} can be obtained from the general rate equations:

$$\frac{dC_{11}}{dt} = -(k_{12} + h_{12})C_{11} + k_{21}C_{21} + h_{21}C_{12} + 0 = 0 \tag{5.4}$$

$$\frac{dC_{21}}{dt} = k_{12}C_{11} - (k_{21} + h_{12}^\dagger)C_{21} + 0 + h_{21}^\dagger C_{22} = 0 \tag{5.5}$$

$$\frac{dC_{12}}{dt} = h_{12}C_{11} + 0 - (h_{21} + k_{12})C_{12} + k_{21}C_{22} = 0 \tag{5.6}$$

$$C = C_{11} + C_{21} + C_{12} + C_{22} = 0 \tag{5.7}$$

Here C indicates the total concentration is constant. These four linearly independent equations can be solved simultaneously for C_{ij}.

Starting from equal gas pressures on either side of the membrane, assuming an average number of molecules (W) adsorb at each channel, and employing (3), the gas transfer rate (R) across the membrane is found to be:

$$R = W(k_{12}C_{11} - k_{21}C_{21}) = W\frac{C}{\Delta}k_{12}k_{21}h_{12}h_{21}^\dagger(N - 1), \tag{5.8}$$

where Δ is the determinant of the coupled linear equations' matrix.

Since $R \neq 0$, there is a net, spontaneous flux of B across the membrane. The net flux ceases when the pressure ratio across the membrane is $N : 1$. The flux is driven entirely by thermal energy. For n moles of ideal gas B, the net entropy decrease is $\Delta S = -nR\ln(N) < 0$.

As a precursor to his membrane engine, Gordon [1] described a purely mechanical engine based on the works of Szilard [10] and Popper [11]. Comparing Gordon's membrane engine with his mechanical one, membrane states X and Y correspond to the trapdoor and piston, respectively. Molecule M breaks the symmetry in the chemical dynamics, thereby allowing the system to evolve irreversibly.

One can estimate the performance of Gordon's engine. If the membrane channel and clathrate cover surface area L^2 and can execute a molecular transaction in average time τ at temperature T, then the areal heat flux $\dot{Q}$ should scale as $\dot{Q} \simeq -\frac{1}{L^2\tau N_A}RT\ln[N]$, where N_A is Avogadro's number. For pore of scale size $L \sim 10^{-8}$m, $N = 6$, and transaction time $\tau \sim 10^{-3}$s at $T = 300$K, one has $\dot{Q} \simeq -0.1$W/m^2. Since chemical membranes are thin, the volume heat flux could be substantial; for stacked arrays of membranes 10^{-7}m thick, the volume heat flux $\dot{Q}$ the immediate example would be $\dot{Q} \simeq -10^6$W/m^3. Clearly, without compensation, this would cool the system very quickly such that this pumping rate could not be sustained. A more streamlined version of the membrane engine was given by Gordon in 1994 [3].

On the surface, Gordon's engine appears superior to its mechanical counterpart in that it relies upon known and plausible biochemical properties, behaviors and structures, rather an on an idealized micro- or nano-sized machine. It retreats from Szilard's classical mechanical modeling, which is suspect at the size scale at which the engine must operate, and instead employs statistical assumptions (transition probabilities) which are more appealing. On the other hand, in

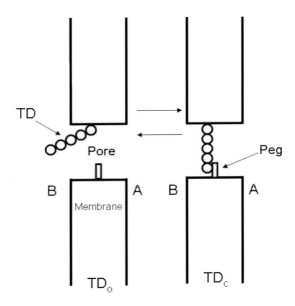

Figure 5.2: Membrane with molecular trapdoor in open and closed states.

light of recent advances in micro- and nanotechnology, especially *micro-* and *nano-electromechanical systems* (MEMS and NEMS) [12], one must question whether Gordon's mechanical engine will ultimately be any less feasibly constructed than his biological one. 'Bottom-up' construction and self-assembly of nanomachines is currently more a dream than a reality, but the rapid progress in the field coupled with sure knowledge of life's success with this approach suggests that, in the end, there may be little difference between the two, aside from descriptive formalism.

5.2.3 Molecular Trapdoor Model

Consider an isolated isothermal system (Figure 5.2) consisting of two chambers of gas G separated by a thin, thermally-conducting membrane containing pores permeable to the gas [2]. The pores are molecularly hinged with a trapdoor (TD) such that it can be in an open (TD_o) or closed state (TD_c). States TD_o and TD_c can be long-lived, for instance if the tip of the door physisorbs on the closing peg or on the pore wall.

This system is claimed to operate asymmetrically in that molecules G moving from compartment B (left) can pass into chamber A only when in the membrane is in the TD_o state since the molecule is unable to force the trapdoor open, against its stopping peg. Molecules convecting from chamber A; however, can pass both when the trapdoor is open (TD_o) and also a fraction of the time when it is closed (X) by forcing the door open with a suitably energetic collision from the right. This is the molecular equivalent of a doggy-door.

The total average flux of G from chamber A to B (J_{AB}) and from B to A (J_{BA})

through the pore can be written

$$J_{AB} = k_{AB}[G_A]\frac{\tau_o + X\tau_c}{\tau} \tag{5.9}$$

and

$$J_{BA} = k_{AB}[G_B]\frac{\tau_o}{\tau}, \tag{5.10}$$

where $\tau = \tau_o + \tau_c$ is the total time of an opening-closing cycle[1]; $[G_{A,B}]$ are the volume concentrations of G in chambers A and B, and k_{AB} and k_{BA} are the rate constants for G entering the channnel from the A and B chambers, respectively. (It is required that $k_{AB} = k_{BA}$, otherwise at equilibrium, with the trapdoor permanently open, one has $J_{AB} \neq J_{BA}$, which violates the principle of microscopic reversibility.)

At equilibrium, it is required that $J_{AB} = J_{BA}$. Setting these equal, one obtains the condition:

$$\frac{[G_B]}{[G_A]} = \frac{\tau_o + X\tau_c}{\tau_o} = 1 + X\frac{\tau_c}{\tau_o} > 1. \tag{5.11}$$

In other words, a pressure difference can spontaneously evolve between the two chambers and, presumably, pressure work can be performed by throttling the gas so as to relieve this difference. In essence, this is an asymmetric molecular valve, a non-sentient Maxwell pressure demon, a chemical version of the Maxwell valve [13].

No explicit chemical design for this device has been proposed for the gas phase, but liquid phase versions have been suggested by K. Varner in the context of "attached osmotic membranes [14, 15]." To date none have been attempted experimentally.

The crucial claim of this paradox is that a gas molecule G can force the trapdoor open from the right, but not from the left. This is questionable. The gas, membrane and trapdoor are at thermal equilibrium. Therefore, the thermal buffeting by the gas molecules is simply another manifestation of the thermal fluctuations that the door undergoes already as a result of being coupled radiatively to the blackbody cavity and conductively to the molecular hinge and peg (intermittantly). In other words, the time the trapdoor spends in state TD_o or TD_c should be unaffected by the presence of the gas. As a result, the $X\frac{\tau_c}{\tau_o}$ term in (5.11) should not be included; thus, (5.11) reduces to $\frac{[G_A]}{[G_B]} = 1$, erasing the paradox[2]. In related work, numerical simulations of the Maxwell demon with a thermally-flapping trapdoor have been conducted by Skordos and Zurek [16]; they find no second law violation.

---

[1]Gordon included additional time scales τ_{oc} and τ_{co} for the times associated with opening-closing and closing-opening. These add detail, but do not add to the essence of the argument, especially since they are expectedly brief, $\tau_{co,op} \sim 10^{-12}$s.

[2]Recently (2003), Gordon has suggested that stereochemical modifications to the trapdoor might overcome this objection by introducing chiral asymmetry into its chemical dynamics.

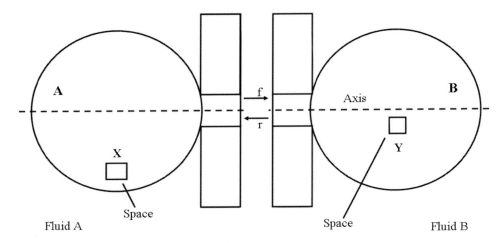

Figure 5.3: Molecular rotor globular protein and membrane in two states (A and B).

Varner's attached osmotic membrane proposal [14] bears strong resemblance to Gordon's trapdoor. It has been suggested on purely thermodynamic grounds that this also does not constitute a second law challenge [17]. The case against Gordon's trapdoor bears resemblance to that raised against Denur's Doppler-demon [18] by Motz [19] and Chardin [20].

5.2.4 Molecular Rotor Model

Combining chemical features of Gordon's first paper [1] with physical forces reminiscent of his second [2], this challenge involves a molecular rotor that pumps either heat or particles across a membrane preferentially in one direction.

Consider a globular protein attached to an adiabatic membrane separating two fluids A and B initially at identical temperatures (Figure 5.3). The protein can exist in two states (A and B) having different moments of inertia (I_A and I_B) with respect to its rotational axis normal to the membrane surface. The protein rotates at angular velocity ω with thermo-statistical probability distribution $p(\omega)$. Via conformational changes, the protein can pass through the membrane adiabatically, preserving both energy and angular momentum during passage. It is claimed that, in principle, net angular momentum (and, therefore, heat) can be conveyed across the membrane by A and B during periods during which their angular momenta are different.

Complex side chain moieties (X and Y) are postulated to reside within the protein's interior volume. Due to a combination of precise steric hinderances and due to differing centrifugal forces experienced by X and Y, it is argued that the A-B transition rates are asymmetrized and detailed balance is destroyed at all ω. Specifically, the forward (A→B) transition rate decreases and the reverse (B→A) transition rate increases with increasing ω. Since the energy transfer rate across the channel depends on ω, net rotational energy (which degrades into heat) is spontaneously transferred from fluid B to fluid A until a sufficient temperature gradient arises to balance the net heat flow. Modifications to this heat pump renders it a particle pump. As described, both violate the second law.

The claimed operation of this molecular rotor is questionable. Issues surrounding angular momentum has not been considered adequately. Presumably, the protein is subject to the principle of equipartition of energy. The salient rotational degree of freedom has an average of $\frac{1}{2}kT$ of thermal energy partitioned to it. For molecules of the size and molecular weight of a globular protein ($m \sim 10^6$amu, $N \sim 10^5$atoms), $\frac{1}{2}kT$ of energy does not buy much angular velocity. For a spherical protein, one expects $\omega \simeq \sqrt{\frac{kT}{I}} \simeq 10^8$rad/s. This is four orders of magnitude smaller than typical rotation frequencies for small molecules. (In fact, the protein is so large that that it behaves much like a classical object in that its thermally-driven angular velocity is small ($\omega \to 0$).) Centripetal forces experienced by moieties X and Y will likewise be small.

Centrifugal distortion of small molecules is well known to affect their vibrational-rotational spectra [21] and can, in theory, affect reaction rates and pathways, especially at high temperature where molecular fragmentation begins, but it is rarely considered significant for low temperature reactions. For the rotor protein under discussion, centrifugally-modified chemical reactions is not a compelling scenario, for several reasons. First, proteins are stable and functional primarily in a narrow, relatively low temperature range ($T \sim 300$K), where angular velocities and centripetal forces are fairly low, probably too low to distort moieties enough to significantly alter or trigger the proposed steric reactions. Second, at a microscopic level, if the steric reactions can be triggered by these small centrifugal forces, it is unclear how well they could be also stabilized against unwanted triggering by random thermal fluctuations. Third, as described by Gordon, X,Y moieties responsible for modifying the protein's moment of inertia appear to account for only a small fraction of the protein's entire mass, perhaps less than 10^{-3}, which would be roughly 1000 atoms. If so, one can expect only a commensurately small change in the moment of inertia and angular velocity. It is questionable whether a 10^{-3} change in angular velocity, which is already 10^4 times smaller than angular velocities associated with molecules that do not typically exhibit centrifugal effects anyway, should have sizable effects on chemical processes. On the other hand, even small asymmetries in competing reaction rates can, in the long term, post a gain. In this sense, Gordon's chemical rotor might be considered a *chemical ratchet*, somewhat akin to the more familiar *thermal ratchets* [22] that rectify thermal fluctuations; however, rather than relying on chemical energy or temperature differences, these chemical ratchets rely on violations of detailed balance.

In all, the mechanical forces associated with rotation are probably too small relative to chemical forces (*e.g.*, those arising from chemical and hydrogen bonds, or van der Waals potentials) to significantly alter chemical behavior of the protein. More detailed calculations, perhaps even fully quantum mechanical ones, should be undertaken to settle this issue.

It is claimed that the membrane is adiabatic with respect to energy and angular momentum transfers from the protein. Inasmuch as rotational and vibrational transition times are relatively short ($\tau \sim 10^{-12} - 10^{-13}$s) compared with protein conformational transition times ($\tau \sim 10^{-3} - 10^{-4}$s), it is unclear how the membrane could remain energetically decoupled so long from the protein during its passage. This problem also applies to the protein's relatively long residence time in fluids A and B. In summary, while the molecular rotor is suggestive of second law challenge in the formal sense, its physical aspects are not compelling.

5.2.5 Discussion

The method underlying Gordon's robust challenges [1, 3] — undermining detailed balance through thoughtful asymmetrization of reaction mechanisms and kinetics — is compelling and finds resonance in the work of later researchers [7, 8]. It remains to be seen whether it can be realized experimentally.

That the membrane engine has not been observed in Nature can be taken as evidence that: a) it is not physically possible (*i.e.*, the second law is absolute or other inherent constraints in chemistry forbid it); b) it is possible, but has not been exploited by life; or c) it is possible, but has not been discovered yet. Without a general formal proof of the second law or complete knowledge of chemistry, option (a) is indeterminate.

The logic of natural selection — "that which survives to reproduce survives" — would seem to demand the existence of thermally-driven biochemistry if natural laws and conditions permit it, and if survival requires it. Under normal terrestrial conditions, where free energy sources are abundant, *thermosynthetic life* would probably be out-competed by standard *free energy life*, but in energy-poor environments, for instance, deep in desert regions the Earth's crust where free energy sources are scarce, thermosynthetic life might have opportunities. Since most microbes and their biochemical processes have not been well characterized — in fact, most microbes cannot even be cultured decently — option (c) remains a possibility. This issue will be considered more deeply in §10.2.

5.3 Denur Challenges

5.3.1 Introduction

Jack Denur is one of the pioneers of the modern second law movement. Over the last 25 years he has advanced two second law challenges relying on the properties of blackbody radiation, in particular, the Doppler shifting of the radiation field. The first [18] is a Maxwell demon that utilizes Doppler-shifted radiation from moving atoms as a means of velocity-sorting them and the second [23, 24] is an advanced, linear version of Feynman's famous pawl and ratchet [25]. While the first [18] appears to have yielded to the second law [19, 20], the second [23, 24] remains a viable challenge even though its second law violating effect is extremely weak and prospects for its experimental realization are dim.

5.3.2 Doppler Demon

Consider a garden-variety Maxwell demon who operates a microscopic gate between two chambers filled with gas at equilibrium [13, 6]. By intelligently observing the atoms and operating the gate, he can establish either: i) a temperature gradient by letting fast molecules pass into one chamber and slow molecules into the other; or ii) a pressure gradient by letting more molecules pass into one chamber rather than the other. In principle, either gradient can be exploited to perform work solely at the expense of heat from a heat bath, in violation of the second law.

A primary difficulty for the Doppler demon is discriminating between fast and slow particles. Brillouin [26] argued that since the entire system is at thermal equilibrium (including the demon himself) and is bathed in the blackbody radiation, everything in the cavity looks the same; therefore, the demon is unable to see incoming molecules. He is blind. In order to see individual molecules, the demon needs a nonequilibrium light source, which generates sufficient entropy to satisfy the second law, regardless of how much work he can extract from the gas.

Denur argues that Brillouin's argument is incomplete. By assumption, the chamber's atoms are in constant thermal motion as they emit and absorb radiation. From his stationary position at the gate, the demon observes Doppler blue-shifted radiation from atoms moving toward him. Thus, by monitoring the subtle blue-shifting of the blackbody radiation, he can anticipate the arrival of high-speed atoms and gate them accordingly. False positives should not foil him since they should occur with equal average frequency from each chamber so their effects cancel out.

The magnitude of the Doppler shifting is roughly

$$\frac{\langle v_x \rangle}{c} \simeq \frac{\langle \delta \nu_m \rangle}{\nu_m}, \tag{5.12}$$

where v_x is the velocity of the atom in the direction of the demon, c is the speed of light, ν_m is the frequency maximum in the Planckian radiation spectrum, and $\delta \nu_m$ is the Doppler frequency shift. Under realistic conditions, the Doppler shift is predicted to be small ($\frac{\langle \delta \nu_m \rangle}{\nu_m} \simeq 10^{-6}$), but within limits of experimental measurement.

Denur's Doppler demon drew critical response. H. Motz [19] pointed out that an increase in intensity of radiation in any particular frequency interval $\nu \rightarrow \nu + \delta \nu$ could be from radiation upshifted or downshifted through scattering by atoms moving toward or away from the demon either in the gas phase or in the cavity walls. Thus, the demon cannot correlate frequency fluctuations with desirable incoming particles. The demon is still blind, as Brillouin predicted [26]. G. Chardin [10] showed that the recoil an atom undergoes when it emits toward the gate reduces the atom's velocity, rendering it useless to the demon's scheme. Chardin's and

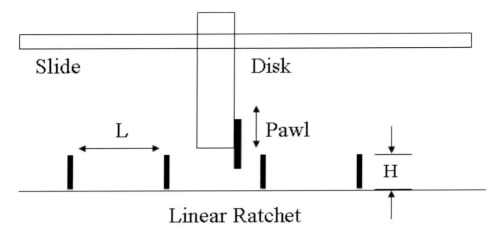

Figure 5.4: Schematic of Denur's linear ratchet and pawl engine.

Motz's arguments reduce to the observation that the gas atoms are an equal component of the blackbody environment just as are the chamber walls. Just as one would not expect the radiation field to betray the existence of individual atoms in the walls, one would not expect it to betray individual atoms in the gas phase.

5.3.3 Ratchet and Pawl Engine

Denur's more recent challenge again relies on the Doppler shifting of the blackbody radiation field, but is immune to previous criticism [19, 20]. Consider the linear ratchet and pawl (RP) system pictured in Figure 5.4, consisting of a disk (mass m'), a rod on which the disk slides frictionlessly, a pawl (mass m), and a linear ratchet with pegs of height H and separation L, satisfying $L \gg H$. Let $m' + m = M \gg m$, so the pawl is a minor mass contribution to the sliding portion of the device; the linear ratchet and slide rod are fixed. Gravity acts downwardly $(-z)$. At temperature T_o the disk/pawl executes Brownian motion in the $\pm x$ direction and the pawl alone also executes Brownian motion in the $\pm z$ direction. The differential probability $dP(V|T_o)$ of finding the disk in the velocity interval between $V - \frac{1}{2}dV$ and $V + \frac{1}{2}dV$ at temperature T_o is given by the one-dimensional Maxwellian distribution:

$$dP(V|T_o) = (\frac{M}{2\pi kT_o})^{1/2} exp[-\frac{MV^2}{2kT_o}]dV. \tag{5.13}$$

This is symmetric in velocity so, in principle, one expects the disk/pawl to randomly move equally in the $+x$ and $-x$ directions. This, however, ignores the effect of Doppler-shifted blackbody radiation caused by the Brownian motion of the disk and the pawl's *asymmetric* location on the disk, facing in the $+x$ direction. The Brownian motion of the disk is random and decorrelated with the fluctuations in the blackbody field. When the disk moves with Brownian velocity V (with $|V| \ll c$), the side facing the direction of travel — for discussion, let this be to the right ($+x$ direction) — experiences a slightly blue-shifted radiation field, while the trailing side of the disk sees a slightly red-shifted field. As a result, to first order in velocity, the leading side of the disk observes a radiation field of temperature

$$T_+ \simeq T_o(1 + \frac{2|V|}{3c}),\qquad(5.14)$$

while the trailing side observes radiation temperature[3]

$$T_- \simeq T_o(1 - \frac{2|V|}{3c}).\qquad(5.15)$$

Denur argues, using realistic physical parameters, that the leading and trailing sides of a microscopic disk can assume the temperature of the radiation field during the characteristic time interval of Brownian fluctuations, while at the same time, the two sides can be sufficiently thermally decoupled to maintain different temperatures. If so, then the vertical ($\pm z$) Brownian motion of the pawl, which is on the right side of the disk, will not fluctuate in a symmetric manner with respect to the disk's horizontal ($\pm x$) velocity. Specifically, when the disk moves to the right ($+x$) the pawl will be at a slightly higher temperature than when it moves to the left ($-x$); therefore, the pawl will on average execute more extreme $\pm z$ Brownian motion when it moves to the right and, as a result, have a higher probability of clearing pegs on the linear ratchet. The upshot of this is that the disk/pawl assembly will preferentially drift to the right ($+x$). This asymmetric net drift is derived solely from thermal energy from the surroundings. As such, since it can be exploited to perform work, it constitutes a violation of the second law.

The probability of ($\pm z$) Brownian motion of the pawl is given similarly to that for the disk/pawl assembly, equation (5.13). The probability $P((z > H)| \pm |V|)$ that its vertical excursions are sufficient to scale the height of the peg ($z > H$) when moving with velocity $\pm|V|$, and thus, to allow passage of the disk (in either direction) depends both on temperature and the disk's instantaneous velocity as:

$$P((z > H)| \pm |V|) = exp[-\frac{mgH}{kT_\pm(|V|)}] = exp[-\frac{mgH}{kT_o(1 \pm 2|V|/3c)}].\qquad(5.16)$$

Under the reasonable condition that $\frac{mgH}{kT_o}\frac{|V|}{c} \ll 1$ and defining $A \equiv \frac{mgH}{kT_o}$, one obtains the simple expression

$$P((z > H)| \pm |V|) = (1 \pm \frac{2A|V|}{3c})e^{-A}.\qquad(5.17)$$

It is clear from (5.17) that the pawl is more likely to Brownian-jump a peg moving to the right ($+|V|$) than to the left ($-|V|$) and, thus, the attached disk is

---

[3]It can be shown that the form of the Planck distribution survives Doppler shifting; the Maxwellian velocity distribution does not.

more likely to move to the right than to the left, *i.e.*, $V_{net} > 0$. To first order, the
ratio of these probabilities $\frac{P(+|V|)}{P(-|V|)} \equiv R_1$ is given by

$$R_1(|V|) = 1 + 2A\frac{|V|}{c}. \tag{5.18}$$

Since the pawl has a net drift it also approaches more pegs from the left than
from the right, and hence it also makes more jump attempts from the left by the
same ratio, $R_1(|V|)$. Thus, the pawl makes more jumps over pegs from the left to
right than vice versa, in the ratio $[R_1(|V|)]^2$. To first order, this is

$$[R_1(|V|)]^2 = 1 + 4A\frac{|V|}{c}. \tag{5.19}$$

The net drift velocity of the disk for Doppler-shifted temperature at velocity
$|V|$ is $V_{net}(|V|)$:

$$V_{net}(|V|) = |V|\{(R_1(|V|))^2 - 1\}e^{-A} \simeq 4A\frac{V^2}{c}e^{-A} \tag{5.20}$$

Averaging this over all V gives

$$V_{net} = 4A\frac{\langle V^2 \rangle}{c}e^{-A}, \tag{5.21}$$

and maximizing V_{net} with respect to A (*i.e.*, $\frac{dV_{net}}{dA} = 0$), one obtains $A_{opt} = 1$ and

$$V_{net,max} = \frac{4\langle V^2 \rangle}{ec} = \frac{4kT_o}{Mec}. \tag{5.22}$$

Denur calculates the maximum power output for this pawl and ratchet to be

$$(\frac{dW}{dt})_{max} = \frac{1}{Lec^2} \cdot [\frac{(2kT_o)^5}{\pi M^3}]^{1/2}. \tag{5.23}$$

Dimensional analysis (with $M \propto L^3$, where L is the characteristic length scale
of the system) shows power output for individual devices scales as $L^{-\frac{11}{2}}$ and the
power density scales as $L^{-\frac{17}{2}}$; both scale as $T_o^{5/2}$. Clearly, small devices and
high temperatures optimize performance. The Denur RP, however, cannot be
made arbitrarily small or run at arbitrarily high temperatures. Implicit in his
analysis is that the disk must be large compared with the typical wavelength of
the blackbody radiation – *i.e.*, $L \gg \lambda \sim \frac{hc}{kT_o}$ – otherwise, diffraction will reduce
the effective opacity of the disk by allowing scattering around it, thus degrading
the temperature difference $(T_+ - T_-)$, and consequently, V_{net} and $\frac{dW}{dt}$. Also,
the device cannot operate at arbitrarily high temperature. The most refractory
materials have upper limit temperatures $T_{max} \leq 4000K$, and even for $T < 3000K$,
evaporation is a significant problem. Taking $T_{max} = 2500K$ as the maximum
operating temperature, taking $L \simeq 10\frac{hc}{kT_{max}} \simeq 6 \times 10^{-5}$m as the minimum device
scale size and letting it be made of low mass density building materials ($\rho \sim$
10^3kg/m^3), one obtains the maximum power per device as $\frac{dW}{dt} \simeq 10^{-47}$W/device
and maximum power density $P^*_{max} \simeq 10^{-34}$W/m^3. The net drift velocity of the

disk is found to be $V_{net,max} \simeq 2 \times 10^{-18}$m/s. These are very small and cast doubt on the practicality of the RP. To put these numbers in perspective, consider the following:

1) Given $V_{net,max}$ it would take roughly $\frac{L}{V_{net,max}} \simeq 10^6$ years for the device to drift its own scale length, or roughly the age of the universe to drift 1 meter.

2) The optimized power densities ($P^* \simeq 10^{-34}$W/m^3) are so low that if the mass of the Sun were devoted to these devices, in total, they would generate only 10^{-7}W of power.

Denur suggests that the power output and velocity of the RP can be improved by substituting the blackbody radiation field with a suitable Maxwellian gas [23]. This would, of course, entail additional caveats and difficulties. The power could be increased by roughly a factor of $\frac{c^2}{4\langle|U|\rangle^2}$ and velocity by roughly $\frac{c}{2\langle|U|\rangle}$, where $\langle|U|\rangle$ is the thermal velocity of the gas $\langle|U|\rangle = (\frac{2kT}{\pi m_{gas}})^{1/2}$. If $\langle|U|\rangle = 1500$m/s, one expects the improvement to V_{net} and P^* to be by factors of roughly 10^5 and 10^{10}, respectively. Despite these significant improvements, V_{net} and P^* are still quite small.

The minuteness of V_{net} and P^* do not make the RP an attractive candidate for an experimentally realizable second law challenge. Compounding these, the practicalities of constructing a working model are daunting. The current art of microfabrication may be adequate to construct such a device, but friction and stiction are likely to be problematic. Stiction commonly seizes up devices similar to the RP. Even in the current best case scenario, using nested multiwalled carbon nanotubes, which have the lowest frictional constants yet measured, friction would probably doom the device. If technical hurdles can be overcome, it is conceivable that Denur's effect might be observable with an ensemble of RPs using advanced interferometric techniques. Long observation times — perhaps multi-year — might be necessary to discern movement along the ratchet.

As a realizable second law violator, Denur's ratchet and pawl is an unlikely candidate, but as a clear demonstration of spontaneous, thermally-driven, velocity-space symmetry breaking, it succeeds. Its main contribution could be its fertile and original insights from which more robust challenges might emerge.

5.4 Crosignani-Di Porto Adiabatic Piston

5.4.1 Theory

B. Crosignani, P. Di Porto and C. Conti, have theoretically and numerically investigated the dynamical evolution of a frictionless, adiabatic piston in a gas-filled adiabatic cylinder subject to the Langevin force [27-32]. They find that in the mesoscopic regime the piston can undergo sizable fluctuations in position and display entropy decreases up to two orders of magnitude greater than those predicted from thermal fluctuations. Furthermore, the system exhibits the disquieting property of failing to settle down to an equilibrium configuration. Although not yet fully developed as an experimentally testable *perpetuum mobile* of the second kind,

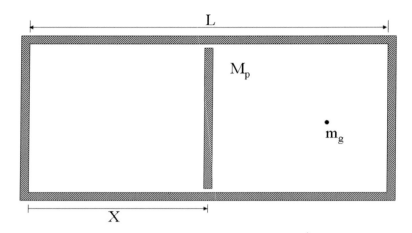

Figure 5.5: Schematic of adiabatic piston and cylinder.

the theoretical behavior of this system raises serious doubts about the validity of the second law in the mesoscopic regime. (Here *mesoscopic* refers to the regime bridging *classical microscopic* and true *quantum mechanical* length scales.) We follow the development of Crosignani, et al.

Consider an adiabatic piston of mass M_p free to move one-dimensionally in an adiabatic cylinder of length L filled with an ideal gas (molecular mass m_g). (Here *adiabatic* means that the piston's and walls' interior degrees of freedom cannot be excited, but that they can receive and transmit impulses to the gas.) The piston divides the cylinder into two regions of lengths X and $L-X$, as indicated in Figure 5.5, and each holds N atoms ($n = N/N_A$ moles) of total mass M_g. At time $t = 0$, let the system be at thermal equilibrium, with $X = L/2$ and with equal initial pressures P, temperatures T_o, and particle numbers N in both regions. Let the piston be released from rest at the center of the cylinder ($X(0) = \frac{L}{2}$, $\dot{X}(0) = 0$). With no outside forces, the piston is driven by the rapidly fluctuating Langevin force [41]. Between the two regions T may vary, but P and N remain fixed.

As derived elsewhere [29], the nonlinear equation for stochastic motion of the piston can be written

$$\ddot{X}(t) = -\sqrt{C}\left[\sqrt{\frac{1}{X}} + \sqrt{\frac{1}{L-X}}\right]\dot{X} + \frac{M_g}{M_p}\left[\frac{1}{X} - \frac{1}{L-X}\right]\dot{X}^2 + A(t) \quad (5.24)$$

Here $C = \frac{16nRT_oM_g}{\pi LM_p}$, with R the ideal gas constant, and $A(t)$ is the Langevin acceleration. Using a piston in a torus model, it can be shown [30] that $A(t)$ should have a white noise spectrum, defined through the standard relation

$$\langle A(t)A(t')\rangle = \frac{8\left(\frac{2m_gkT}{\pi}\right)^{1/2}PS_p}{M_p^2}\delta(t - t'). \quad (5.25)$$

S_p is the cross sectional area of the piston. Clearly, $A(t)$ is frequency independent.

As shown by van Kampen [35], the Langevin approach fails, in general, for nonlinear dynamical systems such as this. To proceed, Crosignani, et al. examine two regimes of operation, corresponding to the conditions $\mu \equiv \frac{M_p}{M_g} \ll 1$ (which can be solved analytically) and the less stringent condition $\mu < 1$ (which they solve numerically). We summarize each.

($\mu \ll 1$): Equation (5.24) is linearized by assuming small displacements of the piston ($\frac{|(X-L/2)|}{L/2} \ll 1$) and also by approximating the square of the piston's velocity by its thermal mean square velocity ($\dot{X} = \sqrt{\frac{kT_o}{M_p}}$). This renders

$$\ddot{X} + 8\sqrt{\frac{2NkT_o}{\pi\mu M_p L^2}}\dot{X} + \frac{8kT_o}{\mu M_p L^2}X = A(t), \qquad (5.26)$$

which is formally identical to the Brownian motion of a 1-D harmonically-bound oscillator of mass M_p:

$$\ddot{X} + \beta\dot{X} + \omega^2 X = A(t) \qquad (5.27)$$

The Brownian harmonic oscillator is well-studied [36].

Once released from its mid-cylinder position, where its thermodynamic entropy is maximum, the piston does not settle down to a final position, but instead fluctuates about $X = \frac{L}{2}$ with a large mean square displacement $\langle x^2 \rangle = \frac{\mu}{2}(\frac{L}{2})^2$. This anomalous behavior does not arise from pressure differences between the two sub-chambers — in fact, their pressures are the same — but rather it is due to temperature variations. The entropy variations associated with these displacements are at odds with a standard corollary of the second law, namely [27],

A closed system in an equilibrium state, once an internal constraint is removed, eventually reaches a new equilibrium state characterized by a larger value of the entropy.

The ensemble average entropy variation about the system's maximum entropy state can be shown to be:

$$\frac{\langle \Delta S \rangle}{k} = -\frac{C_p}{2R}(N\mu), \qquad (5.28)$$

where C_p is the molar heat capacity of the gas at constant pressure. Whenever $N\mu \gg 1$, this implies large entropy decreases. Crosignani, et al. find that for realistic physical parameters at mesoscopic scales, in fact one can have $\frac{\langle \Delta S \rangle}{k} \ll -1$, which is in disagreement with classical thermodynamics.

($\mu < 1$): The primary constraint above (i.e., $\mu \ll 1$) can be relaxed to $\mu < 1$ assuming the Langevin approach validly extends into the mildly nonlinear regime. (This was checked a posteriori and found to be reasonable [32].) One can introduce a characteristic timescale for the system $\left(\tau_p \equiv \left[M_p L(\frac{\pi}{16nRT_oM_g})^{1/2}\right]\right)$ and dimensionless spatial and temporal variables $\xi \equiv \frac{X}{L}$ and $\tau \equiv \frac{t}{\tau_p}$. With these, (5.24) can be recast as the dimensionless equation

$$\ddot{\xi} + \left[\frac{1}{\sqrt{\xi}} + \frac{1}{\sqrt{1-\xi}}\right]\dot{\xi} - \frac{1}{\mu}(\frac{1}{\xi} - \frac{1}{1-\xi})\dot{\xi}^2 = \sigma\alpha(\tau), \qquad (5.29)$$

where $\sigma^2 = (\frac{\pi}{4\sqrt{2}})\frac{\mu}{nN_A} \simeq \frac{\mu}{2N}$ and $\alpha(\tau)$ is the white noise spectrum of unitary power. Derivatives are taken with respect to τ. Equation (5.29) can be solved analytically in the appropriate limits to retrieve established results. For instance, in assuming $|\xi - \frac{1}{2}| \ll 1$ (e.g., the piston does not stray far from the cylinder center) and again approximating the piston's velocity by its formal thermal velocity $\dot{X} = \sqrt{\frac{kT_o}{M_p}}$, (5.29) reduces approximately to (5.27) describing the Brownian motion of a harmonically bound mass.

Numerical simulations carried out on (5.29) are described elsewhere [30, 31]. These confirm that the piston undergoes sizable random fluctuations about $\xi = 1/2$, with time asymptotic rms values of $\xi_{rms} = \sqrt{\langle(\xi - 1/2)^2\rangle}$ extending to a sizable fraction of the cylinder length (e.g., in Figure 5.6 [32], $\xi_{rms} \simeq 0.2$ for $\xi(0) = 1/2$, $\mu = 0.5$, $N = 3 \times 10^4$, for ensemble size = 1000 cases).

The ensemble average entropy deviation for the adiabatic piston (from its $\xi = 1/2$ maximum) is given by

$$\frac{\langle \Delta S \rangle}{k} = \frac{NC_p}{R}\langle \ln\left[1 - 4(\xi - 1/2)^2\right]\rangle \qquad (5.30)$$

Numerical studies indicate that asymptotically this can be of the order of $\frac{S}{k} \simeq -10^4$. In other words, when released from rest in a state of theoretical maximum entropy (thermal equilibrium), the mesoscopic adiabatic piston evolves to very large ensemble average entropy reductions.

Preliminary results from molecular dynamic simulations of point masses (model gas) inside an adiabatic cylinder with a frictionless piston [37, 38] appear to corroborate independently aspects of the Crosignani-Di Porto piston, specifically, that temperature differences spontaneously evolve on opposite sides of the piston and that it undergoes spatial oscillations of the form predicted [29, 30, 32].

Importantly, dimensional analysis shows that this second law challenge — subject to the linearizing approximations above — is extremely sensitive to system scale lengths and that it is viable only for scale lengths on the order of about a micron. Within this range, however, one can envision work cycles that have the net effect of extracting work from its thermal surroundings [33].

Consider the cycle depicted in Figure 5.6. At $t = 0$ (Step 1), the piston halves the cylinder into two equivalent states (N, V_o, T_o). When its latch is removed, the piston moves freely and the system evolves into a new, lower-entropy state, at which time the latch is re-inserted (Step 2). Since the system preserves volume and is thermally insulated from its environment up to this point, one has the conditions: $V_1 + V_2 = 2V_o$ and $T_1 + T_2 = 2T_o$. In Step 3 the sections are physically separated and in Step 4 they undergo reversible adiabatic processes that vary

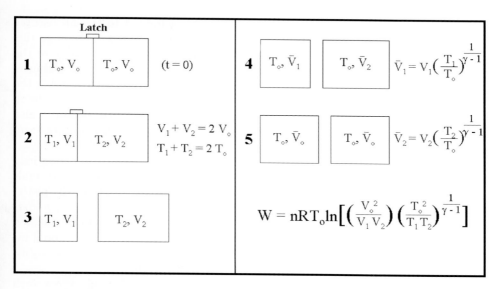

Figure 5.6: Thermodynamic cycle for Crosignani-Di Porto engine.

their volumes and temperatures according to: $\bar{V}_{1,2} = V_{1,2}(T_{1,2})^{\frac{1}{\gamma-1}}$, where γ is the ratio of specific heats ($\gamma = \frac{C_p}{C_v}$). The two subsections are then brought into thermal contact with a heat bath at temperature T_o, undergo reversible isothermal processes, bringing them back to their original temperature and volumes (Step 5). Finally, they are rejoined (Step 6) to complete the cycle.

The total work (W_T) for the cycle is given by

$$W_T = W_1 + W_2 = nRT_o \left[\ln(\frac{V_o}{V_1}) + \ln(\frac{V_o}{V_2}) \right] = nRT_o \ln \left[\frac{V_o^2(T_o^2)^{\frac{1}{\gamma-1}}}{V_1 V_2 (T_1 T_2)^{\frac{1}{\gamma-1}}} \right] \quad (5.31)$$

Since $T_o^2 > T_1 T_2$ and $V_o^2 > V_1 V_2$ (from the Step 2 conditions $T_1 + T_2 = 2T_o$ and $V_1 + V_2 = 2V_o$), it follows that $W_T > 0$. In other words, this cycle performs net positive work, which from consideration of the first law, implies that the work must come at the expense of heat from the surrounding heat bath, a violation of Kelvin-Planck version of the second law. The work is due to the spontaneous and negative entropy variation of the system ($W = T\Delta S$) as it goes from Step 1 to Step 2.

5.4.2 Discussion

The adiabatic piston appears to be a robust theoretical challenge to the second law in the mesoscopic regime. The magnitude of its entropy deviations are significant, as is its failure to settle down to a well-defined equilibrium. It remains to be seen, however, whether a physical embodiment of this *gedanken* experiment can be realized. Prospects for experimental tests are limited to the mesoscopic regime since the chamber cannot exceed roughly a micron in size. Although this is within the current art of MEMS and NEMS technology, it is unclear how the thermomechanical details of its operation will be accomplished — for instance, the application and removal of the thermal insulation between Steps 4 and 5; the operation of the latch; the work extraction mechanism; or the parting and joining of the two subsections. A perfectly-fitting, frictionless, adiabatic piston is also required. Perhaps something along the lines of nested, multi-wall nanotubes will prove suitable since these can slide like nearly frictionless bearings [39].

Crosignani and Di Porto offhandedly speculate that the adiabatic piston may bear on the possibility of life forms being able to extract heat from their surroundings in subversion of the second law [32]. Certainly the scale lengths of the piston and traditional cells coincide well, but the adiabatic piston construction, *per se*, does not seem well-suited to organisms. Perhaps a variant of it might be coupled to some version of a biological Brownian motor or ratchet [?], but this would introduce another damping terms to (5.26) and would require additional analysis. The issue of life and the second law is taken up again in §10.2.

In summary, the Crosignani-Di Porto adiabatic piston qualifies as a theoretical *perpetuum mobile* of the second kind, but its experimental realization appears problematic at this time. It presents a cogent theoretical challenge to standard expectations of the second law in the mesoscopic realm and suggests that further challenges may lurk in this transition region between classical and quantum length scales.

5.5 Trupp Electrocaloric Cycle

5.5.1 Theory

Recently, A. Trupp proposed a simple thermodynamic cycle involving capacitors in liquid dielectrics that appears to challenge the second law [40]. Experiments corroborate key physical processes invoked in the challenge, but no claim of second law violation has been made.

The Trupp cycle involves the charging and discharging, expanding and contracting of a capacitor immersed in a liquid dielectric. Before introducing the cycle, let us review some pertinent aspects of dielectrics:

- A dielectric heats as it is polarized in an imposed electric field; conversely, it cools as the electric field is reduced. This is the electrocaloric effect. Common analogs are found in low-temperature, magnetic cooling and standard gas cycles [41]. One can demonstrate an analogous effect with a rubber band: stretch it quickly and it heats; let it return to room temperature, then slacken it and it cools.

- Coulomb's force law between point charges is modified in the presence of dielectrics, reading,

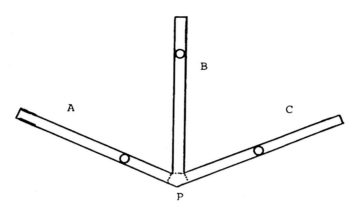

Figure 5.8: Tri-channel with three billiards. When partition P is removed, all billiards scatter into Channel B and remain there.

5.6 Liboff Tri-Channel

In classifying second law challenges, a distinction should be made between closed systems that fail to come to equilibrium (systems for which $\frac{dS}{dt} = 0$ and $S_{system} < S_{equil}$) and systems which demonstrate net entropy reduction ($\frac{dS}{dt} < 0$). Idealized systems of the former type are well known, for instance, billiards reflecting elastically and specularly at small angles from the inner walls of a square box such that the inner region of the box is never traversed, or a linear, one-dimensional gas of elastic colliders that exchange momenta, but which never globally relax to equilibrium (Tonk's gas). Mattis gives a nice collection of idealized classical and quantum systems that come to pseudo-equilibrium locally, but fail to come to equilibrium globally [46].

Liboff [47] discusses an idealized, purely classical irreversible system that violates the second law in the sense ($\frac{dS}{dt} < 0$). Consider three, smooth-walled channels (width $2a$, length $L \gg 6a$) that mutually intersect at 60° angles and share one common vertex, as shown in Figure 5.8. Identical disks ($r = a$) slide frictionlessly in each channel with equal kinetic energies and rebound perfectly elastically from all surfaces.

Let the partition P at the vertex be removed and let the disks arrive at the vertex non-simultaneously. Because of the particular vertex angles, all disks scatter upwardly into Channel B. Once there, they cannot escape, each rebounding ceaselessly between the other disks, the top endcap and bottom vertex. This spontaneous congregation of disks is an irreversible process.

The Boltzmann entropy change for this process is calculated by noting that the configuration space occupied by the disks has collapsed from $3L$ to L. (Since the disk kinetic energies are constant, the velocity space volume is unaffected.) One has for the change in entropy between initial (i) and final state (f):

$$\Delta S = S_f - S_i = k[\ln(\Omega_f) - \ln(\Omega_i)] = k[\ln(1) - \ln(3)] = -k\ln(3) < 0 \qquad (5.37)$$

This irreversible behavior is linked to the vertex, which acts like Maxwell's one-way valve (Maxwell demon). From this purely classical example of "an irreversible orbit in a non-dissipative system," Liboff concludes that "the second law of thermodynamics is not valid in general for idealized irreversible systems" [47].

As a clear example of an idealized, classical, non-dissipative system that demonstrates counterintuitive irreversible behavior, the tri-channel succeeds, but its extreme requisite idealities and boundary conditions doom it as a legitimate, realizable second law challenge.

First, no realistic macroscopic system displays perfectly frictionless motion, perfectly elastic collisions and has perfectly precise angles. Even if the system were reduced to atomic dimensions so as to achieve frictionless, elastic motion, then the walls could no longer be smooth, nor the vertex perfectly angled. Thermal agitation (unless $T = 0$) would further ruin its ideal behavior. The tri-channel appears beyond realistic test at any size scale.

More fundamentally, the idealities of the boundary conditions and constraints (disk fitting perfectly into channels having perfectly smooth walls) lock the system into a phase space of measure zero. Such systems are known to give counterintuitive, second law violating results, but are deemed irrelevant to realistic thermodynamics both because of their zero-measure volumes and because their phase space trajectories are sensitive to infinitesimal perturbations.

For a finite, bounded, two-dimensional plane with three disks (accepting the idealities of frictionless and elastic motion), the 11-dimensional constant-energy phase space manifold, which the system's phase space point explores, is infinitely greater in relative classical area than that which the tri-channel explores. In other words, the constraints and boundary conditions of the tri-channel excises away virtually all of the standard manifold, leaving a subspace of zero measure, for which the special initial conditions give an irreversible result.

The tri-channel is not properly a second law challenge any more than scenarios that can be imagined for almost any thermodynamic system. For example, consider a particular set of initial and boundary conditions among molecules in a lecture hall that result in a pressure fluctuation for which all air molecules move to one side of the hall and remain there. Although many such initial configurations exist, they are so outnumbered by full-room configurations and they are so sensitive to small perturbations that they could not remain on the half-room part of the phase space manifold, that they are deemed irrelevant to the realistic behavior of the system[4]. The tri-channel falls into this category.

In summary, Liboff's tri-channel is a counterintuitive classical example of irreversible behavior, but one too extreme in its idealities to constitute a cogent second law challenge.

---

[4]See footnote 5, page 43.

5.7 Thermodynamic Gas Cycles

Several thermodynamic gas cycles have been forwarded as second law challenges, including ones by R. Fulton, K. Rauen, and S. Amin [48-53]. As of this writing, they are disputed theoretically [54, 55]. Some experiments have been conducted [50], but definitive results are not yet available. Among the challenges discussed in this book, these should be considered longshots. They are classical gas systems dominated by forces and randomizing molecular scattering on short distance scales. They appear to be extensive and they do not seem to be governed by long-range collective behaviors. Furthermore, they do not inhabit extreme thermodynamic regimes that characterize other modern gas-based challenges, for example, the low-pressure, high-temperature regimes of the chemical and plasma systems in Chapters (6-8), or the mesoscopic regime explored by Crosignani, et al. (§5.4).

References

[1] Gordon, L.G.M., Found. Phys. **11** 103 (1981).

[2] Gordon, L.G.M., Found. Phys. **13** 989 (1983).

[3] Gordon, L.G.M., J. Coll. Interf. Sci **162** 512 (1994).

[4] Gordon, L.G.M. in *Quantum Limits to the Second Law*, AIP Conference Proceedings, Vol. 643, Sheehan, D.P., Editor, (AIP Press, Melville, NY, 2002) pg. 242.

[5] Gordon, L.G.M., Entropy **6** 38; 87; 96 (2004).

[6] Leff, H.S. and Rex, A.F., Editors, *Maxwell's Demon 2: Entropy, Classical and Quantum Information, Computing* (Institute of Physics, Bristol, 2003).

[7] Čápek, V., J. Phys. A: Math. Gen. **30** 5245 (1997).

[8] Sheehan, D.P., Phys. Rev. E **57** 6660 (1998).

[9] Becker, W.M., Reece, J.B., and Poenie, M.F., *The World of the Cell*, Chapters 7,8; (Benjamin/Cummings, Menlo Park, CA, 1996).

[10] Szilard, L., Z. Phys. **53** 840 (1929).

[11] Popper, K., Brit. J. Phil. **8** 151 (1957).

[12] Timp, G., Editor, *Nanotechnology*, (Springer-Verlag, New York, 1999).

[13] Maxwell, J.C., letter to P.G. Tait, 11 Dec. 1867, in *Life and Work of Peter Guthrie Tait*, Knott, C.G., Editor, (Cambridge University Press, London, 1911) pg. 213.

[14] Varner, K., Infin. Energy **8/43** 26 (2002).

[15] Varner, K., Infin. Energy **9/49** 24 (2003).

[16] Skordos, P.A. and Zurek, W.H., Am. J. Phys. **60** 876 (1992).

[17] Homel, M., Infin. Energy **9/49** 4 (2003).

[18] Denur, J., Am. J. Phys. **49** 352 (1981).

[19] Motz, H., Am. J. Phys. **51** 72 (1983).

[20] Chardin, G., Am. J. Phys. **52** 252 (1984).

[21] Karplus, M. and Porter, R.N., *Atoms and Molecules* (Benjamin/Cummings, Menlo Park, CA 1970).

[22] Reimann, P. and Hänggi, P., Appl. Phys. A **75** 169; Parrondo, J.M.R. and De Cisneros, B.J., *ibid.* 179; Astumian, R.D., *ibid.* 193 (2002).

[23] Denur, J., Phys. Rev. A **40** 5390 (1989).

[24] Denur, J., in *Quantum Limits to the Second Law*, AIP Conf. Proc. Vol. 643, (AIP Press, Melville, NY, 2002) pg. 326; Entropy **6** 76 (2004).

[25] Feynman, R.P., R.B. Leighton, and M. Sands, *The Feynman Lectures on Physics*, Vol. 1, Chapter 46, (Addison-Wesley, Reading, MA, 1963).

[26] Brillouin, L., *Science and Information Theory*, 2nd Ed., (Academic, New York, 1962) pp. 148-51 and Ch. 13.

[27] Callen, H.B., *Thermodynamics*, (Wiley, New York) (1960) pp. 321-323.

[28] Crosignani, B. and Di Porto, P., Am. J. Phys. **64** 610 (1996).

[29] Crosignani, B. and Di Porto, P., Europhys. Lett **53** 290 (2001).

[30] Crosignani, B., Di Porto, P., and Conti, C., in *Quantum Limits to the Second Law*, AIP Conf. Proc., Vol. 643, (AIP Press, Melville, NY, 2002) pg. 267.

[31] Crosignani, B., Di Porto, P., and Conti, C., arXiv:physics/0305072v2 (2003).

[32] Crosignani, B., Di Porto, P., and Conti C., Entropy **6** 50 (2004).

[33] Crosignani, B., private communication (2004).

[34] Reif, F. *Fundamentals of Statistical and Thermal Physics*, (McGraw-Hill, New York, 1965) pp. 560-565.

[35] Van Kampen, N., *Stochastic Processes in Physics and Chemistry*, (North-Holland, Amsterdam, 1992).

[36] Chandrasekhar, S., Rev. Mod. Phys. **15** 1 (1943).

[37] Renne, M.J., Ruijgrok, M., and Ruijgrok, Th.W., Acta Phys. Polon. B **32** 4183 (2001).

[38] White, J.A., Roman, F.L., Gonzalez, A., and Velasco, S., Europhys. Lett. **59** 479 (2002).

[39] Min-Feng, Y., Yakobson, B.I., and Ruoff, R.S., J. Phys. Chem. **104** 8764 (2000).

[40] Trupp, A. in *Quantum Limits to the Second Law*, AIP Conf. Proc., Vol. 643, (AIP Press, Melville, NY, 2002) pg. 201.

[41] Reif, F., *Foundations of Statistical and Thermal Physics*, Chapters 5, 11, (McGraw-Hill, New York, 1965).

[42] Feynman, R.P., R.B. Leighton, and M. Sands, *The Feynman Lectures on Physics*, Vol. II, §10.5 (Addison-Wesley, Reading, MA, 1963).

[43] Griffiths, D.J. *Introduction to Electrodynamics*, 3rd Ed., Chapter 4 (Prentice Hall, Upper Saddle River, NJ, 1999).

[44] Trupp, A. unpublish., (2002).

[45] Silow, P., Ann. der Phys. Chem. **232** 389 (1875).

[46] D.C. Mattis, in *Quantum Limits to the Second Law*, AIP Conf. Proc., Vol. 643, (AIP Press, Melville, NY, 2002) pg. 125.

[47] R.L. Liboff, Found. Phys. Lett. **10** 89 (1997).

[48] Fulton, R., in *Quantum Limits to the Second Law*, AIP Conf. Proc., Vol. 643, (AIP Press, Melville, NY, 2002) pg. 338.

[49] Rauen, K., Infin. Energy **3/15,16** 109 (1997).

[50] Rauen, K., Infin. Energy **9** 1 (2003).

[51] Rauen, K., Infinite Energy **5/29** 47; **5/26** 17 (1999).

[52] Amin, S., Science **285** 2148 (1999).

[53] Amin, S., Physics Today **52** 77 (1999).

[54] Wheeler, J.C., in *Quantum Limits to the Second Law*, AIP Conf. Proc., Vol. 643, (AIP Press, Melville, NY, 2002) pgs. 345, 352.

[55] Wheeler, J.C., private communications (2003).

6

Gravitational Challenges

Two modern classical gravitational paradoxes are reviewed and critiqued. Each is tracable to 19th-century observations and debates about gravitation and the second law. To varying degrees, each challenge has experimental corroboration. Significant theoretical criticisms have been leveled against each.

6.1 Introduction

Classical gravitational thermodynamics enjoys enduring interest owing in part to surprising, counterintuitive and paradoxical results which regularly emerge from it [1]. Despite its apparent simplicity, it often plays the spoiler in otherwise straightforward thermodynamic discussions. In the Newtonian regime, for instance, the classical Jeans instability [2], the satellite 'paradox' [3], and the 'negative heat capacity' associated with astrophysical self-gravitating systems [4], and gravothermal collapse and oscillation of self-gravitating point-mass systems [5, 6] run counter to ordinary thermodynamic intuition. In the relativistic regime, the Hawking process [7], the related Unruh effect [8] and the deep relationship between the ordinary laws of thermodynamics and the laws of black hole physics stand out as some of the most surprising and significant theoretical physics results of the past 30 years [9].

It has long been recognized that gravitation fits uneasily into the standard thermodynamic paradigm, owing principally to its non-extensivity and scale-free coupling. An *extensive* system is one for which energy and entropy are additive in the sense that they are the sum of the energies and entropies of the system's macroscopic sub-systems [10, 11]. Most thermodynamic systems are extensive, but gravity and electrostatics provide prominent counter-examples. The potential energy Φ of a gravitational system of mass M and size R scales as $\Phi \sim \frac{M^2}{R}$. Since M typically scales as $M \sim R^3$, one has $\Phi \sim M^{5/3} \sim R^5$. Unlike extensive thermodynamic variables that scale linearly with system mass and volume ($\sim R^3$) — and quite unlike *intensive* variables (like pressure and temperature) which are scale invariant — gravitational potential energy is *super-extensive*, growing faster than linearly.

The stability of a self-gravitating system, as prescribed by the virial theorem, places constraints on the system's ratio of kinetic and potential energies. Since gravitational potential energy can grow without bound with the number of particles N as one enlarges a system, one is faced with a thermodynamic choice. Either the temperature (kinetic energy) must increase without bound to offset the potential energy, or the system becomes unstable to local gravitational collapse. The formation of galaxies and galactic clusters from a more homogeneous mass distribution directly after the Big Bang is an example of such "curdling.[1]"

Thus, because of its super-extensivity, gravitational systems cannot usually be carried to the traditional thermodynamic limit ($N \to \infty$ and $V \to \infty$, with N/V or chemical potential μ finite) and traditional approaches to common phenomena such as phase transitions fail. It has been claimed by some [12] that thermodynamics fails for non-extensive systems, while others have shown that by approaching them through microcanonical formalism one can retrieve modified but recognizable thermodynamic behavior [10]. Non-extensivity is a common feature of systems involving long-range forces (*e.g.*, gravitational, electromagnetic).

Whereas on the macroscale gravity exhibits super-extensivity and, thus, in the parlance of particle physics, it has an infrared divergence, on the microscale point-mass systems have an ultraviolet divergence arising from their scale-free coupling ($\sim 1/r$); that is, point masses can approach arbitrarily close to one another and release an arbitrarily large amount of energy. This is also related to the negative heat capacity displayed by gravitating systems. Consider a box containing point-mass gravitators. Via three-body interactions, bound pairs will spontaneously develop. Statistically, these fall into tighter and tighter orbits through additional scattering with third bodies. As the pair falls into tighter orbits it releases increasingly more energy (heat). The more heat that is drawn away from the box, the hotter it becomes — a signature of negative heat capacity.

Super-extensivity is also exhibited by electrostatic systems by analogy with gravity, letting mass become charge ($M \to Q$). Systems with charge imbalance and electric fields (whose energy density ρ_E scales as $\rho_E \sim E^2$) can be super-extensive. The USD solid-state paradox utilizes the non-extensive nature of capacitive electric fields, for example.

In this chapter two second law challenges are unpacked. Historically separated by 125 years, they are thematically joined by the statistical mechanical behav-

---

[1]In the long term, gravitational curdling is the principal source of universal entropy production (§10.3.2).

ior of gases in gravitational fields. The first is a recent challenge involving a gas thermally cycling between an asymmetrically-inelastic gravitator and the walls of its confining blackbody cavity [13, 14, 15, 16]. The second resurrects a historical debate between Maxwell, Boltzmann and Loschmidt regarding temperature stratification of a gas (or solid) in a gravitational field [17, 18].

6.2 Asymmetric Gravitator Model

6.2.1 Introduction

In 1885, during some of the seminal discussions on the subject, H. Whiting pointed out that the molecular velocity sorting performed by the original, sentient Maxwell demon [20, 21] can be accomplished by non-sentient, natural processes [19]:

> When the motion of a molecule in the surface of a body happens to exceed a certain limit, it may be thrown off completely from that surface, as in ordinary evaporation. Hence in the case of astronomical bodies, particularly masses of gas, the molecules of greatest velocity may gradually be separated from the remainder as effectually as by the operation of Maxwell's small beings.

Atmospheric evaporation – the loss of gas from an astronomical object when molecular thermal velocities exceed the escape velocity – is well-known and integral to the studies of planets, stars, and galaxies [22].

Although Whiting identified a natural process that velocity selects like a Maxwell demon, he did not specify a physical embodiment and his observation has lain fallow for over a century. In this section, a modern relative to Whiting's demon is developed. It is fully classical mechanical in that it can be understood entirely in terms of 19th century physics. All of its components (heat bath, gravitator, cavity, test atom) are classically defined; it utilizes classical gravity; its gas-surface collisions and trapping, momentum transfers, and net pressure can be understood in purely classical mechanical terms; and its thermodynamic challenge can be phrased in terms of traditional heat and work.

This gravitational challenge is based on four observations: (1) thermal particles falling through a gravitational potential are rendered suprathermal; (2) suprathermal gas-surface impacts can be nonequilibrium events; (3) the trapping probability for an atom or molecule striking a surface can vary strongly with perpendicular impact velocity and surface type; and (4) in general, atoms trapped on a surface will quickly (within relatively few lattice vibrational periods) reach thermal equilibrium with the surface and desorb in thermal equilibrium with that surface. Experimental and theoretical support for each of these observations is substantial [23, 24, 25, 26]. (In this paper, *suprathermal* refers to particles with average speeds greater than the thermal speed, up to several times the thermal speed.)

Briefly, the paradoxical system consists of a blackbody cavity housing a gravitator of comparable dimensions and a tenuous gas Figure 6.1. The gas in the cavity is tenuous and has a mean free path comparable or greater than the cavity scale length; therefore, gas phase collisions are insufficient to establish standard gas phase equilibrium, and gas velocity distributions are determined primarily by gas-surface interactions. Gas infalling from the walls of a blackbody cavity suprathermally strikes the gravitator which is asymmetrically inelastic on different surfaces and, thereby, asymmetric with respect to trapping probabilities for the gas. The gas rebounds to different degrees depending on the surfaces' inelasticities. A macroscopic analog to gas atoms striking S1 and S2 would be dropping a child's glass marble on two distinct surfaces, say, steel and a soft rug. The marble rebounds more elastically from the steel than from the rug. At the microscopic level, suprathermal collisions (collisions with greater than thermal energies) also display differences in their elasticities depending on the nature of the gas and surfaces.

The resultant asymmetries in the net momentum flux delivered to the different sides of the gravitator by the gas (in essence, a steady-state pressure differential) results in a net unidirectional force on the gravitator. The force persists when the gravitator moves with a velocity small compared with the gas thermal speed. Since a steady-state force can be maintained at a steady-state velocity, in principle, the moving gravitator can be harnessed to do steady-state work. Since the gravitator-cavity system is in a stationary thermodynamic state performing steady-state work, if the first law of thermodynamics is satisfied, then the work performed must come solely from heat from the heat bath, in violation of the second law.

6.2.2 Model Specifications

Formally, the system consists of: (i) an infinite, isothermal heat bath; (ii) a large, spherical blackbody cavity; (iii) a low-density gas in the cavity; and (iv) a spherical gravitator (Figure 6.1). The walls of the cavity are physically and thermally anchored to the heat bath. The gravitator is spherically symmetric both geometrically and with respect to its mass distribution. The high degree of symmetry assumed here is not required, but is analytically convenient. If the heat bath is uniform in its mass density and if the cavity gas has negligible self-gravity, then the spherical cavity interior may be considered a gravitational equipotential save for the gravitational field of the gravitator; in other words, even over long time scales, the gravitator may be considered the sole source of gravitational force within the cavity (This can also be shown from Birkoff's theorem of general relativity.). The gravitator may be at rest (Figure 6.1a), in rectilinear motion, or in synchronous rotation about the cavity center (Figure 6.1b).

The gravitator's two hemispheres are composed of two materials (S1 and S2) distinct in their inelastic responses to the gas, in that one can write their surface trapping probability functions as distinct (e.g., S1 is stiff like a semiconductor and S2 is deformable like a metal lattice with respect to gas collisions). An atom can scatter from a surface in several ways. It may elastically or inelastically rebound or, if it loses enough translational energy, it may be trapped. A trapped atom remains in a weakly bound mobile state until it is either converted to a more tightly

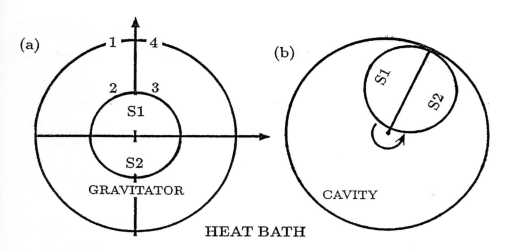

Figure 6.1: Schematic of idealized model system. Points 1-4 in Figure 1a will apply to discussion of Figure 2.

bound physisorbed or chemisorbed immobile state (sticking) or it desorbs due to random thermal excitation from the surface. Experiment and theory indicate that typical trapping times range from on the order of a few nanoseconds to hundreds of microseconds. This corresponds to roughly $10^4 - 10^9$ lattice vibrational periods of a typical solid. As a result of their relatively long and intimate contact with the surface, trapped atoms is taken to desorb in thermal equilibrium with the surface.

A number of models have been developed to describe trapping; most are qualitatively similar. Here we use the "ion cores in jellium model" (ICJM) which is an extension of the Baule-Weinberg-Merrill model [24]. In the ICJM, the trapping probability for atom A on surface j, $P_{trap}(j)$, is given by

$$P_{trap}(j) = [1 - \frac{(1 + \frac{m_A}{m_s(j)})^2 KE_\perp}{4(\frac{m_A}{m_s(j)})(KE_\perp - E_s(j) + W(j))}] \qquad (6.1)$$

provided that $P_{trap}(j) > 0$, otherwise $P_{trap}(j) = 0$. Here m_A and $m_s(j)$ are the masses of the gas atom A and surface j atoms ($j = 1, 2$); $KE_\perp$ is the perpendicular kinetic energy component of A with respect to the surface; $E_s(j)$ is the energy of trapped A on surface j; and $W(j)$ is the depth of the potential energy well associated with surface j. Normal energy scaling is assumed; that is, there is little conversion of surface-parallel momentum to perpendicular momentum during scattering. The ICJM describes well numerous experimental results [24]. Other standard trapping models will yield similar results. In fact, chemically distinct surfaces are unnecessary; mere morphological difference can suffice since what is essential here is not trapping probability, *per se*, but thermalization probability.

Nine parameters specify the system; they are: $T \equiv$ blackbody cavity temperature; $m_A \equiv$ mass of gas atom A; $m_G \equiv$ mass of spherical gravitator; $r_A \equiv$ radius of gas atom A; $r_G \equiv$ radius of spherical gravitator; $r_c \equiv$ radius of spherical blackbody cavity; $n_{cav} \equiv$ average number density of gas atoms A in the cavity volume; $P_{trap}(1)$ and $P_{trap}(2)$. By defining the following scaling parameters – thermal speed ($v_{th} \sim \sqrt{\frac{kT}{m_A}}$), incident surface impact speed for atoms falling from the cavity wall ($v_{inc} \sim \sqrt{(\frac{2Gm_G(r_c-r_G)}{r_c r_G}) + \frac{kT}{m_A}}$), mean free path for gas phase collisions ($\lambda \sim \frac{1}{5.6\pi n_{cav} r_A^2}$), and the first velocity moment of the trapping probability ($<p(j)> \equiv m_A \int_0^\infty v_\perp P_{trap}(j) dv_\perp$) – one can circumscribe the paradox's viable parameter regime in terms of six dimensionless variables as: $\alpha \equiv \frac{v_{inc}}{v_{th}} \geq 1$; $\beta \equiv \frac{<p(1)>}{<p(2)>} \neq 1$; $\gamma \equiv \frac{r_c}{r_G} > 1$; $\delta \equiv \frac{\lambda}{r_c - r_G} \geq 1$; $\epsilon \equiv \frac{m_A n_{cav} v_{th} v_{inc}^2}{\sigma T^4} \ll 1$; and $\zeta \equiv \frac{m_A n_{cav} r_c^3}{m_G} \ll 1$. Here k, σ, and G are the Boltzmann constant, the Stefan-Boltzmann constant, and the universal gravitational constant.

The two primary requirements for the paradox are that the gas-surface impacts are, on average, suprathermal ($\alpha \geq 1$) and differentially inelastic between S1 and S2 ($\beta \neq 1$). The gravitator must fit in the cavity ($\gamma > 1$) and the gas must fall through a sufficiently large gravitational potential for the impacts to be suprathermal ($\gamma > 1$ for $\alpha \geq 1$), but not so large that the gravitator accumulates a dense (collisional) atmosphere or that reflected or desorbed atoms cannot reach the cavity walls. Infalling atoms must reach the gravitator without excessively thermalizing via gas phase collisions ($\delta \geq 1$). For simplicity of analysis – but not strictly required for the challenge – two additional constraints are made. First, gas kinetic energy fluxes are much smaller than radiative energy fluxes ($\epsilon \ll 1$); in other words, blackbody radiation dominates the system's energy transfers. In this way, small surface temperature variations arising from inelastic collisions are quickly smoothed out by compensating radiative influxes or effluxes, and surfaces may be taken to be near the blackbody cavity temperature. Second, the self-gravity of the atmosphere of A is negligible compared with the gravity of the gravitator ($\zeta \ll 1$). These six constraints are physically reasonable and mutually attainable over a broad parameter space. Outside the prescribed inequalities, the system analysis becomes more complex (e.g. $\epsilon \geq 1$, $\zeta \geq 1$, $\delta < 1$) or the paradox fails altogether (e.g. α, β, $\gamma = 1$).

This model is approximated well by a moon-sized gravitator in a low-density gas housed in blackbody cavity of comparable dimensions. The following system self-consistently meets all the above physical assumptions and constraints: $m_A = 4$ amu (He atom), $m_G = 2 \times 10^{23}$ kg, $r_A = 10^{-10}$ m, $r_G = 1.6 \times 10^6$ m, $r_c = 3.21 \times 10^6$ m, $T = 2000$ K, $n_{cav} = 5 \times 10^{10}$ m^{-3}, $E_s(1) = E_s(2) = 2kT$, $W(1) = W(2) = 4kT$, $m_s(1) = 7$ amu, and $m_s(2) = 96$ amu. For this system, $\alpha = 1.7 > 1$, $\beta \neq 1$, $\gamma = 2 > 1$, $\delta = 70 \gg 1$, $\epsilon \sim 10^{-12} \ll 1$, and $\zeta \sim 10^{-19} \ll 1$. Surfaces S1 and S2 are identical except for the masses of their surface atoms. For the discussion to follow this will be called the *standard gravitator*.

6.2.3 1-D One-Dimensional Analysis

(Note: The following 1-D analysis should not be taken at face value. It has been cogently criticized by Wheeler (§6.2.5) [27].)

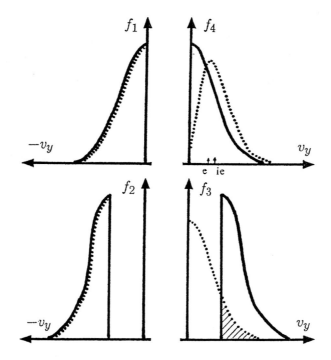

Figure 6.2: One dimensional velocity distributions $f(v_y)$ for gas atoms at locations 1-4 in Figure 1a. Solid (dashed) lines indicate distributions interacting elastically (inelastically) with gravitator surfaces. f_1 taken at instant of desorption from wall; f_2 taken just before surface impact; f_3 taken just after reflection/desorption from surface; and f_4 taken just before collision with cavity wall. Distributions depicted for the case $\frac{Gm_A m_G(r_c - r_G)}{r_c r_G} = kT$. Average velocities of f_4 distributions indicated on abscissa. (See Wheeler (§6.2.5) for conflicting analysis [27].)

Let the system settle into a stationary state. When steady state is reached gas is thermally cycled continuously between the gravitator and the cavity walls: gas falls suprathermally onto the gravitator, inelastically rebounds to different degrees from the hemispheres, and is rethermalized in the cavity. Since the surface trapping probabilities for S1 are S2 are different, the fluxes of gas atoms striking S1 and S2 are partitioned differently between those which are reflected suprathermally back to the cavity walls and those which are trapped and desorbed thermally. As a result of this differential partitioning, the momentum fluxes to the hemispheres are different; these amount to a net pressure (and force) on the gravitator. This pressure can be understood qualitatively by following a velocity distribution of test particles through the cavity in light of conservation of linear momentum. In this analysis, the gravitator will be treated one dimensionally; however, the numerical analyses which follow indicate this argument generalizes to three dimensions.

Let the gravitator begin at rest at the center of the cavity and consider the gas atom velocity distributions $f_1 - f_4$ in Figure 6.2 corresponding to the cavity positions 1-4 in Figure 6.1a. For discussion, the solid-lined distributions correspond to those undergoing purely elastic (e) collisions and dashed-lined distributions to those undergoing purely inelastic (ie) collisions; real distributions will be somewhere in between these. Distribution f_1 is a half-Maxwellian velocity distribution of atoms just desorbed from the cavity wall, starting its fall toward the gravitator. A differential of particle density dN can be expressed as

$$dN_1 = f_1(v)dv = (\frac{A}{\pi})^{1/2}exp[-Av^2]dv, \tag{6.2}$$

where $A = (\frac{m}{2kT}) = v_{th}^{-2}$. During its fall through the gravitational potential to point 2 in Figure 6.2a, distribution f_1 evolves into suprathermal distribution f_2 (depicted just before impact). Notice f_2 has been accelerated en masse such that it is velocity space compressed and has a net drift; its minimum surface-impact velocity is $v_{inc} = \sqrt{\frac{2Gm_G(r_c-r_G)}{r_G r_c}}$. At impact the development of the elastic and inelastic cases diverge. In the purely elastic case, the atoms reflect spectrally ($f_3(v_y) = f_2(-v_y)$), decelerate against gravity, and return to the walls as a thermal distribution f_4, the velocity inverse of f_1; that is, $f_4(v_y) = f_1(-v_y)$. At the wall, f_4 is trapped, rethermalized and recycled as f_1 again. (By an appropriate choice of $E_s(wall)$, $W_s(wall)$ and $m_s(wall)$ and or by increasing its effective surface area with a dendritic or zeolitic structure, the trapping/thermalization probability for the wall can be made essentially unity ($P_{trap}(wall) = 1$).)

By conservation of linear momentum, the net impulse to the gravitator in the elastic case from f_1 to f_4 is: $(\Delta p)_{grav,e} = m_A N((v_y)_{f_1} - (v_y)_{f_4}) = 2m_A N(v_y)_{f_1}$, where N is the total number of gas atoms in the original distribution f_1 and $(v_y)_{f_1}$ is the average velocity of f_1. In the purely inelastic case, f_2 is trapped on the surface and is thermalized when it leaves. Only the slashed portion of f_3, the tail of the distribution ($v \geq v_{inc}$), escapes directly the gravitational potential well and returns to the cavity wall to be recycled as f_1; the rest ($f_3(v < v_{inc})$) falls back to the surface, where it is retrapped, rethermalized and tries again until it leaves the gravitator in the ($v \geq v_{inc}$)-tail. The inelastic $f_4(v)$ is

$$f_4(v) = [\frac{A}{v^2 + B}]^{1/2}v \cdot exp[-A(v^2 + B)], \tag{6.3}$$

where $B = (2Gm_g(\frac{1}{r_g} - \frac{1}{r_c}))$. Here $r_{c,g}$ are the cavity and gravitator radii, m_g is the gravitator mass, and G is the universal gravitational constant. The critical feature is that $(f_4)_{ie}$ is not thermal and has a larger average velocity than $(f_4)_e$, i.e. $(v_y)_{f_4,ie} > (v_y)_{f_4,e}$ (See f_4 in Figure 6.2.), which leads to larger than average momentum and kinetic energy fluxes, as shown below.

The net impulse to the gravitator due to the inelastic collisions of f_1-f_4 is $(\Delta p)_{grav,ie} = m_A N[(v_y)_{f_1} - (v_y)_{f_4}]$. Since $((v_y)_{f_4,ie} \neq (v_y)_{f_4,e}$, one has $(\Delta p)_{grav,e} \neq (\Delta p)_{grav,ie}$. In other words, there is a net impulse imparted to the gravitator by the gas due to the asymmetry in inelasticity between S1 and S2. If there is a steady-state flux of gas to and from the walls and gravitator (Figure 6.1a), then there is a steady-state net force on the gravitator.

Kinetic analysis indicates that the pressure difference between hemispheres can be on the order of 25% of the average gas pressure near the gravitator. Analysis also indicates this pressure asymmetry should persist when the gravitator is moved off center and placed in synchronous rotation about the cavity center Figure 6.1b with a tangential speed up to about 20% of the gas thermal speed. In this case, the gravitator can generate a steady-state torque while maintaining a steady-state angular velocity about the cavity center.

The maximum theoretical power achievable by the synchronously rotating gravitator scales as $P \simeq \chi \frac{m_A r_G v_{th}^3}{r_A^2}$, where $\chi \simeq 10^{-1}$ is a dimensionless constant incorporating system constraints (α - ζ) above. For the *standard gravitator*, the maximum steady-state power output should be on the order of 10^9 W. At the temperature of the universal microwave background radiation field ($T = 2.73$ K) it should generate, in principle, a maximum of roughly 10^4 W.

This system *is not in* and *cannot reach* thermal equilibrium; rather, it is in a steady-state nonequilibrium. The irreversibility of this process can be traced to the inherently irreversible nature of the suprathermal surface impacts coupled with the asymmetry in the inelastic responses of the two hemispheres. Regardless of whether work is extracted from this system, this is a peculiar state of affairs since, in a simple blackbody cavity, one would expect a single velocity distribution to evolve, but it does not. Contrary to normal expectations, however, such nonequilibrium stationary states are not forbidden. For instance, confined two-component plasmas (one with both positive and negative species) with non-zero plasma potentials – in principle, virtually all known plasma types – are inherently nonequilibrium systems. It has been shown that electrostatically confined one-component plasmas (say, electrons in a Penning trap) can reach thermal equilibrium, but when the second species (say, positive ions) is present, the very sign of the plasma potential (that might confine the first species) guarantees that the second species *is not* confined, and therefore, not able to reach thermal equilibrium. (By itself, the Debye sheath at the edge of almost all plasmas is a highly nonequilibrium structure that guarantees the plasma is a nonequilibrium entity.) Plasma paradoxes involving the second law are taken up in Chapter 8.

6.2.4 Numerical Simulations

Three-dimensional numerical test particle simulations of this system were performed. A simulation consisted of integrating the equations of motion of a single test gas atom in the cavity with the gravitator at rest in a background gas and recording the test atom's dynamical variables.

The motion of the test atom was subject to the following principles and processes: (a) Newton's laws of motion; (b) universal law of gravitation; (c) conservation of linear and angular momenta and energy; (d) trapping and desorption probabilities that follow (1); and (e) surface trapping and gas phase collisions that thermalized the test atom; otherwise all collisions were elastic. Each test atom was tracked for on the order of 1-5 years of physical time. During this interval, the test atom traversed the cavity interior thousands of times and underwent tens of thousands of surface and gas-phase collisions, roughly half of which were thermalizing. Collisions included those on the cavity walls and gravitator and in the gas phase; they occurred in various ratios depending on the system parameters, but in typical rough ratios of about $1:1:10^{-2}$, respectively. Newton's equations of motion were integrated via a standard fourth-order Runge-Kutta scheme. Solution convergence was verified via a series of simulations that independently varied the length of the time steps, the number of test atom collisions, and the initial conditions of the test atom. Through these it was verified that the system had settled into a stationary state and that initial conditions did not appreciably affect final results. Code accuracy was checked via known solutions and limiting cases. It is emphasized that in no way was the paradoxical effect 'programmed' into the simulations; it is a robust, emergent phenomenon.

6.2.4.1 Velocity Distributions

Numerical simulations provided velocity distributions for the test atom in the cavity volume between the gravitator and cavity wall at positions of azimuthally symmetric ring- or dome-shaped test patches (See Figure 6.3) located on spherical shells nested within the cavity walls and concentric with the gravitator. Velocity distributions were built up by following the test atom through the cavity and recording its velocity components as it crossed a test patch from 1,500 - 3,000 times. Velocity components v_x, v_y, and v_z were measured directly and $v_r = \frac{x}{r}v_x + \frac{y}{r}v_y + \frac{z}{r}v_z$ was computed from these. The radial component of velocity v_r is normal to the test patch surface.

Velocity distributions $f(v_x)$, $f(v_y)$, $f(v_z)$, and $f(v_r)$ were obtained for test patches specified by radius $r_G < r_{patch} < r_c = 2.01r_G$) and delimited by polar angular extent ($0 \leq \theta_{low} \leq \theta_{patch} \leq \theta_{high} \leq \pi$ rad). The azimuthally-symmetric, spherical gravitator was at rest at the center of the spherical blackbody cavity, therefore, all test patches were azimuthally symmetric. Velocity distributions were normalized with respect to patch area, total number of patch crossings, and total time the test atom spent in the cavity. The *fiduciary velocity distribution*, the distribution against which all other distributions were calibrated, was taken to be that for the *standard gravitator* for 3000 test patch crossings at radius $r_{patch} = 1.605 \times 10^6$m $= 1.003r_G = 0.5r_c$ and polar angular extent $\frac{5}{6}\pi \leq \theta_{patch} \leq \pi$ rad; this corresponds to a 30° solid dome (0.842 steradians) located just above the gravitator surface over the bottom (S2) hemispheric pole. This 0.842 steradian dome (hereafter, a 30° dome) covers 6.7% of the surface area of its sphere and, therefore, gives a relatively local measure of system dynamics.

In Figure 6.4, velocity distributions are presented for the *standard gravitator* at roughly midway between the gravitator surface and cavity wall for 30° domes over the top (S1) and bottom (S2) poles. These display several characteristics which will be useful in interpreting the phase space diagrams to follow. First, $f(v_x)$ and $f(v_z)$ are nearly identical for S1 and S2 – within the limits of statistical fluctuation – and all are roughly Gaussian. Consistent with the azimuthal symmetry of the system, $f(v_x)$ and $f(v_z)$ are roughly the same. The inferred temperature of S1 distributions (T(S1)$\simeq$ 1850K) is greater than S2 distributions (T(S2)$\simeq$1500K); all are cooler than the cavity temperature (2000K). The distributions' reduced temperatures are consistent with the velocity space compression predicted for their

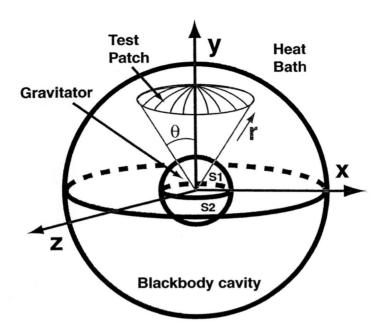

Figure 6.3: Schematic of gravitational system with test patch.

fall through the gravitational potential; that T(S1)>T(S2) is attributable to S1 being more inelastic than S2.

Velocity distributions $f(v_y)$ and $f(v_r)$ are highly non-thermal, non-Gaussian distributions, and are similar to the theoretical ones in Figure 6.2. The distinctive effects of gravity are more evident in these distributions, as well as the critical asymmetries between S1 and S2 gas-surface interactions, which are the basis of the challenge. The similarities between $f(v_y)$ and $f(v_r)$ occur because, over the poles, the r- and y- directions are roughly aligned. They are not, however, perfectly aligned except for $\theta = 0$, so the distributions are, in fact, distinct; the most noticeable differences occur at low velocities, where the depth of their local minima is greater for $f(v_r)$. The distributions display twin peaks; these are evidence of net drift velocities arising from the acceleration of the distributions en masse by the gravitational field of the gravitator. (At other radii the peaks of the distributions are shifted in accord with gravitational effects.) The apparent magnitude of the drift ($\sim$ 1500 m/sec) is about a factor of two below predictions for thermal, half-Maxwellian velocity distributions infalling from the walls and rebounding from the surface, however, this is understandable since the distributions are amalgams of many types of test atoms which camouflage the drift, e.g., inflow from the walls, reflection and outflow from the gravitator, thermalizing atom-atom collisions in the gas phase, and atoms just traversing the cavity volume randomly.

Comparing $f(v_y)$ and $f(v_r)$ between S1 and S2, one observes two main differences. First, the twin peaks of the S2 distributions are fairly sharp and symmetric

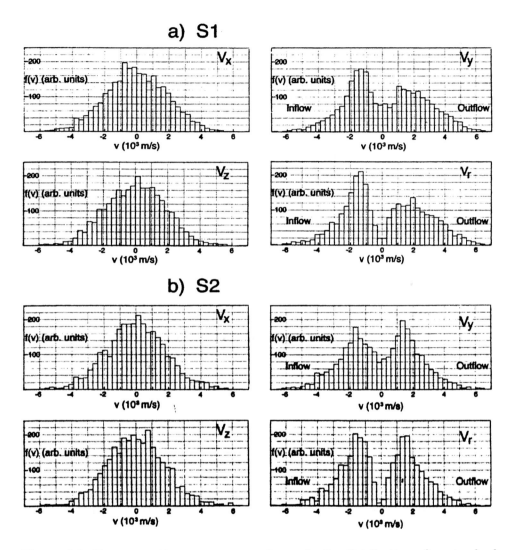

Figure 6.4: Representative x-, y-, z-, and r- velocity distributions for *standard gravitator* for 3000 atomic crossings of 30° dome test patches over S1 (a) and S2 (b), roughly midway between gravitator surface and cavity walls ($r_{patch} = 2.475 \times 10^6$m).

with respect to velocity inversion ($v \rightarrow -v$), while the S1 peaks are more rounded and asymmetric. This is consistent with the different elasticities (surface trapping probabilities) of the surfaces. The bottom hemisphere (S2) is paved with high mass atoms ($m_s(2) = 96$ amu), which renders suprathermal collisions by the test atom ($m_A = 4$ amu) fairly elastic (See (1)); therefore, one expects near (but not exact) velocity inversion symmetry. S1, on the other hand, is paved with low mass atoms ($m_s(1) = 7$ amu), making suprathermal collisions fairly inelastic. Many of these gas atoms are trapped and thermalize on S1 and must evaporate back to the walls. As the data demonstrate, this should make $f(v_y)$ and $f(v_r)$ asymmetric with respect to velocity inversion. Also, as predicted above ((6.3) and Figure 6.2), such surface inelasticity should give rise to additional high-velocity particles in the tail of $f(v)$ outflowing from the gravitator surface. This counterintuitive result is also evident in the data. Tails are evident in both $f(v_y)$ and $f(v_r)$ for both S1 and S2, with the strongest tails evident in S1, the more inelastic of the surfaces. Because the average velocity of the inflow $f(v_y)$ for S1 and S2 are comparable, while their average outflow velocities are different, by conservation of linear momentum, there should be a marked difference in the net momentum flux densities (pressures) exerted on the two gravitator hemispheres by the gas.

The velocity distributions in Figure 6.4 evidence four primary results. First, velocity distributions emerge from the long time average behavior of a single test atom. Additional simulations verified that ensemble average distributions are identical to the time average distributions such that this system meets conditions of a statistical ergodic ensemble. Second, these velocity distributions are stationary, nonequilibrium entities. They are "stationary" because, once they emerge from the time-average statistical behavior of the test atom, they are temporally unchanging thereafter. They are "nonequilibrium" because of the irreversible nature of the suprathermal gas-surface interactions, coupled with the fact that gas phase collisions are too rare to establish standard gas phase equilibrium; gas velocity distributions are determined primarily by gas-surface interactions. (Suprathermal gas-surface collisions must, on average, be nonequilibrium events, otherwise the second law would be violable.) Third, both $f(v_y)$ and $f(v_r)$ display gravitational drift that is responsible for the distributions being suprathermal and, therefore, subject to nonequilibrium, irreversible gas-surface collision processes. In contrast, $f(v_x)$ and $f(v_z)$ do not show drift and are more nearly thermal; they do, however, show the effects of velocity space compression. Fourth, $f(v_y)$ is distinct between S1 and S2. S1's distributions display both greater velocity asymmetry than S2's and also display more high-velocity particles in their outflow. Together, these latter three points provide strong support for the fundamental physical process proposed to explain the paradoxical behavior of system. Compare, for instance, $f(v_r)$ to the hypothesized one-dimensional distributions in Figure 6.2 and equation (6.3).

6.2.4.2 Phase Space Portraits
(Note: What would otherwise be Figures 6.5-6.7 of this section appear as Color Plates I-III.) Phase space diagrams were constructed from multiple, sequential velocity distributions (similar to those in Figure 6.2) for which one system parameter (e.g. test patch radius, polar angle, cavity temperature) was discretely and sys-

tematically varied. Contours in the phase space diagrams are contours of constant phase space density. They are analogous to contours of constant altitude used in topographic maps; interpreting them follows similar principles. Phase space diagrams are useful in that they condense the salient features from multiple velocity distributions and allow one to see both local details as well as global patterns within a system. Integration of the local $f(v)$ or phase space diagrams, particularly to obtain velocity moments, allows one to assess local and global thermodynamic quantities such as number density, particle flux, pressure, kinetic energy density, and entropy density.

The maximum phase space density was arbitrarily assigned value 100, corresponding to the maximum value of $f(v)$ for the *fiduciary velocity distribution* discussed above. For a given velocity distribution, velocities were attributed to each fraction of this maximum phase space density – 90%, 80%, ... 10%, 5%, 2.5%. Each distribution was thus reduced to discrete set of velocities corresponding to its set of fractional phase space densities. These were arranged sequentially, graphically plotted, and then smooth phase space contours were drawn by interpolating between the points.

Color Plate I depicts the phase space diagram v_x versus r (radius from cavity and gravitator centers) for 30° domes over the top (S1) and bottom (S2) poles of the *standard gravitator*. Owing to the azimuthal symmetry of the system, the v_z versus r diagram should be the same; comparison of their velocity distributions verified this. Plate I was constructed from 24 equidistantly radially-spaced velocity distributions. For reference, the two $f(v_x)$ shown in Figure 6.4 are incorporated into Plate I at $r = 2.475 \times 10^6$m. The vertical extent of the error bars in Plate I indicate the coordinate separation of the discrete velocity distributions from which the diagrams were constructed; the horizontal error bars indicate the uncertainty in a contour's velocity due to statistical fluctuations in the data.

In broad terms, the diagram consists of two types of contours: i) a series of open, slanted, nearly linear/parallel, relatively high-velocity, low phase space density contours (2.5% - 30%); and (ii) a series of nested loop contours (40% - 80%) at lesser velocities and lesser radii. The loss of the higher order loop contours with increasing radius corresponds to the loss of gas number density with increasing radius from the gravitator; this is akin to the radial exponential decay of typical planetary atmospheres. (Here, however, the situation is complicated by two caveats: a) the atmosphere is distended beyond the radius of curvature of the planet; and b) the cavity walls act as an additional boundary condition.) Velocity integration gives a roughly exponential decay of particle number density with radius. From the law of atmospheres, $P = P_o exp[\frac{m_A g h}{kT}] \simeq P_o exp[\frac{G m_A m_G}{rkT}]$, one estimates the theoretical e-folding radius to be $_tr_e = 3 \times 10^6$m. Integration of the phase space plot gives a numerical estimate of $_nr_e = 1.5 \times 10^6$m; these are in reasonable agreement, given the caveats.

The slant in the series of open, nearly parallel contours (2.5% - 30%) are consistent with two effects: radial density loss and gravitation acceleration and deceleration. If the cavity atmosphere were composed of uniform density, non-accelerated Maxwellian velocity distributions, then the contours would have no slant and would be aligned vertically, perpendicular to the abscissa.

Several global characteristics of the plots are apparent. First, there is a high-degree of symmetry with respect to velocity inversion ($v \rightarrow -v$). This is expected, given the azimuthal rotational symmetry of the system. There is, however, only a fair degree of symmetry with respect to mirror reflection across the abscissa. Were the gas isotropic in density and temperature throughout the cavity, then one would expect no slant and perfect reflection symmetry, but this is not the case. Because of the location of the test patches, there is some mixing of x and r directions, and as a result, there is evidence of the radial gravitational acceleration in the $v_x - r$ phase space plot as well as hints of the S1-S2 asymmetry with respect to gas-surface interactions. Specifically, there is a difference in radial extent of the low-velocity, looped contours, in that the nested loops for S2 extend further in radius than those for S1. Also, there is a minor mismatch in velocities (~ 100 m/sec) in the high-velocity, open contours (2.5% - 10%) between the S1 and S2 cases, these indicating more high-velocity tail particles over S1 than over S2. As explained earlier, this is consistent with S1 being less elastic than S2 with respect to gas-surface collisions. (A phase space diagram not shown here, v_r versus $m_s(1)$ for the *standard gravitator* (with $m_s(2)$ fixed at 96amu), substantiates the claim that low velocity particles in the bulk of $f(v)$ migrate to the high velocity contours as the elasticity of the collision surface (in this case, S1) is reduced by reducing the mass of gravitator surface atoms.) Overall, Plate I suggests only weakly the dynamic asymmetry responsible for paradox. Plate II, which follows, evidences it more ostensively.

Plate II presents the v_r versus r phase space diagram for the *standard gravitator* constructed from the same 24 test patches used for Plate I: radially spaced, 30° polar domes over S1 and S2. Again, $f(v_r)$ from Figure 6.4 may be used as reference. Recall that gravitational acceleration is solely in the radial direction and that, given the patch location over the poles, distributions $f(v_r)$ and $f(v_y)$ are similar. Three types of serial contours are apparent: i) open, slanted, roughly linear and parallel, high-velocity, low phase space density contours (2.5% - 30%); ii) slanted, nested chevron contours of medium- to high-velocity and high density (40% - 100%); and iii) nested finger-shaped loops with low velocity and low density (5% - 20%). The finger-shaped contours are due to the local minima between the twin peaks in $f(v_r)$ (See Figure 6.4). They are absent in Plate I because $f(v_x)$ are single peaked distributions. These finger contours carry little energy or momentum in their particles.

The chevron contours represent the peaks in the local velocity distributions, as indicated by their large phase space densities (40 - 100%). Their progressive radial disappearance signals the radial decline in gas density within the cavity. Their slants relative to the axes – they match those of the open, linear contours – are evidence both of radial gas density decline and en masse gravitational acceleration and deceleration of $f(v_r)$. For inflow to the gravitator (Quadrant II and III, hereafter, Q-II and Q-III), the slant indicates acceleration with gravity, while for outflow from the gravitator (Q-I and Q-IV) indicates deceleration against gravity. The magnitudes and signs of the slope are in reasonable quantitative agreement with the increase (decrease) in drift velocity expected for gravitational acceleration (deceleration). The open, roughly linear contours have similar interpretation as for those in Plate I.

The asymmetries in Plate II are more pronounced than those in Plate I and support the physical mechanism of the challenge. Quadrants II and III are fairly symmetric with respect to mirror reflection about the abscissa; this indicates that the radial inflow of atoms from the cavity walls to S1 and S2 are comparable. The outflow from S2 (Q-IV) is almost the velocity inverse of S2 inflow (Q-III), but with two differences: (1) Q-IV's intermediate contours (40% - 60%) are less pronounced than in Q-III, indicating fewer medium velocity atoms; while (2) extra atoms appear in Q-IV as high-velocity tail particles, indicated by the relatively higher velocity of the 2.5% - 20% linear slanted contours. Overall, the approximate velocity inversion symmetry of Q-III and Q-IV is consistent with the fairly elastic nature of the S2 surface. This is also borne out by statistics on atom-surface collisions. Of the 2.1×10^5 gas-surface collisions on S2 recorded over all 24 contributing simulations, 90% resulted in elastic rebound (no trapping), while only 10% were inelastic (trapping, thermalizing) events.

The most noticeable asymmetry in Plate II is in the outflow from S1 (Q-I); it is different from that of the other three quadrants in several significant ways. First, Q-I entirely lacks 80% and 90% chevron contours, unlike Q-II,-III, and -IV. Moreover, the Q-I chevrons that are present are significantly smaller than those in other quadrants. Second, Q-II, -III, and -IV have open 30% contours, while Q-I has a chevron. Taken together, these indicate a dearth of low- and medium-velocity particles in the outflow from S1 relative to S2. Offsetting this deficit, however, the open linear contours (2.5% - 20%) in Q-I are noticeably upshifted in velocity relative to the other quadrants, roughly 500 m/sec with respect to Q-IV and nearly 1000 m/sec with respect to Q-II,III. The spreading of these contours in velocity space indicates more high velocity particles in Q-I than elsewhere. Together these indicate two major effects. First, S1 collisions are more inelastic than S2 collisions. This is also borne out by collisions statistics: of the 2.2×10^5 S1 surface collisions, 28% were elastic (non-trapping), while 72% were inelastic (trapping, thermalizing) – the rough inverse of the S2 case above. Second, the contours at large radii near the cavity walls ($r = 3.2 \times 10^6$m) suggest the average velocity of the gas and the momentum flux density (pressure) delivered to the wall is greater for S1 than for S2. Velocity integration of Plate II bears this out. First velocity moments were taken over S1 and S2 near the cavity wall ($r = 3.205 \times 10^6$m $= 0.998 r_c$); these indicated that, at the cavity walls over their respective surfaces, the average velocity of gas from S1 exceeds that from S2: $(v_r)_{S1} = 2.0 \times 10^3$m/sec and $(v_r)_{S2} = 1.7 \times 10^3$ m/sec. Similarly, taking the ratio of the second velocity moments, the ratio of momentum flux densities delivered to (i.e., gas pressure exerted on) the cavity walls by S1 and S2 gas fluxes is $\frac{P(S1)}{P(S2)} \simeq 1.3$. By conservation of linear momentum, a commensurately greater pressure is exerted on the gravitator by gas on the S1 hemisphere than on the S2 hemisphere. Moment integrations of these phase space diagrams quantitatively agree with previous numerical estimates of gross force imparted to the gravitator discussed above, and both of these agree quantitatively with the previous theoretical estimates based on kinetic considerations. Overall, Plate II qualitatively and quantitatively supports the fundamental microscopic processes hypothesized to explain the paradoxical behavior of the system.

Gravitator mass and cavity temperature play reciprocal roles in the operation of the system. At low temperature, the gas freezes out on the gravitator and cannot cycle to the walls, while at high temperatures the effective surface trapping probabilities on S1 and S2 approach one another; i.e., they become symmetrized. In either case, the paradox is lost. Conversely, at high gravitator mass the gas is strongly bound to the gravitator and does not cycle to the walls, while at low mass the gas surface collisions are effectively thermal (rather than suprathermal); i.e., they are symmetric for S1 and S2. Again, the paradox fails. It is only in the narrow regime where the gas atom's gravitational potential energy $(P.E._g)$ and its thermal energy (kT) are comparable that the paradox will succeed. Specifically, it will be shown in Plate III that the viability range is roughly $0.25P.E._g \leq kT \leq 4P.E._g$. Because of the reciprocal roles played by temperature and gravitator mass in the operation of the demon, it suffices to study just one of them.

With this in mind, Plate III presents the v_r versus T (cavity temperature) phase space diagram for the *standard gravitator* for (30°) polar domes over S1 and S2, midway between the cavity walls and gravitator. It is constructed from 20 individual velocity distributions over the temperature range 500K - 8,000 K. (This covers the temperature range over which the *standard gravitator* is most viable.) The diagram consists of three types of curves: i) high- and medium-velocity, open, nested parabolas (2.5% - 30%); ii) nested closed loops (30% - 50%); and iii) low-velocity, nested line/parabolas (5% - 20%). The nested line/parabolas arise due to the local minima between the twin peaks in $f(v_r)$ near $v_r = 0$. Like as for Plate II, these contours are of little dynamical importance to the system. The 30%-50% closed loops are identified with the local maxima in the twin peaks for low temperatures (500K - 3000K). At higher temperature, as expected, velocity distributions broaden, so that these high-density contours are lost as particles move to higher velocity, lower phase space density contours. The open, nested curves (2.5% - 30%) are approximately parabolic; this consistent with the quadratic relationship between thermal energy and velocity $(kT \sim mv^2)$. All curves converge at the origin (T=0, v=0), as expected for a frozen gas.

Several features in Plate III are noteworthy. Quadrants II and III are nearly mirror images of each other – except for a slightly more pronounced 40% loop in Q-III – indicating that the inflow of gas to S1 and S2 are similar. Quadrants III and IV are similar, except for the following minor differences: i) the 30% contour is open in Q-III, but closed in Q-IV, and ii) the 40% contour is less pronounced in Q-IV than in Q-III. The overall similarity of Q-III and Q-IV is, again, consistent with nearly elastic atom-surface interactions on S2 and, again, this is borne out by collision statistics.

Similarly as for Plate II, Q-I is distinct from the other quadrants. It lacks a 50% loop contour entirely; its 40% contour is vestigial; and its 30% contour extends only to $T < 3000$K, whereas in other quadrants they are open out to 8000K. As for the other phase space diagrams, what Q-I lacks in low velocity particles it makes up for in high velocity ones. The 2.5%-10% contours of Q-I are significantly more expansive in velocity space than their counterparts in the other quadrants – over 1000 m/sec faster at the higher temperatures. As before, these carry disproportionately large momentum and kinetic energy to the walls.

In summary, Figure 6.4 and Plates I-III support at a microscopic level, both qualitatively and quantitatively, the physical basis of the challenge. Specifically, these phase space diagrams verify that: a) stationary-state velocity distributions emerge from the long time average behavior of a single test atom; b) the system maintains a cavity-wide, steady-state gas phase nonequilibrium; c) the velocity-space symmetries (asymmetries) of the velocity distributions reflect the geometric and parameter space symmetries (asymmetries) imposed on it; and d) the gas phase number density decrease radially from the gravitator surface to the cavity wall, as expected for a planetary atmosphere. Thirdly, these results support a more general conclusion: that two distinct, steady-state gas phase nonequilibria can interact with one another and still be simultaneously maintained in a single blackbody cavity. Even aside from any second law challenge these two simultaneously maintained distinct gas phase equilibria are peculiar and run counter to normal thermodynamic expectations. Analogous steady-state, nonequilibria have been proposed for plasma, solid state, and chemical systems and will be discussed in later chapters.

6.2.4.3 Gas-Gravitator Dynamics

In addition to the velocity distributions and phase space diagrams, explicit calculations were made of the force and torque exerted on the gravitator by the test atom while the gravitator was at rest and in motion, both at the center of the cavity (Figure 6.1a) and offset from the cavity center so as to circulate within the cavity (Figure 6.1b). These numerical simulations corroborate the analysis above, indicating a net force (and pressure) can persist on an asymmetric gravitator within the range of the system constraints (α - ζ). The values of all system parameters (masses, densities, temperatures, radii, etc.) were varied through and beyond the limits of the viability constraints α - ζ. As in previous sections, test atom collisions were of several types: on S1 and S2 (incident collisions followed by either elastic rebound or by trapping and thermal desorption), on the cavity wall (incident collisions followed by trapping and desorption) and in the gas phase (gravitational scattering or thermalizing collisions). The net impulse on the gravitator at rest (Figure 6.1a) or in rectilinear motion was calculated from conservation of linear momentum; the net torque on the gravitator in synchronous rotation (Figure 6.1b) was calculated via conservation of angular momentum.

Over a wide range of physically realistic parameters ($1 \leq m_A \leq 10$ amu; $4 \leq m_s(j) \leq 200$ amu; 2.73 K$\leq T \leq 2 \times 10^4$ K; 10^5 m$\leq r_G \leq 1.6 \times 10^6$ m; 2×10^5 m$\leq r_c \leq 6 \times 10^6$ m; 3×10^{19} kg$\leq m_G \leq 10^{24}$ kg; -120 m/s$\leq v_{tangential} \leq 120$ m/s; $n_{cav} \leq 2 \times 10^{12}$ m^{-3}), numerical simulations indicated that significant net forces were maintained on the gravitator – either at rest or in motion – as a result of asymmetric trapping (thermalization) probabilities on S1 and S2. Under all circumstances tested, the force was reversed in direction by reversing S1 and S2 surface parameters; the effect was lost when S1 and S2 were symmetrized. For the *standard gravitator*, steady-state, non-zero torques were maintained up to gravitator speeds of roughly 5% of the gas thermal speed – this represents the gravitator's 'terminal velocity.' (The *standard gravitator* was not optimized for power output, but is merely representative.) Outside the regime circumscribed by

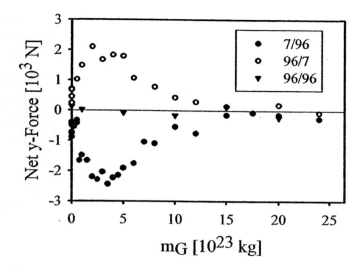

Figure 6.5: Net y-force on symmetric and asymmetric *standard gravitator* versus gravitator mass. Gravitator centered and at rest as in Figure 1a. (30,000 - 65,000 atom collisions per simulation; $n_{cav} = 5 \times 10^{10} \text{m}^{-3}$.)

the constraints ($\alpha - \zeta$) forces (and torques) declined to values within the range of the numerical fluctuations; that is, the challenge was lifted. On the other hand, within the viability regime, the effect was lifted only by the most contrived (e.g. symmetrical) choices for S1 and S2 parameters.

In Figure 6.5 is plotted the net y-force exerted on the *standard gravitator* by all atoms in the cavity ($N_{cav} = \frac{3}{4}\pi(r_c^3 - r_G^3)n_{cav}$) versus gravitator mass. (It is assumed that all gas atoms behave on average like the test atom. This is supported by its previously demonstrated ergodic behavior (Figure 6.4).) Gravitator mass can be expressed (in units of kT) in terms of the gravitational potential energy of a gas atom at the wall ($(P.E.)_{wall}[m_G] = \frac{Gm_A(r_c-r_G)}{r_c r_G}m_G$); for reference, $(P.E.)_{wall}[2 \times 10^{23}\text{kg}] = kT$. Three cases are shown: two asymmetric $\frac{m_s(1)}{m_s(2)} = \frac{7amu}{96amu}$, $\frac{m_s(1)}{m_s(2)} = \frac{96amu}{7amu}$ and one symmetric $\frac{m_s(1)}{m_s(2)} = \frac{96amu}{96amu} = 1$. The symmetric case shows little net force while the two asymmetric cases display significant net y-forces and are mirror images of each other within the limits of the noise. Both at low values of m_G (investigated down to 10^{12} kg $\equiv 5 \times 10^{-13}$ kT in $(P.E.)_{wall}$) and at high values (investigated up to 2×10^{25} kg $\equiv 100$ kT in $(P.E.)_{wall}$) the net y-force is small. At low values of m_G, gas-surface collisions are nearly thermal so S1 and S2 behave nearly identically and little force is seen, or expected. At high values of m_G the gas collisions may be suprathermal, but the gas is held so near the gravitator that it is unable to reach, thermalize and recycle on the cavity walls – again the paradox is lifted. Only in the midrange values (10^{23} kg $\leq m_G \leq 10^{24}$ kg, or $0.25kT \leq (P.E.)_{wall} \leq 4kT$) are the gas-surface collisions appreciably suprathermal and are gas atoms readily able to cycle between the gravitator and walls. It is here that the challenge is most evident. The

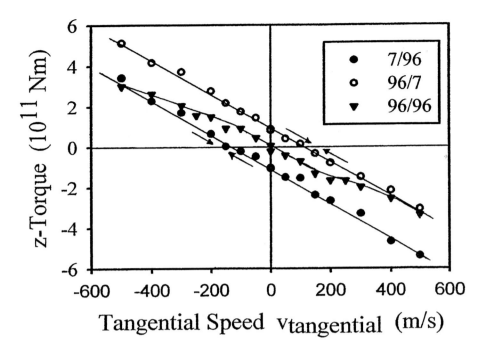

Figure 6.6: Net z-torque on symmetric and asymmetric synchronously rotating *standard gravitator* versus gravitator tangential speed. System attractors indicated by arrows. (25,000 - 50,000 collisions per simulation; $n_{cav} = 10^{12} \text{m}^{-3}$; $r_A = 5 \times 10^{-11}$ m)

maximum net y-force for the asymmetric cases is roughly 13% of the total y-force exerted by all gas-surface collisions to a single hemisphere, in rough agreement with previous theoretical predictions ($\sim 25\%$). At their maxima ($m_G \simeq 3 \times 10^{23}$ kg, $(P.E.)_{wall} \simeq 1.5kT$), the net y-forces are about 10 times larger than either the rms net y-force for the symmetrical case or for the net x- and z-forces for either symmetric or asymmetric cases. The relative sizes of fluctuations in numerically calculated dynamical quantities (for example, in the net linear and angular x-,y-, and z-momenta) scaled inversely with the number of gas-gravitator collisions (N_{coll}) in a run, scaling roughly as expected, that is, approximately as $N_{coll}^{-1/3}$.

In Figure 6.6 is plotted the net z-torque exerted on the synchronously rotating *standard gravitator* (Figure 6.1b) by all gas atoms in the cavity versus the tangential speed of the gravitator. For reference, 100 m/s is roughly $0.05v_{th}$. The same three cases are shown as in Figure 6.5. The symmetric case ($\frac{m_s(1)}{m_s(2)} = \frac{96amu}{96amu}$) demonstrates the torque due to gas drag as the gravitator plows through the cavity gas. The arrows near the data line point to the attractor for the symmetric case (the origin); that is; regardless of its initial $v_{tangential}$, gas drag torque will eventually slow the gravitator to rest. As $v_{tangential}$ approaches about $0.20\ v_{th}$, the symmetric case line meets the respective asymmetric case lines for which the 96

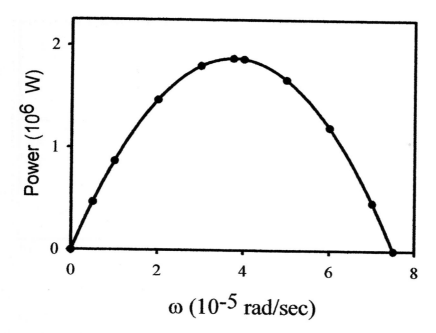

Figure 6.7: Power output from synchronously rotating *standard gravitator* versus angular velocity for case $\frac{m_s(1)}{m_s(2)} = \frac{96amu}{7amu}$. (25,000 - 50,000 collisions per simulation; $n_{cav} = 10^{12}\text{m}^{-3}$; $r_A = 5 \times 10^{-11}\text{m}$)

amu hemisphere faces a headwind. Presumably, this is because, as the gravitator reaches an appreciable fraction of v_{th}, the composition of the leading hemisphere increasingly determines the dynamics of the gravitator.

The attractors for the asymmetric cases ($\frac{m_s(1)}{m_s(2)} = \frac{7amu}{96amu}$ and $\frac{m_s(1)}{m_s(2)} = \frac{96amu}{7amu}$) are at $v_{tangential} \simeq \pm 120$ m/sec. This means the gravitator will move at this tangential speed stably and indefinitely, making it a *perpetuum mobile* of the third kind. On the other hand, if, beyond the torques due to gas drag and asymmetric trapping, an appropriate additional torque is added to the system – say, the torque of an electrical generator located at the axis of rotation – then the gravitator can reside stably and indefinitely at any point along the solid data lines in the first and third quadrants of Figure 6.6. Here, the gravitator can exert a steady-state torque on the generator while maintaining a steady-state angular velocity. Taking r_{lever} as the lever arm (the distance from the gravitator's center to the center of the cavity), if one plots the product $P = \tau \cdot \frac{v_{tangential}}{r_{lever}} = \tau \cdot \omega$ versus ω, one can establish the output power curve for the gravitator/generator, as shown in Figure 6.7 for the case $\frac{m_s(1)}{m_s(2)} = \frac{96amu}{7amu}$ in the first quadrant of Figure 6.6. At either $\omega = 0$ rad/s or at the terminal angular velocity ($\omega_{terminal} \simeq 7.5 \times 10^{-5}$ rad/s) the gravitator/generator produces no power, but for $0 \leq \omega \leq \omega_{terminal}$, the steady-state power output is non-zero and positive, maximizing at $P(\omega \simeq 3.8 \times 10^{-5}\text{rad/s}) \simeq 1.9 \times 10^6$ W. A

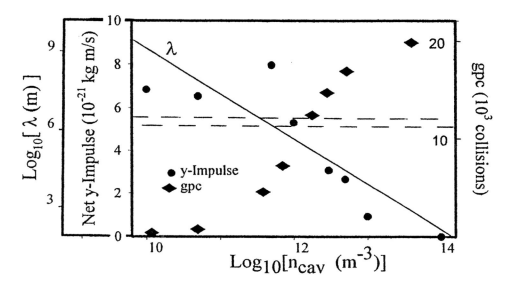

Figure 6.8: Variation of mean free path (λ), number of gas phase collisions (gpc) and the magnitude of the net y-impulse to *standard gravitator* by test atom versus cavity number density, n_{cav}. ($\frac{m_s(1)}{m_s(2)} = \frac{7amu}{96amu}$; 20,000 collisions per simulation)

more realistic estimate also incorporates the counter-torque due to the Doppler-shifted blackbody radiation field. When this is considered, then, at its power maximum ($\omega \simeq 3 \times 10^{-6}$ rad/s or $v_{tangential} \simeq \pm 5$ m/sec) the *standard gravitator* in Figure 6.6 should generate via the generator about 1.6×10^5 W in steady-state power. This is forbidden; this is a *perpetuum mobile* of the second kind.

System behavior is strongly dependent on the gas phase number density and boundary conditions. In Figure 6.8 are coplotted the theoretical gas mean free path (λ), the number of gas phase collisions (gpc) and the net y-impulse to the gravitator by the test atom versus the cavity gas number density, n_{cav}. As expected, as the mean free path becomes comparable to the scale lengths of the cavity and gravitator ($\lambda \sim r_G, r_c$ at roughly $n_{cav} \simeq 10^{12}$ m^{-3}), simultaneously the number of gas phase collisions rapidly increases while the net impulse to the gravitator rapidly decreases, signaling the collapse of the paradoxical effect. (Gas-gravitator and gas-wall collisions also undergo rapid decreases at $n_{cav} \sim 10^{12}$ m^{-3}.) The collapse occurs occurs because thermalizing gas phase collisions reduce the suprathermicity of gas-surface collisions.

The cavity walls play as integral a role to the paradox as does the gravitator itself since it is the walls: i) that act as a reference point for the application of conservation of linear and angular momenta to the gas influxes and effluxes; ii) where gas is rethermalized and recycled for its return to the gravitator; and iii) where the gravitator makes corporeal connection with the universal heat bath via the gas. In other words, were the gravitator simply in free space – that is, were

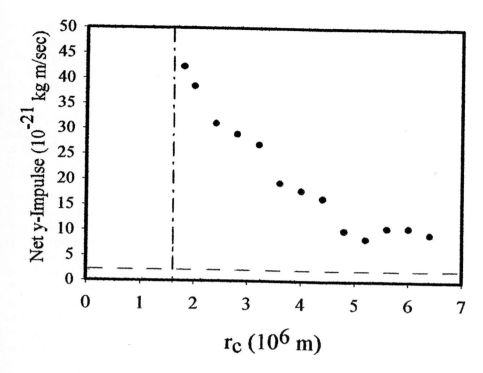

Figure 6.9: Magnitude of net y-impulse imparted to *standard gravitator* by test atom versus cavity radius. (65,000 collisions per simulation)

the cavity walls absent – conservation of momentum dictates that there could be no net force (or torque) applied to the gravitator by the gas – and there would be no paradox. Simulations corroborate this: in Figure 6.9 is plotted the magnitude of the net y-impulse imparted to the gravitator (Figure 6.1a) by the test atom versus radius of the cavity for the case $\frac{m_s(1)}{m_s(2)} = \frac{7amu}{96amu}$. (The vertical dotted line indicates the radius of the gravitator and the horizontal dotted line represents the average magnitude of numerical noise for the x and z components of net impulse.) As evidenced in Figure 6.9, the impulse decreases with increased cavity radius. In the limit $r_c \longrightarrow \infty$, presumably the net y-impulse would decline into the noise – the expected free space value. This critical role of boundary conditions is typical for non-extensive systems [10].

In summary, three-dimensional numerical simulations qualitatively and quantitatively support the challenge to the second law.

6.2.5 Wheeler Resolution

Recently, J.C. Wheeler raised a serious objection against this model (§6.2) [27]; it is not yet clear whether it resolves the paradox. If it does, then inasmuch as the gravitator is closely related to the several other USD challenges (Chapters 6-9), this objection might lead to a resolution of the entire group[2]. Wheeler demonstrates analytically and via numerical simulations that the one-dimensional distribution used by Sheehan (Figure 6.2) is spurious because it is not correctly weighted by velocity. He shows that a 1-D Maxwellian in a gravitational field will retain its form when correctly weighted by velocity so as to reflect particle flux. In this case, distribution f_1 and f_4 should be the same for either hemisphere.

Wheeler goes on to argue that the 1-D distribution can be extended to 3-D and, via Liouville's theorem, the full velocity distributions (arriving at) and (departing from) the gravitator are the same for both hemispheres. By symmetry, then, there is no net force on the gravitator and, thus, the paradox is resolved. No 3-D numerical simulations were conducted to substantiate this claim.

Wheeler's 1-D argument appears sound; however, the 3-D argument, while superficially convincing, is circular. Liouville's theorem, conservation of phase space density, is not valid for systems evolving in a nonequilibrium fashion[3]. Furthermore, 1-D arguments cannot be automatically and legitimately extended to 3-D since the second law has different status in different dimensions. For instance, it is well known that the second law is violable in 1-D [28], but these violations do not extend to the 3-D cases. Likewise, there is no *a priori* reason to believe that a 1-D resolution will naturally extend to the 3-D case; in fact, there are good reasons to believe this one does not. In particular, in attempting to extend from 1-D to 3-D, Wheeler's analysis does not appreciate the fundamental differences between elastic and inelastic suprathermal atomic collisions. An atom that thermalizes instantaneously loses information about its initial conditions, while one that scatters elastically does not. Atoms falling suprathermally onto the hemisphere lose information (increase in entropy) if they thermalize; this is not the case if they scatter elastically or if they were thermal to begin with. S1 and S2 are thermodynamically distinct; in particular, entropy generation is greater on the inelastic side. Moreover, the Wheeler argument does not properly consider the possibility of net particle transport around the hemisphere or the walls of the cavity by surface hopping.

In summary, Wheeler has made an important contribution to understanding the gravitator paradox in one dimension. Unfortunately, the argument does not appear to have been properly extended to three dimensions. Perhaps a more persuasive argument can be devised. Meanwhile, numerical simulations, independent of Sheehan, et al., should be conducted to settle the issue.

6.2.6 Laboratory Experiments [13, 14]

Full three-dimensional analytic expressions for system behaviors are infeasible – after all, even closed-form descriptions of the simpler case of an atmosphere around a spherically symmetric planet without cavity walls are not known – however, 3-D numerical simulations are straightforward and, presumably, accurate. Clearly, full scale experiments of this gravitational system are also infeasible into the forseeable

---

[2]This would be ironic since Sheehan originally proposed the gravitator somewhat as a joke.

[3]Increasing system entropy corresponds to increasing phase space volume visited, and decreasing entropy to decreasing phase space volume.

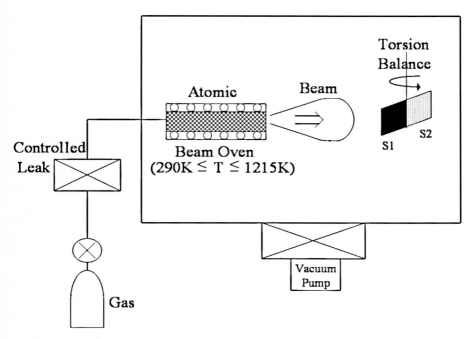

Figure 6.10: Schematic of torsion balance experiment.

future; however, each of the individual underlying physical processes are well-founded and experimentally testable.

A central physical process underlying this challenge is the surface-specific, nonequilibrium suprathermal gas-surface impacts. Their nonequilibrium nature is not doubted theoretically and is documented extensively in the chemical and surface science literature (see references in [23, 24]); it is also amply demonstrated by the ubiquitous examples of gaseous heat convection, upon which countless natural and technological processes rely. In fact, virtually any process in which a gas changes temperature through contact with a surface is an example of a nonequilibrium gas-surface interaction. It is difficult, however, to cite specific examples of two or more distinct surfaces exposed to a single suprathermal gas type in which the surfaces display different inelasticities (i.e. different degrees of momentum transfer with the gas). In theory they can be different (See (1)), but definitive experimental examples have only recently been demonstrated.

High-vacuum torsion balance experiments were conducted in which low-density (i.e. long mean free path), thermal and suprathermal noble gas beams (He and Ar) were directed normally against various target surfaces (metals (Al, W, steel), inorganics (borosilicate glass plate, fiberglass), and organics (acrylate and hydrocarbon films)) that were affixed to the vanes of the torsion balance (Figure 6.10). Gas beam temperatures could be varied continuously from roughly T = 293K (for thermal beams) up to T = 1215K (for suprathermal beams) and were collided on T= 293 K target surfaces on the torsion balance vanes. Surfaces were exposed at either 0 or 45 degrees with respect to beam normal incidence. Argon and helium were chosen as the impact gases since they are chemically inert and monatomic, thereby avoiding the theoretical and experimental intricacies of impact-induced vibrational and rotational modes of the gas and of chemical reactions with the target surfaces. At the suprathermal energies explored, impact energies were well below thresholds for chemical reactions, ionization or sputtering. From the deflection of the torsion balance, which was calibrated, the momentum flux transfer from the gas to the surface could be established with high precision and accuracy, and from this the degree of inelasticity of gas-surface collisions could be determined.

The torsion balance, consisting of thin glass vanes (21 mm square) overlayed with target surfaces attached to a solid copper disc base, was suspended by a 53.4 cm long, 51 micron diameter tungsten torsion fiber over a permanent disc magnet (for magnetic damping of oscillations). This assembly was housed in a 1.12 m tall, vertically aligned, cylindrical 304 stainless steel vacuum vessel (8×10^{-7} torr base pressure, oil diffusion pumped). The fiber's torsion constant (κ) – determined via the balances oscillation period (τ), given its moment of inertia (I) – agreed with theoretical prediction to within 3 % ($\kappa \simeq (\frac{2\pi}{\tau})^2 I = 1.80 \times 10^{-7} \frac{kgm^2}{s^2}$).

Test gases were delivered to the target surfaces via a flow-calibrated gas reservoir system through an open-ended 10 cm long, 0.5 cm diameter cylindrical, ohmically-heated stainless steel blackbody cavity with a centrally-located opaque iron wool diffuser. A type K thermocouple buried in the cavity measured cavity temperatures. Optical pyrometry corroborated thermocouple readings and verified spatial uniformity of temperature with the blackbody cavity. Test gas atoms were estimated to undergo roughly 100 wall surface collisions within the blackbody cavity and so were assumed to be in equilibrium with it upon exiting. Target surfaces were situated a distance $d = 10$ cm directly in front of the exit of the blackbody cavity. The gas mean free path of gas in the beams was longer than d.

Gas temperatures were continuously variable from 293-1215 K and could be maintained within ±3K (well within the 1 % precision of type K thermocouples) during measurements. Gas flow rate through the blackbody cavity as inferred quantitatively (to better than 1%) by monitoring the emptying rate of the fixed volume, isothermal gas reservoir, as established by a capacitance manometer gauge.

Momentum transfer was inferred from the deflection of the torsion balance when exposed to a specific flux of gas atom at a particular temperature. Deflection was inferred from laser reflection from a mirrored side of the balance. Minimum discernable deflections were 2×10^{-4} radians corresponding to net average torques about the balance's axis, forces, and pressures on the vanes of 5×10^{-11} Nm, 3×10^{-9} N, and 5×10^{-6} Pa, respectively.

Since the experiments were conducted at only high-vacuum ($\sim 10^{-6}$ Torr), all surfaces were eventually covered with the same multilayers of residual air, oxides, water, vacuum pump oil, and other standard vacuum contaminants. As might be expected because of similar vacuum surface contaminants, most surfaces behaved similarly with respect to a particular gas – with one exception. In comparing momentum transfer between fiberglass and plate glass it was found that suprathermal He beams transferred up to roughly 15% greater momentum flux density to the plate glass than to the fiberglass at high beam temperatures. The semi-elasticity of the He-glass impacts is supported by the magnitude of the torsion balance deflections. In other words, a greater net pressure was exerted by the He beam on the plate glass than on the fiberglass. On the other hand, for beams at the same temperature as the surfaces, no significant difference in momentum flux density between the two surface types could be inferred. The most reasonable explanation for this is that average collisions of He atoms with a surface are neither fully elastic or inelastic (similarly as to collisions of macroscopic objects), so, whereas on average a gas atom will transfer a fixed amount of momentum during a single collision, with the flat plate glass an atom interacts only once with the surface before leaving and so rebounds not nearly thermalized (i.e. $T \gg 300K$), while with the fiberglass the gas atom can can become trapped in the matrix, undergo multiple collisions and, therefore, can be more nearly thermal (i.e. $T \geq 300K$) when it leaves. In this case, one would expect more momentum to be transferred to the plate glass than to the fiberglass despite the fact that both surfaces are identical chemically. Experimentally, this was observed. It is also noteworthy that one would expect the torsion balance deflection to be the same for both surface types for thermal ($T = 300K$) gas beams. Again, this was observed experimentally.

These experiments corroborate a central physical process involved in the challenge, namely, the different nonequilibrium character of two surfaces with respect to suprathermal gas impacts. In this case, the difference was not due to chemical differences, but due to the morphological differences between the surfaces. (In principle then, one might use plate glass for S1 and fiberglass for S2 and the cavity walls.) The choices of gas, surface type, and temperatures used in the experiments model well several of the parameters presented in the *standard model*. In strictly applying these results to the paradox, however, a number of caveats must be made, chief among them being: (1) that the gravitationally-induced, drifting, cool half-Maxwellian suprathermal beams in the paradox behave comparably to the non-drifting, hot, half-Maxwellian suprathermal beams studied in the experiment; (2) that this behavior will be maintained in a closed, blackbody cavity environment; and (3) that the asymmetry observed experimentally for normal gas-surface collisions also apply to non-normal incident collisions. More sophisticated experiments can be imagined.

In this section several strands of evidence have been woven together in support of a classical mechanical gravitational challenge – analytic arguments, numerical simulations, laboratory experiments – and these mutually support the hypothesis of the challenge. However, none of these strands alone seems able to bear the full burden of proof; in fact, even together they cannot bear this burden because the second law, like all physical laws, is ultimately contingent on empirical verification; and likewise, any purported violation is also contingent. Since it cannot be tested with a full-scale experiment within the foreseeable future, this particular challenge is at most a thought experiment. Its value rests in challenging the common belief that possibilities for classical second law challenges are exhausted and also in its pointing the way to other more cogent and experimentally-testable challenges.

6.3 Loschmidt Gravito-Thermal Effect

The most celebrated gravitational second law challenge revolves around an unresolved dispute between Josef Loschmidt and the two thermodynamic giants, Maxwell and Boltzmann. Loschmidt claimed that the equilibrium temperature of a gas column subject to gravity should be lower at the top of the column and higher at its base [17, 18]. Presumably, one could drive a heat engine with this temperature gradient, thus violating the second law. (Loschmidt's original proposal was for solids, but it can be extended to gases and liquids.) This debate has remained unresolved for over a century. In the last several years, the question has been resurrected by Andreas Trupp [29] and has been explored experimentally by Roderich Gräff [30].

Loschmidt's rationale for a gravitational temperature gradient is straightforward. Consider a vertical column of gas in a uniform gravitational field (acceleration g) of height ($z = h$). Gas molecules (mass, m) at the top of the column possess mgh greater gravitational potential energy than molecules at the base. Thermal motion with (against) the direction of the field increases (decreases) molecular kinetic energy. This net kinetic energy can be transferred from the top to the bottom of the column via collisions. No net particle flow from top to bottom is required since energy transfer is mediated by collision; thus, this is heat conduction, not convection. If this gravitationally directed motion is eventually thermalized, one can write from the first law: $mgz = C_v m \Delta T$, where C_v is the specific heat of the gas at constant volume (Trupp notes that C_v should be replaced by C_p [29]). From this, a vertical temperature gradient can be intuited:

$$\frac{dT}{dz} = -\nabla_z T = \frac{g}{C_v} \qquad (6.4)$$

For a typical gas (e.g., N_2) with $C_v(N_2) \simeq 1100 \text{J/kgK}$, in the Earth's gravitational field, one estimates $\nabla_z T \simeq 10^{-2} \text{K/m}$. This is nearly the well-known standard meteorological adiabatic gradient, where C_p replaces C_v. It is claimed that, in principle, this temperature gradient can be used to drive a heat engine, thereby violating the second law.

The Loschmidt effect was never definitively refuted either by Maxwell or Boltzmann. Maxwell disbelieved it, apparently without mounting a formal proof; rather he appealed to the second law itself [31]. Boltzmann, on the other hand, repeatedly attempted microscopic counterproofs, but was not definitively successful [32, 33]. A review of this debate is given by Trupp [29], who adds his own commentary and analysis. Surprisingly, an experimental measurement of this effect has not been attempted until recently [30].

Loschmidt's argument skates over many crucial thermodynamic and statistical mechanical issues, including:

A) Radiation and convective heat transport, which would counter the conductive energy transport and erase the temperature gradient, are ignored. Heat transport rate is not addressed since (6.4) is derived from equilibrium consideration.

B) The argument ignores microscopic modifications to the gas velocity distribution as it ascends and descends in the field. Notably, it neglects the upwardly flowing, low-velocity particles which are turned back before they can ascend to collide with the molecules above. Sheehan, et al. argue that these are critical to kinetic energy transport in gravitational fields [13-16]; meanwhile, Wheeler argues there should be no net transport at all [27]. Specifically, while Loschmidt and Sheehan agree that there should be spontaneous vertical energy transport in the gas column, they fundamentally disagree on its direction. Loschmidt argues that energy flows downward, while Sheehan claims the net energy flow is upward. Loschmidt applies energy conservation and assumes equilibrium everywhere, while Sheehan examines the full velocity distribution ($f(v_z)$) both analytically and with numerical simulations.

Maxwell's opinion on the Loschmidt effect is the general scientific consensus: that the gas column's temperature must be independent of height [34, 35]. Most proofs can be shown either to analyse the problem incompletely, such as to reach second law compliance before the true problem arises, or else the proof makes *ad hoc* assumptions that deliver the desired result. A good example can be found in Walton [35], where explicit assumptions of equilibrium are made in the derivation of $f(v_z)$ (See (12) in [35]). Not surprisingly, this delivers the Maxwell distribution at all altitudes. Unfortunately, it hides the true development of $f(v_z)$. Trupp [29] considers Garrod's derivation [34].

6.3.1 Gräff Experiments

R.W. Gräff has conducted roughly 50 individual experiments with the goal of putting quantitative limits on the magnitude of the Loschmidt effect [17, 18]. Laboratories in Ithaca (US) and Königsfeld (Germany) simultaneously conduct multiple experiments [30].

Samples in which the Loschmidt effect has been sought include solids (Pb, Fe), liquids (water, oil, carbonated aqueous solutions with dissolved salts and organics), and gases (air, Xe, Ar) at various pressures ($10^{-4} - 10^5$ Torr). Test samples were vertically-aligned long cylinders (aspect ratios 5-10:1) with scale lengths 0.15-0.5m. Temperatures were measured by Type E (Chromomega-Constantan) thermocouples (0.005" dia) situated at various positions in the sample and on the exteriors of its surrounding, nested heat shields. Five thermocouples were strung vertically in series to form a five-leg thermopiles, with the upper and lower terminals taking differential temperature readings (rather than absolute temperature) over 10-20cm heights. Differential temperatures were measured in an attempt to sidestep the thermocouple precision-accuracy limit (so-called *limit of error* [37]) on absolute temperature measurements and also because differential temperatures are sufficient to establish the effect. Voltages were measured to $\pm 10^{-7}$V precision, corresponding to $\pm \Delta T \simeq 3 \times 10^{-4}$K, but the precision claimed by Gräff was only $\Delta T = 2 \times 10^{-3}$K. Type E thermocouple has a ANSI-rated *limit of error* of $\pm 1\%$, suggesting that for absolute temperature measurements at $T = 300$K, one has $\pm \Delta T = 3K$. This over a thousand times worse than the precision claimed by Gräff for differential temperature; however, it is unclear how ANSI specifications apply to differential temperature measurements.

Because the magnitude of the Loschmidt effect is small, the primary experimental challenges involved thermal isolation, uniformity, and stability. Samples were carefully isolated in a nested series of thermal shields and thermal equalizers. Shields consisted of low thermal conductivity layers like styrofoam, plastic wool, plastic fibers, and mirrored glass vacuum dewars. Thermal equalizers were layers of high thermal conductivity materials like iron, aluminum and copper. It can be shown from elementary heat transport theory that the proper alternating of high and low conductivity layers can achieve high degrees of thermal isolation, uniformity, and stability.

In the most sophisticated experiments, small scale thermal inhomogeneities were smoothed by rotating the outer heat shields around the inner shields on timescales much shorter than the system thermal relaxation time ($10s \sim \tau_{rot} \ll \tau_{relax} \sim 10^4 - 10^5 s$). The Loschmidt effect itself is expected to stratify liquids and gas up to their adiabatic convective instability limit in meterology, whereas small scale convection in the sample core could destroy the thermal stratification being sought; therefore, attempts were made to inhibit convection by loading the gas or liquid sample in a matrix of plastic fibers or glass microspheres ($r \simeq 5\mu m$). The volume fraction of liquid to microspheres were roughly 0.4:0.6.

Experiments were conducted in rooms with low levels of vibrations, small seasonal temperature variations ($\Delta T_{season} \simeq \pm 2K$), and minimal vertical temperature stratification between ceiling and floor ($\Delta T_{vert} \simeq 1K$). Experimental runs lasted up to six months each, allowing an apparatus to relax thermally and manifest long-term trends. Typical relaxation times for the core was estimate to be $2 \times 10^4 s$, while for the entire apparatus, typically $2 \times 10^5 s$.

Gräff's experimental results support the existence of Loschmidt's gravito-thermal effect in gases and liquids, but are inconclusive in solids, the phase for which Loschmidt made his original proposal. The most sophisticated long-term experiments reported average vertical temperature gradients of $(\frac{dT}{dz})_{air/exp} = 7 \times 10^{-2} K/m$ for air and $(\frac{dT}{dz})_{water/exp} = 4 \times 10^{-2} K/m$ for water (with glass microspheres), cold at the top and warm at the bottom of the sample columns. Experiment results agree qualitatively with the theoretical estimated temperature gradients from (1) of $(\frac{dT}{dz})_{air/theory} = 9 \times 10^{-3} K/m$ and $(\frac{dT}{dz})_{water/theory} = 2 \times 10^{-3} K/m$. In a recent series of aqueous experiments, the core sample was physically flipped vertically top to bottom up to ten times on time scales of 6-36 hours. The temperature gradient was observed to re-establish itself between flips.

The water/microsphere sample demonstrates a persistent negative temperature gradient while the outer heat shield shows the inverse. The averaging process supresses large, short-term variations in the data (signal-to-noise ratio of order unity). (Controls with the samples held horizontally or at intermediate angles with respect to vertical were not conducted.)

Long-term average voltage (temperature) gradients were unmistakable, however, for all experiments, the signal-to-noise ratios were of order unity, raising the question whether the positive results were the Loschmidt effect or a systematic error. In support of the effect, while the core samples showed persistent negative temperature gradients (cold at top, warm at bottom), the measurements on the heat shields directly surrounding the core consistently exhibited oppositely-directed, positive temperature gradients of greater magnitude. It is difficult to explain this reversal of vertical temperature gradients from inside to outside the apparatus. Aside from the Loschmidt effect, no alternative physical explanation for these results has been advanced.

Arguments against the effect and in favor of systematic error include:
a) The magnitude of the experimental temperature gradients are roughly a factor of 10 larger than theoretical estimates. (It is expected that the temperature gradients should relax somewhat via conduction, convection, and radiation, thus should be less than theoretical estimates, rather than greater.)
b) The data's signal-to-noise ratio is of order unity, but the origin of this noise has not been identified. The impact of 1% *limit of error* for Type E thermocouple on these differential temperature measurements has not been adequately assessed and it is unclear whether the Loschmidt signal can be extracted from noise, even in principle. Adequate control experiments (*e.g.*, varying inclination angle of sample) have not been conducted.
c) Straightforward application of Loschmidt's arguments predict that all material phases (solid, liquid, gas) should exhibit the gravito-thermal effect. While the effect is supported in gases and liquids, it is not yet supported in solids. Gräff notes that experiments with solids thus far have not been as extensive as those in gases and liquids, and more are planned.

The minuteness of the temperature gradient in realistic laboratory settings poses a significant experimental impediment, as can be inferred from Gräff's results [30]. Trupp suggests this difficulty might be overcome by exploring centripetally accelerated systems [36]. Einstein's equivalence principle allows substitution of the centripetal acceleration ($a_c = r\omega^2$) for gravity in (6.4) and state-of-the-art centrifuges can deliver accelerations in excess of $5 \times 10^5 g$'s, one expects commensurately larger temperature gradients than in terrestrial fields.

Consider a centrifuge rotating at angular velocity ω. Substituting the centripetal acceleration for gravity in (6.4), one obtains the temperature gradient, $\frac{dT}{dr} = \frac{\omega^2}{C_v} r$ and an integrated temperature difference between two radii (r_{in} and r_{out}) of

$$\Delta T = T_{out} - T_{in} = \frac{\omega^2}{2C_v}(r_{out}^2 - r_{in}^2) \qquad (6.5)$$

Letting $r_{in} = 0$ and using specifications for commercial ultracentrifuges ($r_{out} \sim 10^{-1}$m, $a_c \simeq 5 \times 10^6$m/s^2), the predicted ΔT for Ar gas radially across a 10cm tube should be $\Delta T \simeq 700$K – in principle, an easily measurable temperature difference. Of course, ultracentrufuges bring a new set of experimental hurdles, including: thermometry and thermal insulation in small, rotating systems; vibration-induced gas convection; and aerodynamic heating of the sample tube. Considering the experimental difficulties experienced by Gräff, Trupp's proposal seems reasonable.

In summary, Gräff's experiments are not yet conclusive and their theoretical underpinnings are disputed by other researchers. Nevertheless, his are the first to test and to support the Loschmidt effect. They deserve serious attention, and ideally, they should be replicated at independent laboratories.

6.3.2 Trupp Experiments

It is natural to attempt to extend the Loschmidt mechanism to gases in other force fields. One obvious application is to evaporation from liquids. Molecules in a liquid are loosely bound to each other, most often by van der Waals forces, ionic or hydrogen bonds [38]. On average in the bulk liquid, these forces cancel, but at the liquid surface they do not. Rather, there is a net force into the liquid — in the case of van der Waals forces, sizably extending out several angstroms — that inhibits evaporation.

Trupp proposes that the forces at the gas-liquid interface act as a highly compressed version of the gravitational field and should, via the Loschmidt mechanism, give rise to a temperature difference between the vapor and liquid phase of suitable solutions [39]. In particular, he predicts that the vapor over a superheated liquid can be significantly cooler than the liquid itself. In principle, this temperature difference could drive a Sterling engine or a heat pump having no moving parts. Trupp cites multiple experimental studies from the 19th century that appear to support this temperature difference [39, 40, 41]. For concentrated aqueous solutions of ionic salts (*e.g.*, aqueous solution of $CaCl_2$) vapor-liquid temperature differences in excess of $40^{o}C$ have been claimed [42].

Experimental subtleties make clean measurements of this temperature gradient more difficult than expected, such that it is still questionable. Experiments of boiling liquids thus far, that indicate temperature gradients, have continuously vented vapor from the confining vessel to the environment; this is required to maintain boiling. Ideally, the liquid and vapor should be held together in a single, sealed vessel, surrounded by an isothermal heat bath. A steady-state temperature gradient under these conditions would be more convincing. (Thermal conduction between the vapor and the walls must be considered carefully in this case.) Additional experimental uncertainty stems from the thermodynamics of contact between the vapor and thermometer. For example, if vapor condenses on a thermometer, it not only releases its latent heat, it also forms a layer of pure liquid unlike the concentrated solution from which it originated, thereby confounding measurements.

Modern non-invasive thermometry should be able to settle this issue. For instance, laser induced fluorescence temperature measurements of gases are possible, using narrow-band, tunable lasers to scan the gas' thermally-Doppler-shifted velocity distributions. This non-perturbing technique has been successfully applied to ions in low-temperature plasmas [43] and should be able to give detailed velocity distributions for vapors from which their temperatures can be inferred.

In summary, the generalized Loschmidt effect applied to evaporation is suspect, but it merits additional theoretical and experimental investigation.

References

[1] Sanmartin, J.R., Eur. J. Phys. **16** 8 (1995).

[2] Shore, S.N. *An Introduction to Astrophysical Hydrodynamics* (Academic Press, San Diego, 1992) pg. 319.

[3] Mills, B.D., Jr., Am. J. Phys. **27** 115 (1959).

[4] Grandy, W.T., *Foundations of Statistical Mechanics* (Dordrecht, Reidel, 1987) pg. 283.

[5] Spitzer, L, *Dynamic Evolution of Globular Clusters* (Princeton University Press, Princeton, 1987).

[6] Hut, P., Complexity **3** 38 (1997).

[7] Hawking, S.W., Commun. Math. Phys., **43** 199 (1975).

[8] Unruh, W.G., Phys Rev. **D 14** 870 (1976).

[9] Wald, R.M.,*Quantum Field Theory in Curved Spacetime and Black Hole Thermodynamics* (University of Chicago Press, Chicago, 1992).

[10] Gross, D.H.E., *Microcanonical Thermodynamics, Phase Transitions in "Small" Systems,* (World Scientific, Singapore, 2001).

[11] Gross, D.H.E., in *Quantum Limits to the Second Law,* AIP Conference Proceedings, Vol. 643, Sheehan, D.P., Editor, (AIP Press, Melville, NY, 2002) pg. 131.

[12] Lieb, E.H. and Yngvason, J., Physics Reports **310** 1 (1999); Phys. Today **53** 32 (2000).

[13] Sheehan, D.P., Glick, J. and Means, J.D., Found. Phys. **30** 1227 (2000).

[14] Sheehan, D.P. and Glick, J., Phys. Scripta **61** 635 (2000).

[15] Sheehan, D.P., Glick, J., Duncan, T., Langton, J.A., Gagliardi, M.J., and Tobe, R., Found. Phys. **32** 441 (2002).

[16] Sheehan, D.P., Glick, J., Duncan, T., Langton, J.A., Gagliardi, M.J., and Tobe, R., Phys. Scripta **65** 430 (2002).

[17] Loschmidt, J., *Sitzungsberichte der mathematisch-naturwissenschaftlichen Classe der Kaiserlichen Akademie der Wissenschaften zu Wein* **73.2** 135 (1876).

[18] Loschmidt, J., *Sitzungsberichte der Mathematisch-Nature Wisenschaftlichen Classe Kaiserlichen Akademie der Wissenschaften zu Wein* **76.2** 225 (1877).

[19] Whiting, H., Science **6** 83 (1885).

[20] Maxwell, J.C. Letter to P.G. Tait, 11 December 1867. In C.G. Knott, *Life and Scientific Work of Peter Guthrie Tait* (Cambridge University Press, London, 1911) pg. 213.

[21] Maxwell, J.C., *Theory of Heat*, Ch. 12, (Longmans, Green and Co., London, 1871).

[22] Chamberlain, J.W. and Hunten, D.M., *Theory of Planetary Atmospheres* (Academic Press, Orlando, FL, 1987).

[23] Hulpke, E. Ed., *Helium Atom Scattering from Surfaces* (Springer-Verlag, Berlin, 1992).

[24] Masel, R.I, *Principles of Adsorption and Reaction on Solid Surfaces* (Wiley, New York, 1995).

[25] Stickney, R.E. and Chanoch Beder, E., in *Advances in Atom. Molec. Phys.* **3**, Bates, D.R. and Estermann, I., Editors, (Academic Press, New York, 1967).

[26] Parilis, E.S., Kishinevsky, L.M., Turaev, N.Yu., Baklitzky, B.E., Umarov, F.F., Verleger, V.Kh., Nizhnaya, S.L., and Bitensky, I.S. *Atomic Collisions on Solid Surfaces* (North Holland, Amsterdam, 1993).

[27] Wheeler, J.C., Found. Phys., in press (2004).

[28] Mattis, D.C., in *Quantum Limits to the Second Law*, AIP Conf. Proc., Vol. 643, (AIP Press, Melville, NY, 2002) pg. 125.

[29] Trupp, A., *Physics Essays* **12** 614 (1999).

[30] Gräff, R., in *Quantum Limits to the Second Law*, AIP Conf. Proc., Vol. 643, (AIP Press, Melville, NY, 2002) pg. 225.

[31] Maxwell, J.C., The London, Edinburgh, and Dublin Philosophical Magazine and Journal of Science **35** 215 (1868).

[32] Boltzmann, L., *Wissenschaftliche Abhandlungen*, Vol. 2, Hasenöhrl, F., Editor, (Leipzig, Barth, 1909) pg. 56.

[33] Boltzmann, L., *Lectures on Gas Theory* (Dover Publications, New York, 1964).

[34] Garrod, C., *Statistical Mechanics and Thermodynamics*, (Oxford University Press, New York, 1995).

[35] Walton, A.J., Contemp. Phys. **10** 2 (1969).

[36] Trupp, A., private communication (2002).

[37] *The Temperature Handbook*, Vol. 29, (Omega Engineering, Inc, 1995).

[38] Sheehan, W.F. *Physical Chemistry*, 2nd Ed., Chpts. 7,8 (Allyn and Bacon, Boston, 1970).

[39] Trupp, A. in *Quantum Limits to the Second Law*, AIP Conf. Proc., Vol. 643, (AIP Press, Melville, NY, 2002) pg. 201.

[40] Sakurai, J. Chem. Soc. London **61** 989 (1892).

[41] Regnault, M.V., *Memoires de l'Academie des Sciences de l'Institut Imperial de France* Tome **XXVI** Paris, 84 (1862).

[42] Trupp, A., private communications (2003).

[43] Sheehan, D.P., Koslover, R., and McWilliams, R., J. Geophys. Res. **96** 14,107 (1991).

7

Chemical Nonequilibrium Steady States

Differential rates of gas-surface reactions under low-pressure blackbody cavity conditions can undermine common notions of detailed balance and, by suitable construction, challenge the second law. Laboratory experiments involving low-pressure hydrogen reactions on high-temperature refractory metal surfaces corroborate these claims.

7.1 Introduction

In this chapter the theory and experimental details of a laboratory-testable challenge to the second law based on heterogenous gas-surface reactions are developed [1, 2, 3]. Theory is developed from the general constraints of detailed balance and then for more specific conditions using primitive rate equations. Experimental support for this effect is presented and prospects for more definitive experiments are discussed.

Under sealed blackbody cavity conditions, gas pressure gradients commonly take three forms: (a) statistical fluctuations; (b) transients associated with the relaxation toward equilibrium; and (c) equilibrium pressure gradients associated with potential gradients (such as with gravity). In the low-density (collisionless) regime, a fourth type of pressure gradient can also arise, this due to steady-state differential thermal desorption of surface species from chemically active surfaces. This gas phase is inherently nonequilibrium in character and can lead to steady-state pressure gradients that can be exploited to perform work solely at the expense of heat from a heat bath, in violation of the second law.

It is considered almost axiomatic that gas phase equilibrium under blackbody cavity conditions is: i) unique; ii) solely determined by chemical species type, temperature, and number density; and iii) independent of the chemical composition of the cavity walls. The gas phase equilibrium constant $K(g)$ is simply a function of temperature T and the Gibbs free energy of the reaction [4]: $K(g) = exp[-\frac{\Delta G}{kT}]$. This can be argued forcefully theoretically and also experimentally by appealing to countless examples. This 'axiom,' however, tacitly assumes that gas phase collisions are sufficiently frequent so as to establish equilibrium without account of gas-surface reactions. Under most everyday circumstances, this assumption is valid. For instance, air molecules at STP have a mean free path of $\lambda \simeq 5 \times 10^{-8}$m and a collision frequency of $\nu \simeq 10^{10}\text{sec}^{-1}$; therefore, regardless of what specific chemical reactions may occur at a boundary surface, *standard gas phase equilibrium* (hereafter, SGPE) is established on very short time and distance scales from any boundary. Furthermore, surfaces typically have sufficient adsorbate coverages to mask gas-surface reactions that might otherwise be manifest.

However, when gas phase collisions are rare compared with gas-surface collisions, SGPE cannot be taken for granted and serious account must be taken of chemical reactions of the gas with the confining walls. In the low pressure regime where surface coverages are low (perhaps less than a monolayer) and surface effects are important, where gas phase collisions are rare compared with gas-surface collisions, a novel steady-state nonequilibrium gas phase can arise which depends primarily on gas-surface reactions, rather than on gas-phase reactions; this results in a *dynamically-maintained, steady-state pressure gradient* (hereafter, DSPG). The DSPG involves two coupled steady-state chemical nonequilibria. Numerical simulations, starting from the primitive rate relations for species concentrations and using realistic physical parameters, support these phenomena and indicate they should be observable in the laboratory. Corroborative laboratory experiments also support this hypothesis [2].

The DSPG represents a new type of pressure gradient. It appears in physical regimes that have not been explored carefully either theoretically or experimentally. Numerous gas-surface interaction studies have been performed, but most of these have been carried out i) at relatively high pressures where standard gas phase equilibrium can be assumed or where sub-monolayer surface coverages cannot be assumed; or ii) in a geometry which does not approximate a blackbody; or iii) where only a single chemically-active surface is involved.

In this chapter, the DSPG will be derived from two perspectives: (i) from the principle of detailed balance, and (ii) from reaction rate theory in the low-gas-pressure and low-surface-coverage regime. In the latter, realistic physical parameters are applied, indicating that the DSPG should be both robust and laboratory-testable. Once the DSPG is established, it is only a minor extension to a second law challenge [3].

The physical origin of the DSPG is straightforward. Consider two surfaces, S1

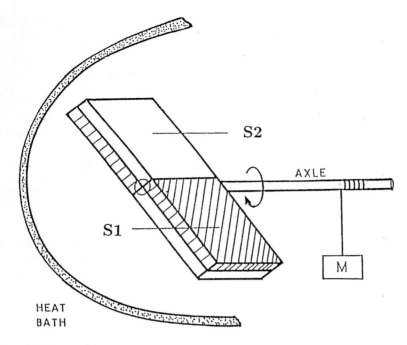

Figure 7.1: Duncan's radiometer. Force from dynamically-maintained, steady-state pressure gradient (DSPG) between S1 and S2 raises weight, M.

and S2, which have different chemical activities with respect to the dissociation of diatomic gas ($2A \rightleftharpoons A_2$) in that, under identical temperatures and pressures, S1 dissociates A_2 and desorbs A more effectively than S2. Since, at a single temperature, the thermal velocity of A is $\sqrt{2}$ greater than that of A_2, if the gaseous influxes onto S1 and S2 are the same and if all effluxing species leave in thermal equilibrium with S1 and S2, then conservation of linear momentum demands that a greater pressure is exerted on S1 than on S2. This pressure difference implies the DSPG.

If the DSPG obtains, then a simple geometric rearrangement can bring it into a form which challenges the second law. Such proposals date back to the early 1990's, but were most clearly enunciated in 1998 by Duncan [3] who proposed a simple radiometer-style second law violator in which S1 and S2 pave alternate faces of turbine was housed in a blackbody cavity surfaced with S2 and housing A_2 and A gas (Figure 7.1). The DSPG operates across the vane faces, creating a steady-state pressure differential on par with the average gas pressure in the cavity. This is used to lift a weight or run an electrical generator continuously. Since this is a nonequilibrium steady state (NESS), the standard NESS literature could be illuminating (§1.4).

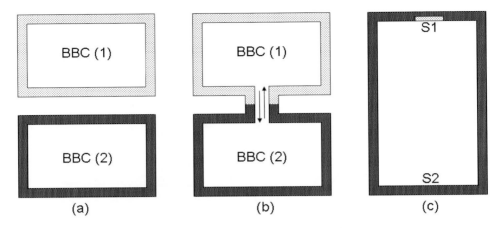

Figure 7.2: Blackbody cavities for establishment of DSPG: (a) separate BBC(1) and BBC(2); (b) Cavities joined by small hole; (c) Single cavity with S1 and S2 surface types; S2 dominates cavity reactions.

7.2 Chemical Paradox and Detailed Balance

Nomenclature from [1] will be used. The initial i refers to surface type, j to chemical species; subscripts ads, des, diss, and recomb refer to the processes of adsorption, desorption, dissociation, and recombination of atomic or molecular species (e.g., $R_{ads}(i, A_j) = R_{ads}(1, A_2)$ is the adsorption rate of A_2 molecules from surface 1.)

Species chemisorbed on a surface can be chemically distinct from what they are in the gas phase. By the very definition of chemisorption, molecular electronic, vibrational, rotational, and conformational states can be substantially modified either from surface to surface, or from surface phase to gas phase. Furthermore, because of the reduced dimensionality of surfaces compared with the gas phase, a species chemical identity can have directionality on a surface; that is, taking the x-y plane to be the surface plane and z-direction perpendicular to the surface, the chemical properties of a chemisorbed species in the (x-y)-direction can be different than its properties in the z-direction. Thus, for two materially distinct species, say monatomic A and diatomic A_2, their chemical identities on surfaces S1 and S2 can not only be different between the two surfaces in the x-y planes, they can also be distinct in the z-direction. It follows, then, that their equilibrium surface concentrations for the reaction $(2 A \rightleftharpoons A_2)$ can be distinct between S1 and S2; i.e., their surface phase equilibrium constants can be distinct: $K(1) \neq K(2)$. It also follows that their surface adsorption and desorption probabilities (P_{ads} and P_{des}) can be distinct for A and A_2 between the two surfaces; i.e., $P_{ads}(1, A) =$

$P_{des}(1, A) \neq P_{ads}(2, A) = P_{des}(2, A)$, and likewise for A_2.[1]

With this in mind, consider two sealed blackbody cavities (BBC(1) and BBC(2)), one tiled with S1 and the other tiled with S2 (Figure 7.2a). Both are held at the same temperature by an isothermal heat bath. Let the gas-surface reactions be such that for S1 the x-y surface equilibrium favors A over A_2 and that for S2 the surface equilibrium favors A_2 over A. Let us consider the surface and gas phase equilibria in each cavity in the limit that: i) species' surface residence times are long enough for surface phase equilibria to be established; ii) surface residence times are long compared to gas phase residence times; and iii) that gas phase concentrations are sufficiently low that gas phase collisions (especially three-body collisions) are too rare to establish SGPE. Under these conditions, gas phase equilibria are set by surface reactions, rather than by gas phase reactions.

One dimensional potential energy diagrams that might be appropriate to S1 and S2 are depicted in Figure 7.3. Figure 7.3a shows a potential energy diagram conducive to dissociative chemisorption of A_2 into A, modeling surface S1. Figure 7.3b, on the other hand, shows a potential conducive to the associative chemisorption of A_2 and recombinative chemisorption of A, thereby modeling surface S2.

At equilibrium for the reaction ($2A \rightleftharpoons A_2$), gas phase concentrations in a given cavity are fixed by the following six constraints:

$$K(s) = \frac{n(s, A_2)}{n^2(s, A)}; \quad K(g) = \frac{n(g, A_2)}{n^2(g, A)} \tag{7.1}$$

$$R_{des}(A) = n(s, A)\nu_{att}(A)P_{des}(A) = \frac{1}{4}n(g, A)v_A P_{ads}(A) = R_{ads}(A) \tag{7.2}$$

$$R_{des}(A_2) = n(s, A_2)\nu_{att}(A_2)P_{des}(A_2) = \frac{1}{4}n(g, A_2)v_{A_2}P_{ads}(A_2) = R_{ads}(A_2) \tag{7.3}$$

$$P_{des}(A) = P_{ads}(A); \quad P_{des}(A_2) = P_{ads}(A_2) \tag{7.4}$$

Equations (7.1) gives equilibrium constants for surface and gas phases, respectively. Equations 7.2 and 7.3 are equilibrium constraints on adsorption and desorption of A and A_2 following the principle of detailed balance. Lastly, (7.4) are statements of microscopic reversibility applied to adsorption and desorption. (Without loss of generality, this constraint can be sharpened to summing discrete quantum mechanical state-to-state transitions between gas and surface phases.) Here v_{A_j} is the gaseous thermal velocity of species A_j; ν_{att} is the attempt frequency for desorption, which is roughly equal to the lattice vibration frequency, $\nu_{att} \simeq \nu_o \sim 10^{13}$Hz for typical surfaces.

Equations (7.1-7.4) can be solved simultaneously to yield the formal gas phase equilibrium constant for each cavity:

---

[1] Microscopic reversibility guarantees that $P_{ads}(i, A_j) = P_{des}(i, A_j)$. Presumably, this should be true both for the average macroscopic behavior and for microscopic state-to-state transitions.

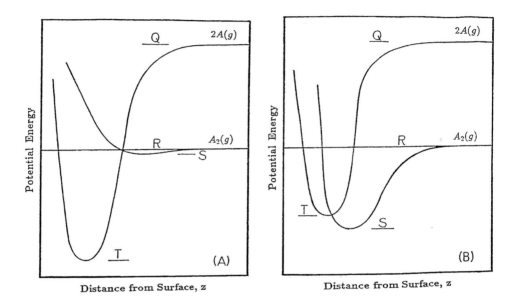

Figure 7.3: One dimensional potential energy diagrams for A and A_2 interacting with surfaces S1 and S2. The zero in potential energy is that of free, gaseous A_2. (a) Surface 1: Dissociative chemisorption of A_2. QR $\equiv$ Dissociation energy of gaseous A_2 in gaseous A, E(A-A); RS $\equiv$ Weak physisorption enthalpy for A_2; RT $\equiv$ Dissociation enthalpy of surface A_2, $\Delta H_{diss,act}(1)$ and also desorption enthalpy of A_2, $\Delta H_{des}(1, A_2)$; $\frac{QH}{2} \equiv$ Desorption enthalpy of A, $\Delta H_{des}(1, A)$). (b) Surface 2: Associative chemisorption of A_2 and recombinative chemisorption of A. QR same as in (a); RS $\equiv$ Desorption enthalpy of A_2, $\Delta H_{des}(2, A_2)$; $\frac{QT}{2} \equiv$ Desorption enthalpy of A, $\Delta H_{des}(2, A)$; TS $\equiv$ Dissociation enthalpy of A_2, $\Delta H_{diss,act}(2)$.

$$K(g) \equiv \frac{n(g, A_2)}{n^2(g, A)} = \frac{\left[\frac{v_A}{\nu_{att}(A)}\right]^2}{4\left[\frac{v_{A_2}}{\nu_{att}(A_2)}\right]} \frac{n(s, A_2)}{n^2(s, A)} = \frac{1}{4} \frac{v_A^2}{v_{A_2}} \frac{\nu_{att}(A_2)}{\nu_{att}^2(A)} K(s) \qquad (7.5)$$

Since $v_A^2 = 2v_{A_2}$ and since $\nu_{att}(A) \simeq \nu_{att}(A_2) \simeq \nu_o$, this can be simplied to

$$K(g) \simeq \left[\frac{v_{A_2}}{2\nu_o}\right] K(s) \qquad (7.6)$$

Given that v_{A_2} is independent of surface effects and $\nu_{att} \simeq \nu_o$ is relatively independent of surface type, but given that $K(s)$ can be strongly dependent on surface type, (7.6) implies that the gas phase equilibrium in a cavity $K(g)$ can be set by the surface phase equilibrium. Thus, at a single temperature, sealed BBC(1) and BBC(2) can harbor distinct gas phase equilibria. This is not particularly surprising considering that if gas phase collsions are rare, then there is essentially no mechanisms to set the gas phase equilibrium. In other words, the surface phase Gibbs free energy trumps the gas phase Gibbs free energy for the reaction ($2A \rightleftharpoons A_2$).

The stage is now set to form a steady-state chemical nonequilibrium. Let the two cavities be connected by a small hole, as depicted in Figure 7.2b. The hole appears to each cavity like a material wall of the opposite chamber. If the gas phase equilibria in the two cavities are distinct, then, under the additional constraint that the total flux of A atoms (the sum in the forms of monatomic A and diatomic A_2) passing through the hole in each direction is the same, there can obtain a steady-state nonequilibrium in each cavity. The total A-flux constraint for the two isothermal cavities is written: $n(1, A) + \frac{1}{\sqrt{2}}n(1, A_2) = n(2, A) + \frac{1}{\sqrt{2}}n(2, A_2)$. For the limiting case $K(1) \ll 1$ and $K(2) \gg 1$, one has from (7.5,7.6) that $n(1, A) \gg n(1, A_2)$ and $n(2, A_2) \gg n(2, A)$, in which case the A-flux constraint becomes $n(1, A) \simeq \frac{1}{\sqrt{2}}n(2, A_2)$.

The total A-flux constraint guarantees that neither cavity gains nor loses atoms. If the hole is small, then the gas phase concentrations in each cavity are shifted only slightly from their respective equilibria. Both gas phases now represent steady-state nonequilibria since they each must continuously process what for each is a nonequilibrium gas flux: the distinctive flux from the other cavity. The magnitude of the nonequilibrium shift is proportional to the size of the hole. If the hole is kept small neither cavity will dominate the other's gas phase and each will process the small nonequilibrium flux from the other. This can also be argued persuasively from Le Châtelier's principle. Also, regarding temperature variations, if the particle numbers are kept low and the cavities are thermally well-coupled to the surrounding heat bath (well thermostatted), then the unbalanced, steady-state, nonequilibrium dissociation and recombination will not significantly alter surface temperatures away from the heat bath temperature. Thus, two distinct chemical equilibria can be conjoined to form a steady-state chemical nonequilibrium. The DSPG follows directly from here by appealing to ideal gas behavior.

The ideal gas law does not discriminate between monatomic and polyatomic gases with respect to gas pressure; gas pressure depends only on temperature and number density. Since the gas phase concentrations of A and A_2 can be partitioned differently between BBC(1) and BBC(2), owing to the chemical differences of S1 and S2, then, even under the total A-flux constraint, the total number densities of particles can be distinct between the two cavities. If the gas is treated as ideal (which is a reasonable approximation at the low pressures assumed here, but which is not required for the result), it follows that the pressures in the two cavities will be distinct. Specifically, one has:

$$P_1 = [n(1, A) + n(1, A_2)]kT \neq [n(2, A) + n(2, A_2)]kT = P_2. \qquad (7.7)$$

The pressure difference between BBC(1) and BBC(2) ($\Delta P = P_1 - P_2$) can be shown to be the same as (18) in [1]. In the limiting case above ($K(1) \ll 1$ and $K(2) \gg 1$) with its A-flux constraint ($n(1, A) \simeq \frac{1}{\sqrt{2}} n(2, A_2)$), applying (7.7), one finds $P_1 \simeq n(1, A)kT \neq n(2, A_2)kT \simeq \sqrt{2}P_1 \simeq P_2$; that is, $\frac{P_2}{P_1} = \sqrt{2}$ and $\Delta P \neq 0$. In general, ΔP will be non-zero for chemically distinct cavities except for highly contrived initial conditions. Since this pressure difference occurs continuously over a finite distance Δx between the cavities, there is a pressure gradient ($\nabla P \sim \frac{\Delta P}{\Delta x}$). This establishes the DSPG. Even if the gas is less than ideal, the DSPG is unlikely to disappear since ΔP can be made to be of the same order of magnitude as the total gas pressure. The gas would need to be highly non-ideal (with precisely the correct signs and magnitudes of non-ideality) to negate ΔP.

It is counterintuitive that two isolated equilibria (BBC(1) and BBC(2)), when combined, should form a steady-state nonequilibrium, however, this follows logically from treating the gas and surface phases on equal footing with respect to detailed balance. In BBC(1) and BBC(2), the gas phase is essentially chemically inert, therefore, the surface phase concentrations primarily determine the gas phase concentrations. The 'axiomatic' claim that the gas phase is unique and can only be SGPE – regardless of surface type – cannot be justified since it treats the gas phase as a preferred phase relative to the surface phase with respect to detailed balance.

More to the point, chemical equilibrium is the result of the balance of forward and reverse rates of specific chemical reactions. If a specific process is absent or its rate is small compared with the rates of competing reactions, then its effect on the equilibrium is small or zero. In BBC(1) and BBC(2) gas phase reactions are virtually absent, while surface-specific chemical reactions (*e.g.*, adsorption and desorption) dominate, therefore, it is they that primarily determine the gas phase equilibria in BBC(1) and BBC(2). From the connected cavities in Figure 7.2b it is a small step to a single cavity with two surfaces (Figure 7.2c), and from there on to Duncan's radiometer (Figure 7.1).

7.3 Pressure Gradients and Reaction Rates

The DSPG is supported by the principle of detailed balance, but it is not sufficiently specific to point to laboratory tests. In this section, the DSPG is derived from reaction rate theory under standard approximations used in systems with low gas pressure and low surface coverage. When realistic physical parameters are ascribed to this hypothetical chemical system, it is found that the DSPG should be a robust, laboratory-testable effect.

Consider a sealed blackbody cavity into which is introduced a small quantity of dimeric gas, A_2. The cavity walls are made from a single chemically active material (S2), except for a small patch of a different material (S1). By definition, in steady state the average numbers of A and A_2 on any surface and in the cavity volume are time invariant, *i.e.*,

$$\frac{dN(i, A_j)}{dt} = 0 \qquad (7.8)$$

where the subscripts i = 1,2,or c stand for surfaces 1 or 2 or the cavity volume; and N is the average number of either species A or A_2. Equation (7.8) can be expanded in terms of the various sources and sinks of A and A_2:

$$\frac{dN(c, A)}{dt} = 0 = [R_{des}(1, A) - R_{ads}(1, A)](SA)_1 + [R_{des}(2, A) - R_{ads}(2, A)](SA)_2 +$$

$$[2R_{diss}(c, A_2) - R_{recomb}(c, A)]V_{cav} \qquad (7.9)$$

$$\frac{dN(c, A_2)}{dt} = 0 = [R_{des}(1, A_2) - R_{ads}(1, A_2)](SA)_1 + [R_{des}(2, A_2) - R_{ads}(2, A_2)](SA)_2$$

$$+ [\frac{1}{2}R_{recomb}(c, A) - R_{diss}(c, A_2)]V_{cav} \qquad (7.10)$$

$$\frac{dN(1, A)}{dt} = 0 = [R_{ads}(1, A) - R_{des}(1, A) + 2R_{diss}(1, A_2) - R_{recomb}(1, A)](SA)_1$$

$$(7.11)$$

$$\frac{dN(1, A_2)}{dt} = 0 = [R_{ads}(1, A_2) - R_{des}(1, A_2) + \frac{1}{2}R_{recomb}(1, A) - R_{diss}(1, A_2)](SA)_1$$

$$(7.12)$$

$$\frac{dN(2, A)}{dt} = 0 = [R_{ads}(2, A) - R_{des}(2, A) + 2R_{diss}(2, A_2) - R_{recomb}(2, A)](SA)_2$$

$$(7.13)$$

$$\frac{dN(2, A_2)}{dt} = 0 = [R_{ads}(2, A_2) - R_{des}(2, A_2) + \frac{1}{2}R_{recomb}(2, A) - R_{diss}(2, A_2)](SA)_2$$

$$(7.14)$$

Here R refers to adsorption, desorption, dissociation, or recombination rates [$m^{-2}s^{-1}$ for surfaces and $m^{-3}s^{-1}$ for volume]; and $(SA)_1$, $(SA)_2$, and V_{cav} are the surface areas of S1 and S2, and the cavity volume, respectively. Relations (7.9-7.14) are generally applicable to gas-surface systems and, in principle, can be simultaneously solved if given adequate thermodynamic information. Number densities $n(i, A_j)$ can vary spatially due to local differences in reaction rates.

The following chemical constraints (a-f) are assumed for the cavity system discussed above. These constraints are commonly assumed in gas-surface studies [5, 6, 7, 8] and are easily shown to be both valid and self-consistent within a broad parameter space. Further discussion of these constraints can be found elsewhere [1].

a) The gas phase density is low such that gas phase collisions are rare compared with gas-surface collisions. (In other words, the mean free path of gas atoms is very long compared with cavity scale lengths; *i.e.*, $\lambda \gg L_{cav}$.) However, the average pressure is much greater than the rms pressure fluctuations; *i.e.*, $P_{cav} \gg \delta P_{rms}$.
b) All species contacting a surface stick and later leave in thermal equilibrium with the surface.
c) The only relevant surface processes are adsorption, desorption, dissociation, and recombination.
d) Fractional surface coverage is low so adsorption and desorption are first order processes.
e) A_2 and A are highly mobile on all surfaces and may be treated as two-dimensional gases.
f) All species spend much more time in the surface phases than in the gas phase. In other words, the characteristic time any species spends on a surface before desorbing (its desorption time, τ_{des}) is much longer than its thermal-velocity transit time across the cavity, τ_{trans}. Also, for S1 the time scales for dissociation of A_2 and recombination of A is short compared with the desorption time. (These allow the surface concentrations of A and A_2 to be in chemical equilibrium.)

For this chemical model, the six general rate relations (7.9-7.14) can be solved simultaneously or they can be recast into the following five equations in the six variables, $n(i, A_j)$, with one variable taken as independent.

$$n(c, A) \simeq \frac{4}{v_A \tau_{des}(2, A)} n(2, A) \qquad (7.15)$$

$$n(c, A_2) \simeq \frac{4}{v_{A_2} \tau_{des}(2, A_2)} n(2, A_2) \qquad (7.16)$$

$$K(1) \simeq \frac{n(1, A_2)}{n^2(1, A)} \qquad (7.17)$$

$$K(2) \simeq \frac{n(2, A_2)}{n^2(2, A)} \qquad (7.18)$$

$$\frac{v_A}{4} n(c, A) + \frac{2v_{A_2}}{4} n(c, A_2) \simeq \frac{1}{\tau_{des}(1, A)} n(1, A) + \frac{2}{\tau_{des}(1, A_2)} n(1, A_2). \qquad (7.19)$$

Here τ_{des} is given by:

$$\tau_{des}(i, A_j) \simeq \frac{1}{\nu_0} F(i, A_j) exp[\frac{\Delta E_{des}(i, A_j)}{kT}] \qquad (7.20)$$

In (7.15-7.19), $n(i, A_j)$ is the surface or volume number density of A_j; v_{A_j} is the thermal speed of A_j (v_{A_j} is taken to be the same for gas and surface phases); and $F(i, A_j) \equiv (\frac{f}{f^*})$ is a ratio of partition functions. f is the partition function for the species in equilibrium with the surface, and f^* is the species-surface partition function in its activated states. For real surface reactions, $\frac{f}{f^*}$ typically ranges between roughly $10^{-3} - 10^4$. Here ΔE_{des} is the desorption energy (experimental values typically range from about 1 kJ/mol for weak physisorption up to about 400 kJ/mole for strong chemisorption).

A two-dimensional gas phase can form on a surface, mediated by gas-surface interactions. Because of the myriad of possible contributing physical effects, no single theory completely describes this state, however, at low surface coverages standard phenomenological models give the surface equilibrium constant $K(i)$ for the reaction $(2 \, A \rightleftharpoons A_2)$ to be:

$$K(i) = \frac{n(i, A_2)}{n^2(i, A)} \simeq \frac{r_A v_A}{v_{vib}} \frac{\gamma_{recomb}(i)}{\gamma_{diss}(i)} exp[\frac{\Delta E_{diss,act}(i)}{kT}] \qquad (7.21)$$

Here $\Delta E_{diss,act}$ is the energy of activation for dissociation of A_2 on surface i (typical values are 1-500kJ/mole); r_A is the atomic radius of A; v_{vib} is the vibrational frequency of the molecule; γ_{diss} is the probability of a molecular vibration leading to dissociation on the surface $(0 \leq \gamma_{diss} \leq 1)$; and γ_{recomb} is the recombination probability for A-A surface collisions $(0 \leq \gamma_{recomb} \leq 1)$. By no means is this relation intended to describe all possible surface equilibria. In general, $K(i)$ is unique for a given gas-surface combination, here in (7.21), via the γ and $\Delta E_{diss,act}$ terms. In theory, the surface equilibrium constant, $K(i)$, can vary as $0 \leq K(i) \leq \infty$; experimentally K is well known to vary for different molecules, surfaces, and temperatures [5, 6, 7, 8].

The meaning of (7.15-7.19) can be inferred from inspection: (7.15) and (7.16) are statements of conservation of A and A_2 within the cavity; (7.17) and (7.18) are statements of chemical equilibrium on S1 and S2; and (7.19) states conservation of total A atoms on S1. With these five equations and with particular system parameters (*e.g.*, those in Tables 7.1, 7.2), one can calculate the steady-state surface and volume species densities for this system. Note that (7.17) and (7.18) describe chemical equilibrium on S1 and S2, but again, gas phase equilibrium is not guaranteed in this model.

In addition to recasting the rate relations, the model constraints (a-f) also place the following four limits on surface and volume densities:

Limit 1: The lower limit of cavity density is that at which statistical pressure fluctuations, δP_{rms}, remain negligible compared with the pressure difference, ΔP. A standard relation between rms pressure fluctuations and the number of particles in a system, N, is given by [9]: $\frac{\delta P_{rms}}{P} \sim \frac{1}{N^{1/3}} \sim [\frac{1}{n(c)L^3}]^{1/3}$, where L is the scale size of the system and P is the average gas pressure. A criterion for rms pressure fluctuations to be negligible is: $\delta P_{rms} \sim \frac{P}{n(c)^{1/3}L_{S1}} \ll \Delta P$, where L_{S1} is the scale size of the small S1 patch.

Limit 2: The upper limit cavity density is that density at which the mean free path, λ, remains long compared with the cavity scale lengths. Roughly, it is: $\lambda \sim \frac{1}{\pi r_A^2 n(c)} \gg L_{cav}$.

Limit 3: The upper limit surface species density, $n(i, A_j)$, is that at which the fractional surface coverage, θ, remains much less than unity

Molecular Weight A_2	40 amu
Atomic Weight A (m_A)	20 amu
Atomic Radius, A (r_A)	5×10^{-10} m
RMS Velocity $A_2(v_{A_2})$	790 m/sec
RMS Velocity A (v_A)	1.1×10^3 m/sec
Cavity A_2 Density$(n(2, A_2))$	2×10^{16} m^{-3}
Cavity Temperature (T_3)	1000 K
Cavity Radius (R)	0.1 m
S1 Patch Scale Length	10^{-3} m
Surface Area Ratio, $\frac{(SA)_2}{(SA)_1}$	10^9
E(A-A)	240 kJ/mole
Surface Lattice Frequency, ν_o	10^{13} Hz
A_2 Vibrational Frequency, ν_{vib}	10^{13} Hz
Monolayer Density	10^{19} m^{-2}

Table 7.1: Thermodynamic and operating parameters for a model DSPG system.

	Surface 1	Surface 2
$\Delta E_{des}(A)$	250	200
$\Delta E_{des}(A_2)$	260	190
$\Delta E_{diss,act}$	0	30
$F(A)$	10^{-2}	10^3
$F(A_2)$	10^3	1
γ_{diss}	10^{-1}	10^{-9}
γ_{recomb}	10^{-6}	10^{-1}

Table 7.2: Thermodynamic surface parameters for model DSPG system. All ΔEs are in kJ/mole.

$(\theta \ll 1)$.

Limit 4: The lower limit surface density $n(1, A)$ is set at that density for which the recombination time of A on S1, $\tau_{recomb}(1)$, remains much less than the desorption times, $\tau_{des}(1, A_j)$.

The critical requirement for the DSPG is this: that in steady state, S1 and S2 desorb distinctly in the same environment simultaneously. This will occur if $\alpha(1) \neq \alpha(2)$. For low surface coverage where desorption is a first order process, the desorption rate ratio, $\frac{R_{des}(i, A_2)}{R_{des}(i, A)} \equiv \alpha(i)$, is given by:

$$\alpha(i) \equiv \frac{R_{des}(i, A_2)}{R_{des}(i, A)} = \frac{n(i, A_2)}{n(i, A)} \frac{F(i, A)}{F(i, A_2)} exp\{\frac{\Delta E_{des}(i, A) - \Delta E_{des}(i, A_2)}{kT}\}. \quad (7.22)$$

The ratio α varies as $0 \leq \alpha \leq \infty$. Experimental signatures of differential α's are abundant [2,10-15]. If $\alpha(1) \neq \alpha(2)$, the cavity gas cannot be in standard gas phase equilibrium since this equilibrium must, by definition, be unique while the cavity gas phase is twained by two distinct $\alpha(i)$.

In principle, the DSPG effect can arise in a sealed blackbody cavity where $\alpha(1) \neq \alpha(2)$, regardless of the relative surface areas of S1 and S2. However, a simple case to analyse is one in which the surface area of S1 is much less than that of S2; that is, $(SA)_1 \ll (SA)_2$. In this case, if the total desorptive fluxes of A_2 and A from S2 each far exceed the total fluxes from S1, then S2 will almost completely determine the surface and volume inventories of A and A_2, regardless of the behavior of S1. (This can be argued cogently from Le Châtelier's Principle.) The conditions that the instantaneous fluxes of A and A_2 from S2 each greatly exceed those from S1 can be written:

$$\frac{R_{des}(2, A_2)}{R_{des}(1, A_2)} \gg \frac{(SA)_1}{(SA)_2} \quad (7.23)$$

and

$$\frac{R_{des}(2, A)}{R_{des}(1, A)} \gg \frac{(SA)_1}{(SA)_2}. \quad (7.24)$$

Effectively, S1 is made an arbitrarily small 'impurity' in the chemical dynamics of the cavity.

Under conditions (7.23) and (7.24), and assuming all species leave all surfaces thermally, the pressure difference between S1 and S2, $(\Delta P = P_1 - P_2)$, can be expressed as:

$$\Delta P = m_A v_A R_{des}(1, A) + m_{A_2} v_{A_2} R_{des}(1, A_2) - m_A v_A R_{des}(2, A) - m_{A_2} v_{A_2} R_{des}(2, A_2) \quad (7.25)$$

or it can be written in terms of the desorption ratios α as:

$$\Delta P = (2 - \sqrt{2}) m_A v_A R_T(A) [\frac{\alpha(2) - \alpha(1)}{(2\alpha(1) + 1)(2\alpha(2) + 1)}] \quad (7.26)$$

where $R_T(A)$ is the total flux density of A onto a surface, $R_T(A) = \frac{1}{4}[n(c, A)v_A + 2n(c, A_2)v_{A_2}]$. Notice from (7.26) that so long as $\alpha(1) \neq \alpha(2)$, then $\Delta P \neq 0$. If

	Surface 1	Surface 2
$n(i, A)[m^{-2}]$	8.8×10^{16}	4×10^{12}
$n(i, A_2)[m^{-2}]$	4.2×10^{9}	3.2×10^{15}
$\theta(i, A)$	8.8×10^{-3}	4×10^{-7}
$\theta(i, A_2)$	4.2×10^{-10}	3.2×10^{-4}
$\tau_{des}(i, A)[s]$	0.012	2.9
$\tau_{des}(i, A_2)[s]$	4000	8.7×10^{-4}
$\tau_{diss}(i)[s]$	10^{-12}	3.7×10^{-3}
$\tau_{recomb}(i)[s]$	10^{-5}	2.3×10^{-6}
$R_{des}(i, A)[m^{-2}s^{-1}]$	7.3×10^{18}	1.4×10^{12}
$R_{des}(i, A_2)[m^{-2}s^{-1}]$	1.1×10^{6}	3.7×10^{18}
$\frac{R_{des}(i,A_2)}{R_{des}(i,A)} \equiv \alpha$	1.4×10^{-13}	2.6×10^{6}

Table 7.3: Summary of derived system parameters for starting parameters in Tables 7.1 and 7.2 for the cavity concentration $n(c, A_2) = 2 \times 10^{16} m^{-3}$ and temperature T = 1000 K.

ΔP persists over a distance scale Δx, the pressure gradient is roughly $\nabla P \sim \frac{\Delta P}{\Delta x}$.

7.4 Numerical Simulations

Owing to the many independent variables specifying it – about two dozen in Tables 7.1 and 7.2 – complete parametric analysis of a realistic DSPG system is intractable. However, it will be shown for one particular DSPG system that: a) with physically realistic parameters, a steady-state pressure difference, $\Delta P \gg \delta P_{rms}$, is obtained; and b) the physical constraints of the model are self-consistent.

Let a cavity (scale length $L_{cav} = 0.1$m) be coupled to an 'infinite' 1000 K heat bath. The surface area of S1 (scale length $L_{S1} = 10^{-3}$m) is 10^{-9} times less than that of S2. (Let the cavity have dendritic structure and let S2 be porous.) Other system parameters are given in Tables 7.1 and 7.2. Derived system parameters are summarized in Table 7.3. In Figure 7.4 are plotted the various equilibrium surface and volume species densities versus volume density $n(c, A_2)$. These are calculated from simultaneous solution of (7.15-7.19), given $n(c, A_2)$ as the independent variable. Simultaneous solution of the more general equilibrium relations, (7.9-7.14), under the approximation of surface chemical equilibrium, render the same results as the simplified equations to within about 10 %.

Several features in connection with this system and with Figure 7.4 are noteworthy:

1) As expected, each $n(i, A_j)$ increases linearly (logarithmically) with increasing $n(c, A_2)$.

2) Species A_2 dominates surface 2 and cavity inventories while A dominates surface 1.

3) Inspection of Figure 7.4 and Table 7.3 indicates, surfaces 1 and 2 display different desorption ratios for all values of $n(c, A_2)$. In particular, at $n(c, A_2) = 2 \times 10^{16}$ m^{-3}, one has $1.4 \times 10^{-13} = \alpha(1) \ll \alpha(2) = 2.6 \times 10^6$.

4) The different desorption ratios occur simultaneously and in steady-state in a single cavity.

5) The volume density interval (bounded by the two up arrows on the abscissa in Figure 7.4), $2 \times 10^{14} \leq n(c, A_2) \leq 2 \times 10^{17}$ m^{-3}, satisfies all the constraints and limits described in the main text and indicates the most viable region of operation for this system. The right limit line in Figure 7.4 is set by the condition that $\lambda \gg L_{cav}$. Here it is taken to be $\lambda = 10L_{cav} \simeq$ 1m. The lower limit line is set by the condition that $\tau_{recomb}(1) \ll \tau_{des}(1, A)$. This puts a lower limit on $n(1, A)$. Here it is taken to be $10n(1, A) = 7.6 \times 10^{14}$m^{-2}. The left limit line is set by the condition that the statistical pressure fluctuations, δP_{rms}, over the scale length of the S1 patch be much less than the pressure difference, ΔP. Here the limit is taken to be $\delta P_{rms} \leq 10\Delta P$, rendering a lower limit density, $n(c, A_2) = 4 \times 10^{11}$m^{-3}. The upper limit line is set by the condition that the surface coverage by any species be much less than one monolayer. Here it is taken to be $\theta = 0.1$, or $n(i, A_j) = 10^{18}$m^{-2}. From these limits, it appears this system should display the DSPG effect over about three orders of magnitude in cavity gas density ($2 \times 10^{14} \leq n(c, A_2) \leq 2 \times 10^{17}$ m^{-3}).

6) The pressure difference, ΔP, should range between $8 \times 10^{-7} \leq \Delta P \leq 8 \times 10^{-4}$ Pa over the viable cavity density range. This pressure is significant in the context of the DSPG; i.e., $\Delta P \gg \delta P_{rms}$.

7) It was verified numerically and analytically that the values of any parameter in Tables 7.1 and 7.2 could be varied – in some cases, up to several orders of magnitude from their table-stated values – and the DSPG effect should persist.

In summary, there appears to be a broad range of physical values over which the DSPG effect is viable. Furthermore, for this representative system the DSPG model is self-consistent; in other words, the physical parameters necessary for the validity of the model constraints are generated by the system itself. Details can be found in Appendix B in [1].

At these gas pressures, gas phase populations have little effect on the total cavity inventories of either species. Analysis indicates gas phase collisions, regardless of their products, cannot shift cavity inventories of either species by more than about one part in 10^6 from those values obtained by entirely neglecting those collisions. Furthermore, any compositional changes caused by gas phase collisions are erased during the long surface residence times of both species. Not surprisingly, given its dominant surface area, the cavity wall (S2) is the principal reservoir for both species. For instance, at the cavity concentration, $n(c, A_2) = 2 \times 10^{16}$m^{-3},

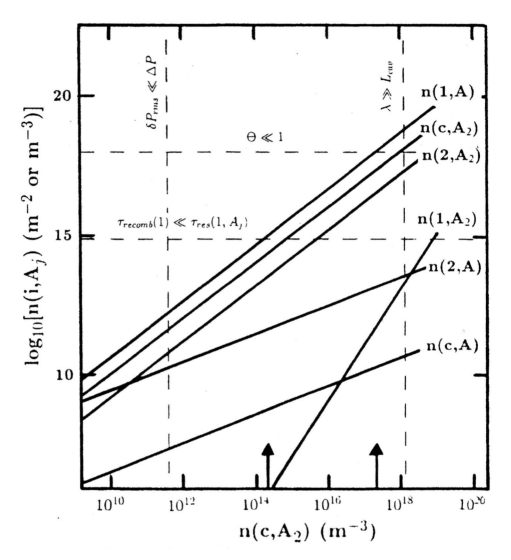

Figure 7.4: Variation of surface and cavity species densities versus cavity density $n(c, A_2)$ for representative system. Model limits are indicated by dotted lines. Up arrows on the abscissa indicate limits for most viable cavity densities of operation.

the combined volume and surface loads of A and A_2 are roughly 4×10^{15} atoms and 3.2×10^{18} molecules. The number fractions of A atoms associated with S1 : S2 : cavity volume are 2.2×10^{-5} : ~ 1 : 1.4×10^{-9}. For A_2 molecules the fractions are 1.3×10^{-15} : ~ 1 : 6.3×10^{-6}. These ratios indicate S2 dominates cavity inventories of both species.

Surface 2 also dominates the fluxes of both species. It was claimed that inequalities (7.23) and (7.24) must be satisfied for S2 effluxes to greatly exceed S1 effluxes. From Tables 7.1 and 7.3, it can be shown that $3.4 \times 10^{12} = \frac{R_{des}(2,A_2)}{R_{des}(1,A_2)} \gg \frac{(SA)_1}{(SA)_2} = 10^{-9}$, and $1.9 \times 10^{-7} = \frac{R_{des}(2,A)}{R_{des}(1,A)} \gg \frac{(SA)_1}{(SA)_2} = 10^{-9}$. Both inequalities are satisfied, so S2 dominates system chemistry.

7.5 Laboratory Experiments

7.5.1 Introduction

In this section, experimental results are presented which corroborate the theoretically predicted, central physical processes of the DSPG. These experiments do not conclusively prove the existence of the DSPG; it can be inferred from the results only with the acceptance of caveats, which will be discussed later. Most of these caveats can be lifted with more sophisticated experiments.

In these experiments, surface dissociation rates were inferred for low-pressure (0 - 90 Torr) hydrogen (H_2) on two high-temperature ($T \leq 2500$ K) surfaces: tungsten (W) and molybdenum (Mo). The primary experimental conclusions bearing on the DSPG were: i) when heated singly (*i.e.*, one surface hot) or when heated simultaneously (*i.e.*, both surfaces hot) to identical temperatures and under identical hydrogen pressures, Mo displayed significantly greater (up to 2.4 times greater) H_2 dissociation rates than W; ii) under low-pressure H_2 atmospheres, gas phase concentrations of H and H_2 over heated Mo and W were not held at gas phase equilibrium but were set by surface-specific reaction rates; and iii) when Mo and W were heated simultaneously in the low-pressure regime and brought into close physical proximity, the net H_2 dissociation rate on both surfaces was reduced – similarly in absolute magnitude for both surfaces, but more in relative magnitude for W than for Mo.

Results (i) and (ii) are not surprising and, in fact, have been studied extensively for these and many other gas-surface combinations — after all, heterogeneous nonequilibrium catalysis undergirds much of the world chemical industry — however, direct comparisons of two metals under identical thermodynamic conditions are less common. Result (iii) especially distinguishes these experiments in that, taken together, results (i-iii) corroborate the central theoretically predicted physical process of the DSPG, namely, that two chemically distinct surfaces can simultaneously and cooperatively maintain different, nonequilibrium, low-density gas phase concentrations over their surfaces [in a closed backbody cavity environment]. From these results, one can infer the existence of a steady-state, nonequilibrium pressure difference (and gradient) between the two surfaces.

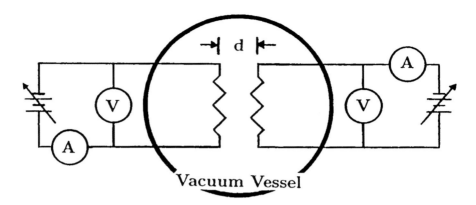

Figure 7.5: Schematic of experimental apparatus. The W and Mo filaments are symbolized by resistors.

7.5.2 Apparatus and Protocol

The experimental apparatus and protocol were straightforward. Centered within a stainless steel cylindrical vacuum vessel (40 cm high, 30 cm diameter)), two slab filaments (W and Mo; 6 cm x 3.3 mm x 0.127 mm) were dc ohmically heated ($300 \leq T \leq 2500$ K) in a constant pressure atmosphere (either vacuum, pure He, or pure H_2; 10^{-6} Torr $\leq \mathcal{P} \leq 90$ Torr), with their broad surfaces facing one another (See Figure 7.5). Their physical separation (d) could be varied over more than a factor of 20 (1.5 mm $\leq d \leq 35$ mm). Electrical power was supplied a filament to achieve a fixed temperature against the several possible energy loss channels (e.g., convection, conduction, radiation, H_2 dissociation). Temperature was measured with an optical pyrometer, was truthed by color-matching the filaments against each other, and also by comparison of power dissipation by the filaments under vacuum and He. The pyrometer was calibrated against the melting points of six pure metallic filaments (W, Ta, Mo, Hf, Zr, Ni). Temperature-dependent spectral emissivities were used (ϵ_{Mo}(2000K - 2250 K) = 0.37-0.36; ϵ_{W}(2000K - 2250K) = 0.43). Gas pressure was measured using calibrated capacitance manometer gauges. At the experimental temperatures, the vapor pressures of Mo and W were low and did not appreciably affect the results.

For typical experimental temperatures and pressures (e.g., T = 2200 K, $\mathcal{P}(H_2)$ = 2.5 Torr), the fraction of power dissipated in the various loss channels was roughly as follows: radiation ($\sim$ 0.40); H_2 dissociation ($\sim$ 0.45); conduction ($\sim$ 0.05); simple convection ($\sim$ 0.05); enhanced convection due to excitation of H_2 rotational/vibration modes ($\leq$ 0.05). These values, of course, depended on the filament temperature, gas type and pressure, and on filament separation. The latter could lead to radiative and convective co-heating of the filaments, as well as co-heating due to surface recombination of H atoms generated by the opposite filament. (The dissociation reaction, $H_2 \longrightarrow 2H$, is highly endothermic; the H-H bond energy is $E_b = 431$ kJ/mole.)

The power consumed by a filament dissociating hydrogen, P_{hd}, as well as other loss channels, could be inferred by running the filament at a fixed temperature first in vacuum, then in He, and lastly in H_2, the latter two at identical pressures. This protocol followed roughly that of Jansen, et al. [11] in their investigation of H_2 dissociation on rhenium; the present results agree well qualitatively and quantitatively with theirs.

The filament power consumption in vacuum, P_{vac}, was due strictly to radiation ($P_{rad} = \epsilon \sigma T^4 A_{filament}$) and to heat conduction from the hot filament to the cool copper electrodes holding it: $P_{vac} = P_{rad} + P_{cond}$. Since the Mo and W filaments had identical dimensions and comparable spectral emissivities and coefficients of thermal conductivity, not unexpectedly, their radiative and conductive losses were similar. When the filaments were run in He another loss channel, simple convection ($P_{conv,s}$), was added. The convective power consumption was inferred by subtracting P_{vac} from the total power consumption in He; that is, $P_{conv,s} = P_{He} - P_{vac}$. Helium was used since its thermal conductivity at low temperatures ($T \leq 1700$ K) is within about 15% of that of H_2 and, therefore, can be used to estimate the simple convective losses due to H_2. Since the convective losses are relatively small in either He or H_2, this fractional difference amounts to only about 1% of the total power consumption of a filament in H_2.

Lastly, from the power loss in a H_2 atmosphere (P_{H_2}), the power consumption due to hydrogen dissociation, P_{hd}, could be estimated from $P_{hd} \simeq P_{H_2} - P_{He}$. Properly included in P_{H_2} for higher temperatures ($T \geq 1700$ K) are additional convection terms, $P_{conv,a}$, associated with excitation of rotational/vibrational modes of H_2 and the additional heat capacity arising from the generation of 2H from a single H_2. Estimates indicate that these contribute 10% or less to P_{hd}. The relative smallness of $P_{conv,a}$ can be estimated by considering the ratio of H-H bond energy ($E_b = 4.47$ eV) to the equipartitioned thermal energy ($\sim$ kT ~ 0.20 eV, for T = 2250 K). This ratio is $\frac{E_b}{kT} = \frac{4.47}{0.20} \simeq 22$. Also, presumably, the vibrational/rotational excitation of H_2 should be nearly the same on both W and Mo, so these power losses should not mask the power losses due to the true chemical differences between the surfaces. Plasma production was not evident in these experiments, nor was it expected since the ionization energy of hydrogen atoms (13.6 eV) and molecules (15.4 eV) far exceeded the thermal energy ($kT \sim 0.2$ eV). Photodissociation of H_2 at the experimental temperatures was negligible.

A high degree of physical symmetry was built into the apparatus in order to assure that conductive, convective, and radiative losses were as similar as possible for both filaments. The filaments, manufactured to identical physical dimensions, were placed symmetrically facing each other at the center of the cylindrical vacuum vessel, oriented parallel to the plane of vessel's own plane of bilateral symmetry (this with respect to its horizontal and vertical ports). The copper electrodes holding the filaments were identical; their power and diagnostic wire leads were symmetrical in design and placement. The power supplies, ammeters, and voltmeters for both filaments were identical models and were calibrated against each other and against a third, fiduciary instrument. Experimental gases were not passed through the chamber on a continuous basis during experiments since dynamic pressure control is not wholly reliable and since forced gas flow could lead to asymmetric convective instabilities; rather, a single gas load was used for a related set of measurements. The experimental measurements to follow were repeatable to within about 10%.

7.5.3 Results and Interpretation

In Figure 7.6 is plotted P_{hd} versus temperature for Mo and W filaments heated individually at 2.5 Torr pressure H_2. (The case where the filaments were heated simultaneously displayed similar results.) The consistently greater power consumed by the Mo establishes that the surface dissociation rate on Mo is greater than on W. It also indicates that the gas phase concentrations of H and H_2 over Mo and W cannot both be at gas phase equilibrium values, since, by the general requirements of equilibrium thermodynamics, the gas phase equilibrium concentrations are unique at a given temperature and pressure. In fact, these experiments show that nonequilibrium gas phase concentrations are maintained over both surfaces. (One would not expect gas phase equilibrium to be established in the low pressure regime ($\mathcal{P}(H_2) \leq 10$ Torr) since the mean free path for the three-body collisions (required for hydrogen recombination reaction $2H \longrightarrow H_2$) is many times longer than the vessel scale length. Rather, gas-surface reactions, either on the filaments or at the vessel wall, should primarily determine the gas phase concentrations.)

In Figure 7.7 is plotted the difference in P_{hd} between Mo and W per Torr H_2 ($i.e.$, $\frac{\Delta P_{hd}}{\mathcal{P}(H_2)} \equiv \frac{P_{hd}(Mo) - P_{hd}(W)}{\mathcal{P}(H_2)}$) versus H_2 pressure, $\mathcal{P}(H_2)$. Filaments were heated individually at T = 2200 K with 33 mm separation. (Similar experiments with the filaments heated simultaneously gave similar results.) At low pressures ($\mathcal{P}(H_2) \leq 10$ Torr), where the mean free path for the three-body, H-atom recombinative collisions is long, one expects surface-specific reaction rates to dominate such that $\frac{\Delta P_{hd}}{\mathcal{P}(H_2)}$ should be relatively large. The data show this. The non-zero values of $\frac{\Delta P_{hd}}{\mathcal{P}(H_2)}$ at low pressure are indications of both nonequilibrium gas phase concentrations of H and H_2 and also of a larger surface dissociation rate for H_2 on Mo than on W. At high pressures, on the other hand, the three-body recombinative mean free paths are relatively short, gas phase reactions begin to dominate, and gas phase equilibrium is approached over both surfaces. As a result, the chemical distinction between the two surfaces vanishes and $\frac{\Delta P_{hd}}{\mathcal{P}(H_2)}$ decreases to zero, also indicated by Figure 7.7.

In Figure 7.8 is plotted P_{hd} versus filament separation distance d (See Figure 7.5). Mo and W were heated simultaneously to T=2200K at $\mathcal{P}(H_2) = 2.5$ Torr. Also plotted, for reference, are the P_{hd} values for the filaments run singly (dotted lines). At large separations, P_{hd} values for the filaments run simultaneously approach the values of P_{hd} for the filaments run singly; this is not surprising. As d is reduced, however, P_{hd} decreases for each filament. The absolute value of the decrease is roughly the same for each filament at all separations, but the relative change is greater for W than for Mo. Also, the power consumption for the fila-

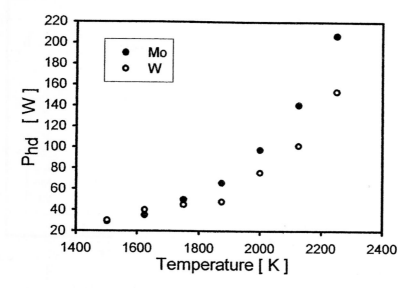

Figure 7.6: Hydrogen dissociation power, P_{hd}, versus filament temperature for W and Mo filaments. ($\mathcal{P}(H_2) = 2.5$ Torr; $d = 6.5$ mm; filaments heated individually)

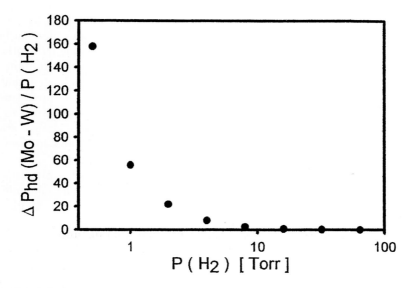

Figure 7.7: Difference in P_{hd} per torr versus hydrogen pressure for W and Mo.($T = 2200$ K; $d = 33$ mm; filaments heated singly)

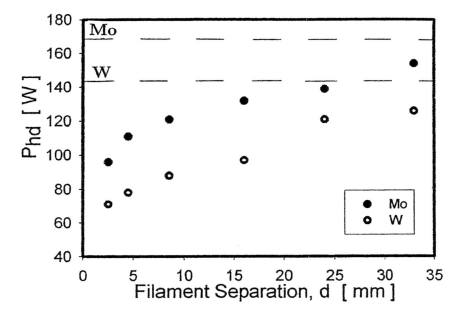

Figure 7.8: Hydrogen dissociation power, P_{hd}, versus filament separation, d, for W and Mo. ($T = 2200$ K; $\mathcal{P}(H_2) = 2.5$ Torr; filaments heated simultaneously). Dotted lines indicate P_{hd} for filaments heated singly.

ments remain distinct in the $d \longrightarrow 0$ limit. The most reasonable explanation for the decrease in P_{hd} with decreased filament separation is that each filament begins to recombine H atoms desorbing from the other filament. The highly exothermic recombination reaction reduces the electrical power necessary to maintain a filament at 2200 K. Additional experiments at other pressures indicated that this proximity effect is most pronounced at low pressure and at high temperature.

This behavior of Mo and W in the hydrogen atmosphere is what one expects of a DSPG system, namely, a stationary nonequilibrium thermodynamic state in which each surface dissociates H_2 at a different rate and maintains its own local and distinct gas phase concentrations at each separation distance.

Under the assumptions that the influx of gas to each surface is the same and that species leave in thermal equilibrium with the surfaces, one can infer the difference in momentum flux densities $\mathcal{F}_p$ (i.e., pressure) between the two filaments in terms of known and experimentally measured quantities:

$$\Delta \mathcal{F}_p \simeq \frac{[P_{hd}(Mo) - P_{hd}(W)]m_A v_A}{E_b A_{filament}} \qquad (7.27)$$

For typical experimental values (e.g., T = 2200K, $\mathcal{F}(H_2) = 2.5$ Torr), one infers a steady-state pressure difference between the surfaces of $\Delta \mathcal{F}_p \simeq 1$ Pa. This is small compared with the background pressure in the vessel ($\frac{\Delta \mathcal{F}_p}{\mathcal{P}(H_2)} \simeq 0.002$), but it is significant in the context of the DSPG. It is two orders of magnitude greater than those predicted for statistical rms pressure fluctuations and is several orders of magnitude greater than that predicted in the reaction rate derivation of the DSPG, but it is consistent with the detailed balance derivation. For comparison, it is also several orders of magnitude greater than pressure fluctuations associated with audible sound.

7.6 Discussion and Outlook

In this chapter, theoretical and experimental developments of the DSPG are presented. The outlook for confirmation of this effect, and its attendant second law challenge, are promising.

The low-pressure, low-surface coverage regime (§7.3), although admitting laboratory tests, probably underestimates the physical limits of the DSPG effect. The first derivation (§7.2), based on detailed balance is more general and permits the DSPG at much higher pressures and surface coverages. With respect to the rate equation approach, many potentially interesting surface effects can be considered, for example, higher order surface reaction rates, multidimensional molecule-surface potential energy surfaces, polyatomic molecular reactions, surface loading effects, tunneling, incorporation, absorption, surface defects, edge effects, side chemical reactions, activation energies of desorption, precursor states, and potential energies of mobility. These details are not required for the detailed balance approach since it deals with only total particle fluxes.

The range in gas number density over which the DSPG is predicted by detailed balance should extend over several orders of magnitude more than the reaction rate model. In the latter, the DSPG is limited to an upper limit density of about 10^{17}m^{-3} ($10^{-3}\text{Pa} \sim 10^{-8}\text{atm}$), while in the former it is limited only by the constraint that gas phase equilibrium is not established via gas phase reactions. For example, given the long mean free paths for three-body collisions and the relaxation of the low-θ constraint, the pressure range for the DSPG might be extended up to $10^{23} - 10^{24}\text{m}^{-3}$ ($10^2 - 10^3\text{Pa} \sim 10^{-3} - 10^{-2}\text{atm}$); that is five to six orders of magnitude greater than predicted in §7.2-7.4. This higher pressure range eases concerns about statistical pressure fluctuations being in competition with the DSPG, and also makes laboratory searches more inviting.

Many laboratory experiments over the last 100 years corroborate the key physical processes of the DSPG , namely, that different surfaces can display different dissociation rates for gases at low pressure, when reaction rates are surface limited [5-8,10-15]. Laboratory searches for the DSPG should be possible. The broadest base of technical knowledge for molecular-surface interactions exists for light diatomic molecules (*e.g.*, H_2, N_2, O_2, CO, Cl_2) with transition metals (*e.g.*, Fe, Ni, Pt, Cu, Pd, Au, Ag) and semiconductors (*e.g.*, Si, Ge, GaAs). These reactions are usually most vigorous at high temperatures ($T > 1000\text{K}$), but, in principle, these effects should also be possible at low temperatures. Surface desorption and dissociation energies can be less than 0.1 kJ/mole for van der Waals interactions, so these effects might occur at or below room temperature, perhaps even below 100K for weakly bound van der Waals molecules such as Ar_2 or He_2. Experiments with noble gases on metals indicate that some species (*e.g.*, He_2) might, in principle, display the DSPG effect down to within a few degrees of absolute zero [16-20]. An experimental signature of this should be a variation in the second virial coefficient for a van der Waals gas depending on the composition or stucture of the confining surface.

The experimental results presented (§7.5) strongly corroborate key behaviors of the DSPG, but they do not conclusively prove its existence and their corroboration requires several caveats. First, it is emphasized that direct measurements of pressure were not made in these experiments; the pressure difference $\Delta\mathcal{F}_p$, was inferred under the assumptions that the gaseous influxes to Mo and W were the same and that effluxes left both surfaces at thermal equilibrium with them. These are reasonable assumptions, but if, for example, Boltzmannian population statistics for vibrational/rotational/electronic states were severely violated for gas collisions on one or both surfaces, then the inferred ΔP_{hd} apparent in Figure 7.8 (and from which $\Delta\mathcal{F}_p$ is inferred) could be largely erased – but this in itself would be quite interesting thermodynamically. Second, in these experiments, the surfaces were heated ohmically, whereas the DSPG calls for them to be heated under blackbody cavity conditions, that is, by blackbody radiation, convection, and conduction, rather than by the passage of electrical current. It is not known if the chemical activity of the Mo and W surfaces were significantly altered by current flow.

Finally, and most importantly, these experiments were not conducted under sealed, high-temperature blackbody cavity conditions. The region between the heated filaments, at best, only approximated a blackbody cavity, while the actual cavity walls (vacuum vessel) were significantly cooler than the filaments (*e.g.*, 330 K versus $\sim$ 2200 K). The species concentrations of the flux from the walls to the filaments was certainly different from the flux from the filaments to the walls; in particular, it was probably enriched in H_2. Despite this, there are good reasons to believe that the nonequilibrium gas phase concentrations in the vicinity of the filaments, especially when their separation d was small, may have fairly well approximated that which would have arisen inside a true high-temperature blackbody cavity. As mentioned above, gas was not passed through the vacuum vessel continuously, but rather, a single load of gas was used in each experiment. This allowed a steady state gas phase concentration to develop in the entire cavity within a few seconds. At the experimental pressures, the mean free path for two-body collisions was much shorter than the scale length of the filaments (10^{-5} m $\sim \lambda \ll l_{filament} \sim 10^{-3}$ m), so the gas could be taken to be diffusive, in which case, one would expect that both surfaces of both filaments would interact strongly with each others' surface effluxes. And, when the filaments were closely spaced ($d \sim 2$ mm $< l_{filament}$), the gas phase concentrations of H and H_2 in this gas diffusive environment were likely to be similar between the front and back of each filament. Therefore, if the distinct P_{hd} values for Mo and W indicated in Figure 7.8 are maintained in the $d \longrightarrow 0$ limit, then one could reasonably expect that that they would remain distinct under sealed, high-temperature blackbody cavity conditions. This behavior would amount to the DSPG.

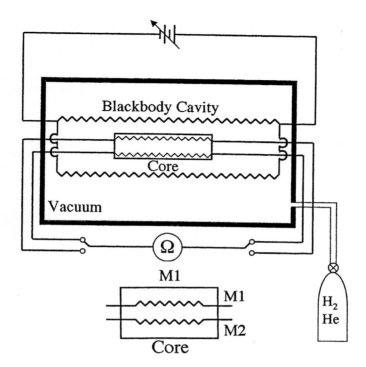

Figure 7.9: Proposed experimental design for improved DSPG experiments.

To lift the above caveats, one might construct Duncan's turbine or idealize it with a torsion balance. In fact, Sheehan conducted torsion balance experiments with a tungsten/molybdenum torsion vane similar to that in Figure 7.1, suspended by a fine tungsten torsion fiber in a high-temperature blackbody cavity. Alternatingly infusing the cavity with helium and hydrogen (as for the previously discussed filament experiments), vane deflections consistent in direction and magnitude with the DSPG effect were observed, but these were confounded by pressure fluctuations and by the slow, hysteretic thermal expansion of material parts, such that a reliable conclusion could not be reported. Given suitable time and resources, however, this experiment could probably be successfully staged.

At the time of this writing, more sophisticated experiments are in progress (Figure 7.9). These will measure the temperature-dependent electrical resistivities of thin filaments of W and Mo inside sealed high-temperature W and Mo core cavities filled with hydrogen. In Figure 7.9, the core will be heated radiatively by the outer, ohmically-heated cavity. High-precision resistance measurements will allow discernment of small absolute temperature variations (on the order of 1 K or less) between W and Mo filaments due to differential hydrogen dissociation/recombination rates. This experimental design should lift most of the caveats associated with the previously described experiments. If steady-state temperature differences between the filaments are observed, this would imply the DSPG effect, but even more directly this temperature difference would imply the possibility of a heat engine operated across the temperature gradient between the W and Mo filaments, but powered ultimately by the single external heat bath. A number of resolutions have been forwarded for this chemical paradox. These are compiled and discussed elsewhere [3].

In summary, the DSPG effect, and by extension a second law challenge, appears to be a robust chemical phenomenon that is open to direct experimental test. Corroborative experiments support it, but more definitive tests are needed.

References

[1] Sheehan, D.P., Phys. Rev. E **57** 6660 (1998); original theory unpublished (1992).

[2] Sheehan, D.P., Phys. Lett. A **280** 185 (2001).

[3] Duncan, T., Phys. Rev. E **61** 4661 (2000); Sheehan, D.P., Phys. Rev. E **61** 4662 (2000).

[4] Pauling, L. *General Chemistry*, 3rd Ed., (W.H. Freeman, San Francisco, 1970); Sheehan, W.F. *Physical Chemistry*, 2nd Ed., (Allyn and Bacon, Boston, 1970).

[5] Hudson, J.P. *Surface Science, An Introduction*, (Butterworth-Heinemann, Boston, 1992).

[6] Tompkins, F.C. *Chemisorption of Gases on Metals*, (Academic Press, London, 1978).

[7] Masel, R.I. *Principles of Adsorption and Reaction on Solid Surfaces*, (John Wiley, New York, 1996) pg. 116.

[8] Raval, R., Harrison, M.A., and King, D.A., in *The Chemical Physics of Solid Surfaces and Heterogeneous Catalysis*, Vol. 3, King, D.A. and Woodruff, D.P., Editors, (Elsevier, Amsterdam, 1990) pg. 39.

[9] Andrews, F.C. *Equilibrium Statistical Mechanics*, (John Wiley and Sons, New York, 1962) pg. 177.

[10] Otsuka, T., Ihara, M., and Komiyama, H., J. Appl. Phys. **77** 893 (1995).

[11] Jansen, F., Chen, I., and Machonkin, M.A., J. Appl. Phys. **66** 5749 (1989).

[12] Eenshuistra, P.J., Bonnie, J.H.M., Los, J., and Hopman, H.J., Phys. Rev. Lett. **60** 341 (1988).

[13] Wise, H. and Wood, B.J., in *Advances in Atomic and Molecular Physics*, Vol. 3, (Academic Press, New York, 1967) pg. 291.

[14] Madix, R.J. and Roberts, J.T., in *Surface Reactions*, Springer Series in Surface Sciences, Vol. 34, Madix, R.J., Editor, (Springer-Verlag, Berlin, 1994).

[15] Langmuir, I. and Kingdom, K.H., Proc. Roy. Soc. A **107** (1925).

[16] Luo, F., McBane, G.C., Kim, G., and Giese, C.F., J. Chem. Phys. **98** 3564 (1993).

[17] Hobza, P. and Zahradnik, R., Chem. Rev. **88** 871 (1988).

[18] Blaney, B.L. and Ewing, G.E., Ann. Rev. Chem. **27** 553 (1976).

[19] Sauer, J., Ugliengo, P., Garrone, E., and Saunders, V.R., Chem. Rev. **94** 2095 (1994).

[20] Chizmeshya, A. and Zaremba, E., Surf. Sci. **268** 432 (1992).

8

Plasma Paradoxes

Second law paradoxes are proposed for two low-temperature ($T \sim 2000$K) plasma systems. Each has theoretical and experimental corroboration.

8.1 Introduction

This chapter is concerned with two challenges based in plasma physics, designated *Plasma I* and *Plasma II* [1, 2, 3, 4]. They arise from: i) the disparate physical behaviors of ions (high mass, positive charge, slow time response) and electrons (low mass, negative charge, fast time response), which allow steady-state macroscopic electrostatic sheaths to form at boundaries; and ii) the manner in which certain low-temperature plasmas are produced — via surface ionization. Although distinct in themselves, these plasma challenges serve as central links between the other USD challenges (§10.1.1). The gravitational and solid state challenges, for example, were inspired by Plasma I, and the chemical challenge can be seen to be an analog of Plasma II.

Both plasmas systems are theoretically motivated and both have laboratory experiments that corroborate their central physical processes. (The classical thermodynamic explanation of Plasma I [1, 3] is buttressed by a fuller quantum mechanical interpretation [4].) They reside at the high-temperature, low particle density limits of currently known second law challenges.

This chapter will be divided into two parts, covering the theoretical and ex-

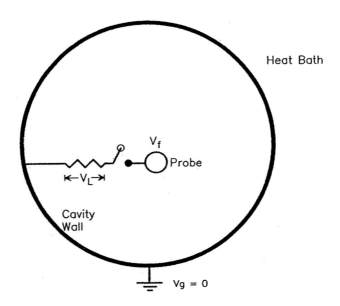

Figure 8.1: Schematic of *Plasma I* system.

perimental support for each system. Resolutions to the paradoxes are reviewed.

8.2 Plasma I System

8.2.1 Theory [1, 3, 4]

Consider an electrically conducting probe suspended in a high-temperature, black-body cavity housing a low-density plasma (Figure 8.1). The cavity walls are electrically and thermally grounded to the heat bath and the probe is connected to the cavity walls through a load. The load can be conservative (*e.g.*, a motor) or dissipative (*e.g.*, a resistor). The probe and load are small enough to represent minor perturbations to the cavity properties.

Since the probe can electrically float with respect to the walls, if the voltage between the probe and walls is non-zero, this voltage can drive a current through the load, doing steady-state work. This can be brought into sharper relief by placing a switch between the probe and the load (Figure 8.1). In this case, when the load is electrically shorted to ground, the probe is physically disconnected from the walls (ground) and will electrically charge as a capacitor to the plasma floating potential. When the switch is reversed, the probe will discharge as a capacitor through the load and plasma. With an ideal switch, this charging and discharging of the probe through the load can be repeated indefinitely, performing work in apparent violation of the second law.

In this discussion, several electrostatic potentials will be introduced; with reference to Figure 1, they are:

$V_g \equiv$ reference potential of cavity walls and heat bath (ground), both taken to be zero ($V_g = 0$).
$V_{pl} \equiv$ potential difference between the bulk plasma and ground.
$V_f \equiv$ potential difference between ground and the probe when it is electrically *floating* in the plasma; *i.e.*, when the probe draws no net current from the plasma.
$V_L \equiv$ potential difference across the load.
$V_{pr} \equiv$ potential difference between the probe and the walls.

This system can be treated through the formalism of Q-machines and self-emissive Langmuir probes [5, 6]. The cavity walls are at ground potential, here taken to be zero. The plasma potential (V_{pl}, the potential between the bulk plasma and the cavity walls (ground)) will be positive or negative depending on the work function and temperature of the walls, and the plasma type and concentration. In the absence of any net current to the plasma or walls, the plasma potential is calculated by equating the current leaving the walls to the current received by the walls. If thermionic electron emission from the walls dominates over plasma production, one has an *electron-rich* plasma in the cavity, in which case one may estimate the plasma potential by equating the Richardson emission from the walls to random electron flow from the plasma into the walls:

$$AT^2 \exp(-\frac{e\Phi}{kT}) \exp(\frac{eV_{pl}}{kT}) = \frac{nev_e}{4}. \qquad (8.1)$$

Here Φ is the wall material's work function, T is temperature, V_{pl} is the plasma potential, v_e is the average electron thermal speed ($v_e = \sqrt{\frac{8kT_e}{\pi m_e}}$ for a Maxwellian distribution), k is the Boltzmann constant, m_e is the electron mass, n is the plasma particle density, and A is the Richardson constant for the material with a value of about $6 - 12 \times 10^5 \frac{A}{m^2 K^2}$ for pure metals. Here V_{pl} is taken to be negative for electron-rich plasmas, however V_{pl} may be positive if the electron current from the plasma to the walls exceeds the Richardson emission from the walls to the plasma; in this case one has an *ion-rich* plasma. This occurs above a *critical* plasma density, n_c:

$$n_c = \frac{4AT^2}{ev_e} \exp(-\frac{e\Phi}{kT}), \qquad (8.2)$$

as can be seen from (8.1). For tantalum ($\Phi_{Ta} = 4.2$ V) at T = 1500 K, the critical density is $n_c \simeq 2 \times 10^{12} m^{-3}$. The plasma potential can be negative (positive) for an electron (ion)- rich plasma, and zero only for very specific cavity conditions.

A probe placed in this plasma will achieve a potential with respect to the plasma and walls depending upon the probe temperature, resistance to ground (load resistance, R_L), and the nature of the current flux to it. The potential difference between the probe and the walls is given by the intersection of the load line

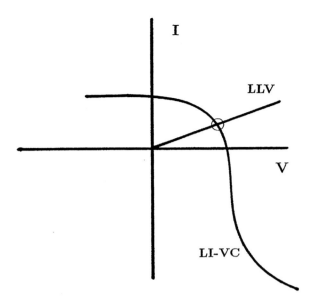

Figure 8.2: Probe load line voltage (LLV) and Langmuir current-voltage char-acteristic (LI-VC). The intersection of the two curves is the probe-wall voltage, V_{PW}.

voltage, $V_L = I_L R_L$, with the probe's Langmuir current-voltage chararacteristic curve, as shown in Figure 8.2. Here V_L and I_L are the load voltage and current, respectively. The probe for this system will be nearly in thermal equilibrium with the walls, therefore, is expected to be self-emissive. As detailed in Hershkowitz et al. [6], self-emissive probes should electrically float near the plasma potential, therefore, so long as its resistance to ground is large, the probe should reside near the plasma potential, V_{pl}, and V_{pl} should be roughly equal to V_L. If the load resistance is zero, the probe is shorted and will reside at ground potential – an uninteresting case. The power consumed by the load may be expressed as $P_L = \frac{V_L^2}{R_L} = I_L^2 R_L$. Ideally, V_L should be the potential difference between the probe and walls, which, for an emissive probe and large R_L, should be roughly the plasma potential, V_{pl}, as given in (8.1) for electron-rich plasmas. The maximum current through the load should be roughly the current intercepted thermally by the probe; that is, $I_L(max) \simeq \frac{n e v_e}{4}(SA)_p$, where $(SA)_p$ is the probe surface area. (See §8.4 for additional discussion on this point.) The rate of entropy production by the load is $\frac{dS_L}{dt} = \frac{P_L}{T} = \frac{V_L^2}{R_L T}$; this will be positive (negative) for a purely dissi-pative (conservative) load. Note that the paradox is not restricted to systems with thermionically emitting walls; any plasma system with a non-zero plasma poten-tial appears viable. Note that the load cannot supply unlimited power by reducing R_L since the probe cannot supply unlimited current to the load – maximally only that which it receives from the plasma – so the load power has an upper limit: $P_L \sim I_L^2 R_L$.

That this probe electrically floats with respect to the walls, that it can be almost unperturbing to the plasma, and that a steady-state current can flow through the load performing work should not be surprising; these behaviors have been observed in many plasma experiments. However, if one admits that this system does work on the load while maintaining a time-averaged steady-state (not necessarily equilibrium) temperature and species concentration profiles, and if one assumes the first law of thermodynamics is satisfied, then a paradox involving the second law naturally develops.

The crux of this challenge is deduced by assuming the second law is upheld by the paradoxical system and then identifying the non-trivial necessary condition(s) that make(s) it so. Specifically, as the paradox is posed, the second law is violated iff the load consumes power; therefore, for the second law to hold, the load does not consume power; *i.e.*, $P_L = 0$. The load's electrical power consumption is: $P_L = \frac{V_L^2}{R_L}$. But $R_L < \infty$, therefore, $V_L = 0$. Thus, the crux of the paradox lies with the potential drop across the load: If $V_L = 0$, there is no paradox; if $V_L \neq 0$, the second law is violable. No other non-trivial, necessary conditions are apparent.

As in Figure 8.1, it is assumed the probe is immersed in an blackbody cavity plasma $(T_e \simeq T_i)$. The general requirement is deduced in the following five points.

    i) $V_L = V_{pr}$.
    ii) The plasma potential satisfies the inequality: $V_f \leq V_{pl}$.
    iii) V_{pr} satisfies the inequality $\mid V_f \mid \geq \mid V_{pr} \mid \geq V_g = 0$, depending on
    the magnitude of R_L. As $R_L \to \infty$, $V_{pr} \to V_f$; as $R_L \to 0$, $V_{pr} \to 0$. If
    $R_L \neq 0$, then $V_{pr} \neq 0$. In particular, if $V_f < 0$, then $V_f \leq V_{pr} \leq 0$.
    iv) If $V_{pl} < V_g = 0$, then by (i-iii) above, V_L satisfies the inequality
    $V_L \leq 0$, and if $R_L \neq 0$, then the strict inequality holds: $V_L < 0$. Hence,
    $V_L \neq 0$.
    v) For any value of V_{pl}, then by (i, iii) above, V_L satisfies the inequality
    $\mid V_f \mid \geq \mid V_L \mid \geq 0$, therefore, if $V_f \neq 0$ and $R_L \neq 0$, then $V_L \neq 0$.

Point (v) establishes the general requirement for second law violation in terms of the floating potential; it is: $V_f \neq 0$. Point (iv) sets a more restricted requirement in terms of the plasma potential; it is: $V_{pl} < 0$. (Other values of V_{pl} may also be viable, but these must be checked on a case by case basis.)

Point (i) can be verified by inspection of Figure 8.1. Points (ii) and (iii) require discussion of V_{pl} and V_f. The floating potential, V_f, is defined as that potential at which the net current is zero to the probe when it electrically floats in the plasma. A probe will achieve a potential with respect to the plasma and walls depending upon the probe temperature, resistance to ground (R_L), and the nature of the current flux to it. The potential difference between the probe and walls, V_{pr}, is given by the intersection of the load line voltage, $V_{pr} = V_L = I_L R_L$, with its Langmuir current-voltage characteristic curve, as shown in Figure 8.2. The slope of the load line voltage curve is $\frac{1}{R_L}$, so as $R_L \to \infty$, the intersection occurs at $V_{pr} = V_f$, and as $R_L \to 0$, the intersection is at $V_{pr} = 0$; this is the case where the probe is shorted to ground. Thus, for $0 \leq R_L \leq \infty$, one has $0 \leq \mid V_{pr} \mid \leq \mid V_f \mid$, and in particular, if $V_f < 0$, then $V_f \leq V_{pr} \leq 0$. Point (iii) is established.

In an isothermal, Maxwellian plasma the electron thermal speed is much greater than the ion thermal speed ($v_e = \sqrt{\frac{m_i}{m_e}} v_i \gg v_i$), so a probe at V_{pl} will collect more electron than ion current. (Here $v_{e,(i)}$ is the average electron (ion) thermal speed and $m_{e,(i)}$ is the electron (ion) mass.) However, if the probe is allowed to float to V_f, it will charge negatively, so as to reflect electron flux, in order to achieve the required condition of zero net current; in other words, $V_f \leq V_{pl}$. (To good approximation, it can be shown that, for isothermal plasma systems such as this, the relationship between V_f and V_{pl} is given by: $V_f - V_{pl} \simeq \frac{kT}{2e}[\ln(2\pi \frac{m_e}{m_i}) - 1]$. This quantity is negative definite. Here T is temperature, k is the Boltzmann constant.) When the probe is self-emissive — as it may be here — then $V_f \simeq V_{pl}$. Point (ii) is established.

The condition for Point (iv) – that $V_{pl} < 0$ – is met ostensibly by many plasma systems. (For steady-state systems such as this, one may often estimate the V_{pl} by equating the current leaving the walls to the current received by the walls.) For example, in the *electron-rich* plasma in the original paradox, V_{pl} is inherently negative. With $V_{pl} < 0$, this model system meets the restricted requirement for second law violation ($V_{pl} < 0$). And, since $V_f \leq V_{pl} < 0$, the general requirement ($V_f \neq 0$) is also met.

Laboratory experiments have corroborated most salient aspects of the paradoxical system. Countless Langmuir probe and emissive probe measurements support $V_f, V_{pl}, V_{pr} \neq 0$ and $V_f \leq V_{pl}$ for most plasmas, many of these at or near local thermal equilibrium and many generated by purely thermal processes. Notable among these are double-ended Q-machine plasmas [7, 5]. These are created by surface ionization of alkali metal (*e.g.*, K, Rb, Cs) or alkali earth metals (*e.g.*, Sr, Ba) on heated ($T \geq 2000K$) high work function refractory metals (*e.g.*, Ta, W, Re). These plasmas are generated thermally and are effectively in equilibrium with their walls. In the experiments to follow, it will be shown that, using purely thermal plasma generation, a non-magnetized plasma, in near thermal equilibrium blackbody cavity environments can support non-zero, steady-state probe-wall potentials occurred over a wide range of temperatures ($T_{cav} \leq 2060$ K) regardless of the composition of the cavity walls, probes, electrical loads or leads; and it is shown that steady-state voltage V_L can be supported across a load — the crux condition for second law violation. Of course, these experiments themselves did not violate the second law because the entropy produced by the dissipative load were minute compared with the entropy produced in conducting the experiments. However, were the load conservative, were the universe at high temperature (~ 2000 K), were no vacuum, ancillary apparatus, or experimenter required, then it appears that the entropy production rate for the experiments should be negative.

8.2.2 Experiment [1]

8.2.2.1 Apparatus and Protocol

Experiments were conducted approximating the paradoxical system; a schematic of the apparatus is given in Figure 8.3. The blackbody cavity and heat bath were approximated by a heated, hollow 5 cm diameter, 5 cm high molybdenum cylinder, with wall thickness 0.64 cm. The diameter and height of the cylinder interior were 3.8 cm and 4.7 cm, respectively. The cavity was heated by a continuous, alumina-insulated tantalum filament interstrung 16 times through the cylinder walls with uniform angular spacings, radially half-way between the cylinder inner and outer walls. Flat, thin molybdenum caps (0.15 cm thick) sealed the cylinder ends; the top cap had a 0.64 cm hole through which probes were inserted into the cavity. For all but one experiment the cavity was lined with thin tantalum foil (0.025 cm thick). From the perspective of the probe, the experimental cavity walls represent both the idealized cavity walls and infinite heat bath.

Cavity temperatures were standardly inferred with a type C thermocouple (5% W-Re/ 26% W-Re) buried at the radial and vertical midpoint of the oven wall, roughly equidistant between two heater elements and the cavity's interior and exterior walls (point A in Figure 8.3a). Spatial resolution was taken to be the sizes of the thermocouple tips, roughly 4mm. Estimates of temperature variations within the cavity were made by comparing simultaneous measurements from three thermocouples located at points A,B, and C in Figure 8.3a, specifically, at A) the standard wall position; B) the geometric center of the blackbody cavity, where probes were typically situated; and C) the entry hole for the probe. Over the heating/cooling cycle, the wall (point A) and cavity center (point B) temperatures were similar, with the center temperature consistently lower than the walls, typically by 5-50 K. Over much of the temperature range where probes displayed non-zero potential with respect to the walls (probe-wall voltage, $V_{PW} \neq 0$) they were in agreement within the rated experimental uncertainty of the thermocouples ($\pm 1\%$). The thermocouple at point C is believed to have measured the coolest region of the cavity, near the probe entry hole. Here heat could escape via conduction along the probe's alumina support rod or via blackbody radiation directly out through the hole. Point C temperatures were 70 - 100 K below wall temperature (point A) over the nonzero-V_{PW} temperature ranges. Thermocouple measurements were corroborated with optical pyrometer measurements and agreed well with electrical power heating requirements, given the thermal insulation used around the cavity.

Cavity temperatures could be varied between 290 K and 2060 K; higher temperatures melted the alumina insulation tubes. Good thermal stability was achieved; e.g., at 1400 K temperature drift could be held to 12 K over 30 minutes for an average temperature variation of 7×10^{-3} K/sec. The cavity was insulated by series of 10 nested heat shields. The probe was guided through the heat shields into the cavity by a small-bore (0.64 cm ID) tantalum tube which was physically, thermally, and electrically anchored to the cavity walls and cavity tantalum inner lining. The entire assembly was housed in a cylindrical vacuum vessel (L = 61 cm, Dia = 28 cm) and operated at base pressure of 1-2 $\times 10^{-6}$ torr. Pressure measurements inside the cavity were not made.

Probes consisted of small, refractory metal strips or wires exposed to the cavity environment. Probe materials tested included tungsten (work function, $\Phi = 4.5$ V), tantalum ($\Phi = 4.2$ V), molybdenum ($\Phi = 4.2$ V), hafnium ($\Phi = 3.5$ V), and lanthanum hexaboride ($\Phi = 2.66$ V) on tantalum. These were chosen for their refractory natures and because their work functions were either greater than, less

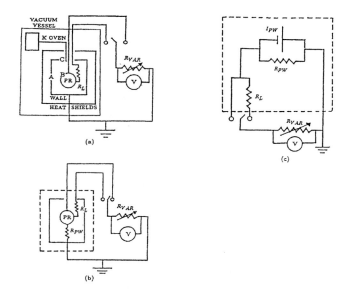

Figure 8.3: Schematic of Plasma I experimental apparatus. a) Physical schematic; b) Equivalent circuit for apparatus; c) Norton-equivalent circuit for apparatus. Dashed lines indicate the high-temperature region of the apparatus. Notation: probe (PR), load resistor (R_L), variable resistor (R_{var}), voltmeter (V).

than, and equal to that of the cavity walls. The probes were supported by multi-hole, 0.25 cm diameter, high-purity alumina rods (Coors AD998). Zirconia cement (Aremco 516) was sometimes used for local insulation. All probe and thermocouple alumina support rods had marginal physical thermal contact with cavity metal surfaces and were heated primarily by radiation. Several probe shapes and sizes were tried, varying from small cylinders (L = 0.15 cm, Dia = 0.025 cm) to large rectangles (L = 1.4 cm, W = 0.3 cm). The ratio of probe surface area, $(SA)_P$, to cavity surface area, $(SA)_{cav}$, was small; $2.1 \times 10^{-4} \leq \frac{(SA)_P}{(SA)_{cav}} \leq 1.5 \times 10^{-2}$.

Simulated loads, R_L, consisting of length or coils of fine wire (either 0.013 cm diameter tantalum or 0.005 cm diameter tungsten) were connected to the probes and housed in the cavity as shown in Figure 8.3a. Surface areas of load resistors, $(SA)_R$, were less than that of the probes; surface area ratios were varied as $0.02 \leq \frac{(SA)_R}{(SA)_P} \leq 0.6$. Load resistances increased with increased cavity temperature due to the temperature dependence of bulk resistivity. *In situ* measurements of R_L agreed within a factor of two with predictions based on resistor dimensions and bulk resistivity.

The probe and diagnostic circuit were simple and passive (See Figure 8.3a). The primary measurements for these experiments were of the potential difference between the probe and wall, called V_{PW}, and the electrical resistance between the probe and wall, called R_{PW}. Both varied with cavity temperature. V_{PW} and R_{PW} could be inferred from voltage drops across the variable resistor, R_{var}. (The value of R_{var} could be varied as $0\Omega \leq R_{var} \leq 2M\Omega$.) From measurements of V_{PW} and R_{PW} the current flowing from the probe or through the load resistor could be calculated via Ohm's law and from these load power drops, $P_L = \frac{V_{PW}^2}{R_L}$, and entropy production rates, $\frac{dS_L}{dt} = \frac{P_L}{T}$, could be estimated. Passive micro-ammeters and active ohmmeters were used regularly to verify the measurements of the passive circuit. The equivalent circuit for the probe, load, diagnostic circuit, and R_{PW} is given in Figure 8.3b.

The probe and wall, if they reside at different potentials, constitute the terminals of a battery whose output voltage is V_{PW}. The battery is current-limited by the net current it receives from the plasma; for an electron-rich plasma, the current limit, I_{lim}, is roughly $I_{lim} \simeq nev_e((SA)_P + (SA)_R)$, where the variables have been defined previously. In light of this, a revised, Norton-equivalent circuit is given for the apparatus in Figure 8.3c.

R_{PW} was strongly dependent on temperature. As the cavity was heated and thermionic electron emission and plasma production began, R_{PW} fell dramatically, from values greater than 300 kΩ for $T \leq 1000$ K to values as low as 40 Ω at T = 2060 K. Electrical conductance through the probe's alumina support rod was excluded as the primary cause of this resistance decrease; plasma formation was probably the primary cause.

Care was taken to avoid the several possible solid state thermoelectric effects – the potentially most serious of which was considered to be the Seebeck effect – that could confound plasma-related voltage and current measurements. For example, all diagnostic and grounding wires leading out the cavity and vacuum vessel were made of pure elemental metals, usually matched sets of tantalum or tungsten. They were brought to a single isothermal reference block outside the vacuum vessel (T = 293 K) before continuing as copper wires to the diagnostic circuit. In an extreme cases, Experiments 7 and 8 (Table 8.1), the probe, load resistor, cavity walls, diagnostic and grounding wires from the cavity to the diagnostic's thermal reservoir were all composed of tantalum. By symmetry, one expected little or no thermoelectric potential to develop between probe and walls, yet still sizable potentials (≥ 700 mV) were observed, similarly to all other metal combinations.

Atomic potassium was introduced into the cavity using an atomic beam oven directed into the cavity as shown in Figure 8.3a. Since the ionization potential of potassium (I.E. = 4.34 V) is close to the work function of the tantalum cavity walls ($\Phi = 4.2$ V), atomic potassium was ionized with high probability in the heated cavity. For example, at 1500 K the ionization probability of potassium on tantalum is roughly 0.4. Thus, introducing potassium was a convenient way to boost cavity plasma density while introducing only a small population of neutrals. Density was roughly estimated by noting when the plasma potential switched polarity, this corresponding to the critical density as given in (8.2).

8.2.2.2 Results and Interpretation[1]
Several different simulated paradoxical systems were investigated. Table 8.1 summarizes by experiment the elemental compositions of the experimental compo-

EXP	C	P	R	L	T_{max}(K)	P T	V_{max}/V_{min}(mV)
1	Ta	Ta	– – –	Ta	2060	– – –	120/ − 65
2	Ta	LaB_6/Ta	– – –	Ta	1980	– – –	140/ − 64
3	Ta	W	– – –	W	1970	– – –	206/ − 100
4	Ta	W	W	W	1780	– – –	47/ − 25
5	Ta	W	W	W	1900	K	405/ − 400
6	Ta	Ta	W	W	1650	K	385/ − 280
7	Ta	Ta	Ta	Ta	1750	K	56/ − 720
8	Ta	Ta	Ta	Ta	1580	K	25/ − 330
9	Ta	Ta, Mo W, Hf	– – –	Ta	1650	– – –	156/ − 348
10	Mo/Ta	Ta, Hf W	– – –	Ta	1860	– – –	410/ − 374

Table 8.1: Summary of experimental parameters for Plasma I: Experiment number (EXP); Chemical composition of cavity (C), probe (P), load resistor (R), and electrical leads (L); Maximum temperature achieved during experiment (T_{max}(K)); Plasma type (P T); Maximum and mininimum V_{PW} achieved during experiment (V_{max}/V_{min}(mV)).

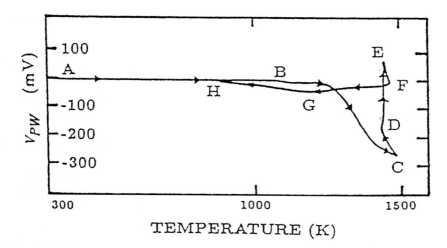

Figure 8.4: Experimental probe-wall voltage, V_{PW}, versus temperature curve for Experiment 7. Arrows indicate direction of increasing time. Explanation of alphabetic labeling given in text.

nents, the maximum achieved cavity temperatures, maximum and minimum V_{PW}, and ionizable species introduced into the cavity. All systems displayed qualitatively similar behaviors, conforming to expectations for the theoretical paradoxical system. As a means by which to discuss typical system behavior, a representative case, Experiment 7 in Table 8.1, will be described in detail.

In Experiment 7, the cavity walls, probe $((SA)_{PR} = 7 \times 10^{-5}$ m^2), load resistor $((SA)_R = 4 \times 10^{-5}$ m^2) were all tantalum, and leads out of the cavity to the isothermal reference reservoir were also all constructed from tantalum. Figure 8.4 depicts the V_{PW} versus cavity temperature curve for Run 1, Day 3. Arrows indicate the direction of increasing time. The following are descriptions and interpretations of the various intervals of the thermal cycle alphabetized in Figure 8.4.

Interval A——→B (Duration 12 minutes): The cavity is electrically heated and the tantalum probe begins to develop a negative potential with respect to the walls at about 1000K, probably due to electron thermionic emission and/or plasma production.

Interval B——→C (Duration 7 minutes): The cavity is heated to 1475 K, monotonically reaching a minimum potential of -260mV.

Interval C——→D (Duration 7 minutes): Heating power to the cavity is reduced from 285 W to 170 W; the cavity cools.

Point D (Duration 25 minutes): The cavity temperature is stabilized at 1430 - 1450 K for roughly 20 minutes. V_{PW} also stabilizes at approximately -150mV, varying only 5 % during this time. (Notably, here $qV_{PW} \simeq kT$.) This interval is significant because its quasi-steady-state condition appears to approximate the critical requirement of the paradoxical system that a voltage be maintained across the load in a steady state fashion. The potassium oven is heated.

Interval D⟶E   (Duration less than 15 seconds): V_{PW} undergoes a sudden and rapid increase at the stable temperature 1450K, varying from -150mV to +56 mV in an interval of less than 15 seconds, probably due to a sudden surge of potassium atoms. The density, estimated to be $n_c \simeq 10^{12} \text{m}^{-3}$, agrees fairly well with the estimated atomic flux output from the potassium oven.

Interval E⟶F   (Duration 10 - 12 minutes): The potassium oven cools. V_{PW} decreases from 56mV to -20mV.

Interval F⟶G⟶H   (Duration 40 minutes): Cavity heating power is ceased, cavity temperature decreases and V_{PW} reaches a minimum of -45mV, then increases to zero.

No experiment in Table 8.1, in itself, violated the second law because for all experiments the power dropped and the entropy produced by the load resistor, R_L, were minute compared with the power required and entropy produced in carrying out the experiments. For example, during Experiment 7 at temperature T = 1710 K, the load resistance and current were measured to be $R_L = 3.3\Omega$ and $I_L = 17.5 \mu A$. The power drop across R_L and the entropy production rate were $P_L = 9.3 \times 10^{-11}$ W and $\frac{dS_L}{dt} = 5.4 \times 10^{-14}$ W/K. Presumably, were the load purely conservative the entropy production rate would have been negative. On the other hand, substantially more power was required just to carry out the experiment; this included the electrical power to heat the cavity (~ 500 W), to operate the vacuum pumps and ancillary apparatus (~ 600 W), and to animate the experimenter (~ 100 W). If the roughly 1200 W used to power the experiment is exhausted ultimately into the 2.73 K universal blackbody, then the net entropy production for the experiment was roughly $\frac{dS_{exp}}{dt} \simeq 400$ W/K. This positive entropy production rate is roughly 10^{16} times greater than the presumed negative production rate from a conservative 3.3 Ω cavity load. However, were the load conservative, were the universe at T = 1710 K, were no vacuum, ancillary apparatus, or experimenter required, then it appears that the entropy production rate for the experiment would have been negative: $\frac{dS_{exp}}{dt} = \frac{dS_L}{dt} = -5.4 \times 10^{-14}$ W/K ≤ 0. This negative entropy production is small, but significant in the context of the paradox.

In summary, non-zero probe-wall potentials, V_{PW}, were observed over a wide range of temperatures ($T_{cav} \leq 2060$ K) for all systems tested, regardless of the composition of the cavity walls (Ta or Mo/Ta), probes (Ta, W, Mo, Hf, or LaB$_6$), load resistors (Ta or W), or electrical leads (Ta, W). All probes demonstrated current flows to ground. The behavior of V_{PW}, R_{PW}, and I_{PW} versus cavity temperature can be explained in terms of plasma effects. Quasi-steady-state V_{PW} and cavity conditions were demonstrated. Estimated power drops in R_L and entropy production rates were small, but significant in the context of the paradox. Finally, it was shown that the experiments, themselves, did not violate the second law.

These experiments did not perfectly simulate the idealized paradoxical system. Experimental non-idealities included: i) extreme temperature differences between the cavity and exterior environment ('universal heat bath'); ii) spatial and temporal temperature variations within the cavity; iii) influx and efflux of particles from the cavity; iv) possibly unaccounted for or unwanted thermoelectric potentials; and v) thermal and chemical degradation of the cavity contents. The experiment has generated controversy (§8.4). Detailed analysis indicates that, although these non-idealities may degrade the comparison of the experimental results to theory, they probably do not invalidate the primary paradoxical effect, namely, quasi-steady-state, non-zero probe-wall potential. It would be helpful if similar experiments could be conducted near room temperature since many experimental non-idealities might be ameliorated. Unfortunately, this would probably require either materials with exceptionally low work functions or plasma species with exceptionally low ionization energies.

8.3 Plasma II System

8.3.1 Theory [2]

The Plasma II challenge is the plasma analog to the chemical challenge (Chapter 7). Both rely on differential gas-surface reactions to create a pressure difference between two surface under blackbody cavity conditions and both operate in the high-temperature, long mean free path regime. However, unlike the chemical system, Plasma II makes use of the potential gradients of the Debye sheath to accelerate ions and electrons to superthermal speeds ($qV \gg kT$) and thereby magnify specific impulses over what can be achieved using neutrals gas alone. In this respect it is like Plasma I, which also relies on sheaths.

The Plasma II system consists of a frictionless, two-sided piston in a high-temperature, plasma-filled blackbody cavity surrounded by a heat bath (Figure 8.5). Apposing piston faces are surfaced with different work function materials. Owing to differential neutral, electronic, and ionic emissions from the different materials, a steady-state pressure difference is sustained between the piston faces which, in principle, can be exploited to do work. In order to satisfy the first law, the work performed by the piston must be derived ultimately from the heat bath, but this leads to an apparent contradiction of the second law. Laboratory experiments testing a critical aspect of the paradox – that different surfaces can simultaneously thermionically emit distinctly in a steady-state fashion in a single blackbody environment – corroborate theoretical predictions. Again, it is strongly emphasized that these experiments did not, themselves, violate the second law; they served only to verify contributing high-temperature thermodynamic behaviors.

Consider a universe consisting of a blackbody cavity surrounded by an infinite, high-temperature heat bath. The cavity interior is bathed in blackbody radiation and a low-density ionizable gas, B. Introduce into the cavity a frictionless, two-sided piston as shown in Figure 8.5. The piston is electrically and thermally grounded to the walls, as are the walls to the heat bath. Particles freely move between sides of the cavity partitioned by the piston (it may be perforated.). The majority of the piston is of identical composition as the walls (surface type 1),

HEAT BATH

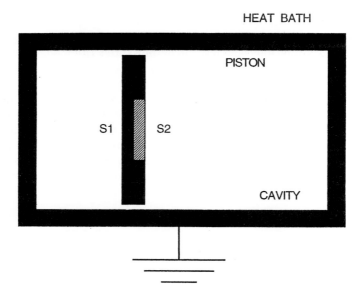

Figure 8.5: Schematic of Plasma II system.

however, on one piston face is a small patch having a different work function (surface type 2). It is small in the sense that it is relatively unperturbing to global plasma properties. The work functions of surface types 1 and 2 and the ionization potential of B are ordered as: $\Phi_2 \gtrsim I.P. > \Phi_1$.

The following plasma model is assumed:

1) Plasma is created by a combination of Richardson emission and surface ionization of B. Surface ionization and recombination are governed by the Langmuir-Saha relation. Richardson emission greatly exceeds ion emission for all surfaces, giving an electron-rich plasma with a negative plasma potential. Secondary electron emission is negligible.

2) Particles come into thermal equilibrium with the surface they contact, leaving as half-Maxwellians.

3) The plasma is effectively collisionless for charged and neutral particle-particle collisions.

4) The Debye length is short compared with plasma dimensions.

5) The plasma is quasi-neutral ($n_e \simeq n_i$).

6) The plasma is in a stationary state.

The following nomenclature from [2] will be used: j is particle flux density ($\mathrm{m^{-2}s^{-1}}$) and F is momentum flux density, pressure, ($\mathrm{Nm^{-2}}$); superscript right arrow ($\rightarrow$) indicates a flux away from a surface and ($\leftarrow$) indicates flux toward a surface (here, arrows do not signify vectors); subscripts $n,i,e,1$, and 2 refer to neutral and ionic B, electrons, and surfaces 1 and 2, respectively; subscript k refers to an arbitrary surface; and S1 and S2 refer to surfaces 1 and 2. The x coordinate is the direction perpendicular to a surface.

Surface-ionization has been extensively studied both theoretically and experimentally [9, 10, 11], as have surface-ionized plasmas, particularly in the context of Q-machines [5, 12, 13] and thermionic power generators [14, 15]. In fact, the present plasma can be roughly considered to be an unmagnetized, three-dimensional Q-plasma with a sliding hot plate. The theory presented here draws heavily from that of Q-plasmas.

If thermal equilibrium holds, the fractional ionization of B leaving a surface (even with a sheath) is given by the Langmuir-Saha relation:

$$\frac{n_i}{n_n} = \frac{g_i}{g_n} \exp[\frac{e(\Phi - I.P.)}{kT}], \tag{8.3}$$

where $\frac{n_i}{n_n}$ and $\frac{g_i}{g_n}$ are the ratio of densities and the ratio of statistical weights for ionized and neutral B, respectively. The ionization probability is:

$$P_i = \frac{n_i}{n_n + n_i} = [1 + \frac{g_n}{g_i} \exp\{\frac{e(I.P. - \Phi)}{kT}\}]^{-1} \tag{8.4}$$

The recombination (neutralization) probability is: $P_n = \frac{n_n}{n_n+n_i} = 1 - P_i$. From the principle of detailed balance, one can infer that, in steady state:

$$\vec{j}_{i,k} = P_{i,k} \overleftarrow{j}_{B,k} \tag{8.5}$$

and

$$\vec{j}_{n,k} = P_{n,k} \overleftarrow{j}_{B,k}, \tag{8.6}$$

where $\overleftarrow{j}_{B,k}$ is the total flux density of B onto surface k: $\overleftarrow{j}_{B,k} = \overleftarrow{j}_{i,k} + \overleftarrow{j}_{n,k}$.

The electron current density from surface k with a plasma is given by:

$$\vec{j}_{e,k} = J_{R,k} \exp(\frac{eV_{pl}}{kT}) = AT^2 \exp(-\frac{e\Phi_k}{kT})\exp(\frac{eV_{pl}}{kT}) = \frac{n_e q v_e}{4} \tag{8.7}$$

Here $J_{R,k}$ is the Richardson current density, V_{pl} is the plasma potential (negative for an electron-rich plasma), T is temperature, k is the Boltzmann constant, v_e is the average electron thermal speed ($\overline{v_e} = \sqrt{8kT/\pi m_e}$) for a Maxwellian distribution), q is an electronic charge (absolute value), m_e is the electron mass, n_e is the electron number density, and A is the Richardson constant for the surface with a value of about 6-12 $\times 10^5$ (A/m²K²) for pure metals.

The physical processes leading to plasma formation are straightforward: electrons are "boiled" out of the metal (Richardson emission) and ions, created by surface ionization, are accelerated off the metal surface by the electron negative space charge. Ions, in turn, ease the electrons' space charge impediment, thus releasing a quasi-neutral plasma from the surface. Actually, if $V_{pl} < 0$, this Q-plasma is essentially a charge-neutralized, low-energy ion beam leaving the surface. The ions are a velocity-space-compressed, drifting, half-Maxwellian and the electrons are an essentially thermal half-Maxwellian. In Q-machines, ion drift speeds can be several time the average ion thermal speed, but still much less than the average electron thermal speed, *i.e.*, $\overline{v}_i < v_d \ll \overline{v}_e$.

The ordering $\Phi_2 \stackrel{>}{\sim} I.P. > \Phi_1$ allows, with appropriate plasma density and temperature, and surface areas $((SA)_1$ and $(SA)_2)$, the following additional assumptions for the plasma system:

  i) Surface 2 ionizes B well and recombines it poorly $(\vec{j}_{n,2} \stackrel{<}{\sim} \vec{j}_{i,2} \sim \overleftarrow{j}_{n,2})$,

  while surface 1 ionizes B poorly, but recombines B well $(\vec{j}_{i,1} \ll \vec{j}_{n,1} \sim \overleftarrow{j}_{n,1})$. In fact, the ion current density from S2 exceeds that from S1 by a factor of $\frac{P_{i,2}}{P_{i,1}}$, that is, $(\vec{j}_{i,2} \simeq \frac{P_{i,2}}{P_{i,1}} \vec{j}_{i,1}.)$

  ii) Since S1 dominates plasma properties by virtue of its greater surface area $((SA)_1 \gg (SA)_2)$, and in light of result (i) above, the net flux of B to any surface is predominantly neutral B: $(\overleftarrow{j}_{B,k} \sim \overleftarrow{j}_{n,k} \gg \overleftarrow{j}_{i,k})$.

  iii) Surface 2 will be relatively unperturbing to cavity plasma conditions if the S2 ion current into the plasma is much less than the total S1 ion current; that is, if $\vec{j}_{i,2}(SA)_2 \ll \vec{j}_{i,1}(SA)_1$. This condition can be stated equivalently as: $\frac{P_{i,2}}{P_{i,1}} \ll \frac{(SA)_1}{(SA)_2}$

  iv) The neutral flux onto S1, onto S2, and off of S1 are all roughly the same: $(\vec{j}_{n,1} \simeq \overleftarrow{j}_{n,1} \simeq \overleftarrow{j}_{n,2})$. Since S2 ionizes well, $\vec{j}_{n,2}$ is smaller than these by the factor $\frac{P_{n,2}}{P_{n,1}}$.

  v) The electron emission off S1 exceeds that off S2 by a factor $\exp[\frac{(\Phi_2-\Phi_1)}{kT}]$: $(\vec{j}_{e,1} = \exp[\frac{(\Phi_2-\Phi_1)}{kT}]\vec{j}_{e,2})$. Since S1 dominates global plasma properties, the thermal electron current to each surface should be roughly the same, that is, $(\overleftarrow{j}_{e,1} \simeq \overleftarrow{j}_{e,2})$.

Because of the differences between electronic, ionic, and neutral masses and the different currents of each leaving S1 and S2, a steady-state pressure difference can be supported between piston faces. A comparison will be made of momentum flux densities to and from surface 2 and the corresponding patch of surface 1 diametrically through on the other side of the piston (See Figure 8.5). If these are different, the pressure difference, $\Delta F = F_2 - F_1$,in principle, can be exploited to do work. Fluxes will be measured through Gaussian surfaces which enclose the piston surfaces and their plasma sheaths out to where their potential gradients vanish in the plasma. Note that the plasma locally outside S2 may be different from that outside S1, but should revert to the S1-type within a few Debye lengths.

In regard to momentum fluxes to any surface, if S2 is effectively unperturbing to plasma properties and if the piston moves slowly $(v_{piston} \ll \overline{v}_{e,i,n})$, then the neutral, electronic, and ionic momentum fluxes onto S1 and S2 should be essentially the same, and therefore, should exert no net pressure difference on the piston, $\Delta \vec{F}$. It is only departing particle currents that contribute to a net pressure difference.

Neutrals: As expected for an ideal gas, the average pressure difference between S1 and S2 patches due to departing neutrals, $\Delta \vec{F}_n$, should be:

$$\Delta \vec{F}_n = \Delta \vec{F}_{n,2} - \Delta \vec{F}_{n,1} = (P_{i,1} - P_{i,2}) \frac{n_n kT}{2} \tag{8.8}$$

where n_n is the average cavity neutral number density. The factor of $\frac{1}{2}$ arises since only departing neutrals contribute to the pressure. In the limit that $(P_{i,1} \ll P_{i,2})$, one can write roughly

$$\Delta \vec{F}_n \simeq -\frac{P_{i,2} n_n kT}{2} \tag{8.9}$$

Electrons: The velocity distribution function for Richardson electrons should be half-Maxwellian, but modified to account for the retarding negative plasma potential. Just outside the Gaussian surface, it is:

$$\vec{f}_{e,k}(v_x) = n_{e,k} \sqrt{\frac{m_e}{2\pi kT}} \exp[-\eta_p] \exp[-\frac{mv_x^2}{2kT}]; \qquad v_x > 0 \tag{8.10}$$

where $\eta_p = |\frac{qV_{pl}}{kT}|$ and $n_{e,k} = 4\frac{J_{R,k}}{q}\sqrt{\frac{\pi m_e}{8kT}}$ and $J_{R,k}$ is given in (8.7). The velocity distributions for S1 and S2 are distinguished only by their work functions which fold into $J_{R,k}$. The electron current leaving each surface may be written:

$$\vec{j}_{e,k} = \frac{J_{R,k}}{q} \exp[-\eta_p] \tag{8.11}$$

The plasma electron density, $n_{e,p}$, should be $n_{e,p} = n_{e,1} \exp[-\eta_p]$, which is twice that calculated from (8.10) since electrons also originate from the opposite wall. The average pressure difference between S1 and S2 due to electron fluxes is

$$\Delta \vec{F}_e = \frac{\pi}{4} \frac{m_e \bar{v}_e}{q} \exp[-\eta_p][J_{R,2} - J_{R,1}] = \frac{\pi}{4} m_e \bar{v}_e (\vec{j}_{e,2} - \vec{j}_{e,1}) \tag{8.12}$$

In the limit that $\vec{j}_{e,1} \gg \vec{j}_{e,2}$, this is roughly

$$\Delta \vec{F}_e = -\frac{\pi}{4} m_e \bar{v}_e \frac{J_{R,1}}{q} \exp[-\eta_p] = -\frac{\pi}{4} m_e \bar{v}_e \vec{j}_{e,1}. \tag{8.13}$$

Ions: The ion velocity distribution function just outside the Gaussian surface in an electron-rich plasma should be:

$$\vec{f}_{i,k}(v_x) = n_{i,k} \sqrt{\frac{m_i}{2\pi kT}} \exp[\eta_p] \exp[-\frac{m_i v_x^2}{2kT}]; \qquad v_x \geq \sqrt{-\frac{2qV_{pl}}{m_i}} \tag{8.14}$$

where $n_{i,k} = P_{i,k} \frac{4\overleftarrow{j}_B}{\overline{v}_B} \simeq P_{i,k} \frac{4\overleftarrow{j}_n}{\overline{v}_n} = P_{i,1} n_n$ since B exists primarily as neutrals in the cavity. Since the ions are accelerated through a negative sheath, the lowest velocity for plasma ions is $v = \sqrt{-\frac{2qV_{pl}}{m_i}}$.

The average ionic pressure difference, $\Delta \vec{F}_i$, is:

$$\Delta \vec{F}_i = m_i \overline{v}_i (\vec{j}_{i,2} - \vec{j}_{i,1}) \simeq m_i \frac{n_n \overline{v}_n}{4} (P_{i,2} - P_{i,1}) \int_{\sqrt{2\eta_p}}^{\infty} \frac{v_x f_{i,k}(v_x)}{n_{i,k}} dv_x \tag{8.15}$$

If $-qV_{pl} \gg kT$ (as is typical for a Q-plasma), then with regard to the momentum flux, the ions can be considered roughly a mono-energetic beam, and if $P_{i,2} \gg P_{i,1}$, then $\Delta \vec{F}_i$ is roughly

$$\Delta \vec{F}_i \simeq m_i \frac{n_n \overline{v_n}}{4} P_{i,2} \sqrt{-\frac{2qV_{pl}}{m_i}} \tag{8.16}$$

Although S2 emits a disproportionate ion current, the excess ions are effectively neutralized on a type-1 surface after a single pass through the plasma. The plasma potential is obtained through the quasi-neutrality condition, $n_{i,p} \simeq n_{e,p}$, or by solving (8.7) for V_{pl}.

Combining (8,12,15), the net pressure difference between S1 and S2 is

$$\Delta F_{net} = \frac{(P_{i,1} - P_{i,2}) n_n kT}{2} + \frac{\pi}{4} \frac{m_e \overline{v_e}}{q} (J_{R,2} - J_{R,1}) \exp[-\eta_p] + \tag{8.17}$$

$$m_i \frac{n_n \overline{v_n}}{4} (P_{i,2} - P_{i,1}) \int_{\sqrt{2}\eta_p}^{\infty} \frac{v_x f_{i,k}(v_x)}{n_{i,k}} dv_x$$

Under the approximations given for (8.9, 8.13, and 8.16), the net pressure difference between the piston patches is roughly

$$\Delta F_{net} \simeq -\frac{P_{i,2} n_n kT}{2} - \frac{\pi}{4} \frac{m_e \overline{v_e}}{q} J_{R,1} \exp[-\eta_p] + m_i \frac{n_n \overline{v_n}}{4} P_{i,2} \sqrt{-\frac{2qV_{pl}}{m_i}} \tag{8.18}$$

Except for very specific plasma parameters, ΔF_{net} should be non-zero. In fact, examining (8.17), it seems that for ΔF_{net} to vanish, both the Richardson current densities and ionization probabilities for S1 and S2 must be identical; this requires work functions for S1 and S2 be identical. Except for very special choices of S1 and S2 (e.g., S1 and S2 the same material), this is unlikely.

This plasma-mechanical system has similarities to the Plasma I system. In the latter, a non-zero plasma potential was exploited to do work via a spontaneous net current flow from a plasma probe, through a load to ground. Spontaneous organization of random current was given as the origin of the entropy decrease. In the present system, again a non-zero plasma potential is exploited, but here to help create a pressure asymmetry on a piston. This asymmetric pressure is the analog of the asymmetric current through the load in Plasma I; in fact, both may be viewed as succeeding via asymmetric momentum fluxes (§10.1.1). In the present system, however, the plasma potential is not essential; even were it absent the system should operate still. Notice, in (8.17) or (8.18), if $V_{pl} = 0$, the net pressure on the piston remains non-zero. At a deeper level, this is because equipartition of energy does not imply equipartition of linear momentum. If a neutral B arrives at a surface, thermally equilibrates and leaves, on average it will carry away $\sim kT$ of energy and average linear momentum $\overline{p} = m_n \overline{v}_n$. On the other hand, if it leaves as a thermal ion and an electron, each independently leaves with an average of $\sim kT$ of energy and with combined average momentum of $\overline{p} = m_i \overline{v}_i + m_e \overline{v}_e$ which will be greater than the momentum of the neutral. So, if a surface (S2) dissociates a species (B) more efficiently than another surface (S1), and if all particles leave thermally, a greater momentum flux density leaves S2; hence by conservation of linear momentum, a greater pressure is exerted on it. In this respect, Plasma II resembles the chemical paradox (Chapter 7) in which chemically distinct surfaces render disproportionate products for the dissociation reaction $(2A \rightleftharpoons A_2)$. As such, Plasma II can be seen a hybrid between Plasma I and chemical systems.

The Plasma II pressure effect should be observable experimentally. Consider a hypothetical system much like that in Figure 8.5 set to the following initial conditions: a hollow hafnium cylinder ($\Phi = 3.5$ V, L = 50 cm, Dia = 50 cm) containing a low-friction, hafnium piston with a small circular patch of tungsten ($\Phi = 4.5$ V, Dia = 1 cm), immersed in a heat bath (T = 2200 K). The neutral density of ionizable gas, potassium (I.P. = 4.3 eV), is established at $n_n = 5.4 \times 10^{15}$ m^{-3}.

Under these initial conditions, the plasma model assumed above is obtained. The plasma is collisionless ($n_i = 4 \times 10^{13}$ m^{-3}), electron-rich ($V_{pl} = -2.1$ V) and its Debye length is short compared with plasma dimensions ($\lambda_D = 5 \times 10^{-2}$ cm). Surface 2 ionizes potassium well and Richardson emits poorly with the reverse true for surface 1; ion and electron current densities are each disparate by factors of at least 80 between S1 and S2. Surface 2 is relatively unperturbing to global plasma properties since its production of ions is far less than that of S1; that is, $2 \times 10^4 \simeq \frac{(SA)_1}{(SA)_2} \gg \frac{P_{i,2}}{P_{i,1}} \simeq 80$. Neutrals dominate the potassium particle flux to any surface ($\frac{\overleftarrow{j_n}}{\overleftarrow{j_i}} \sim 40$). A summary of hypothetical system parameters is given in Table 8.2.

The net pressure difference between S1 and S2 patches should be roughly $\Delta F \simeq 1.3 \times 10^{-4}$ Pa, dominated by ionic pressure, so the piston should move to the left. Were the piston to move at roughly 5% of the neutral thermal speed, it could deliver about 6×10^{-7} W of power, a value comparable to that required for a 1 mg ant to climb a wall with speed 2 cm/sec. This mechanical power is over 10^{12} times less than the blackbody radiative power from cavity surfaces, so the mechanical work drawn should have negligible effect of the cavity properties.

This paradoxical effect is small but robust; many system constraints can be relaxed without destroying it. For instance, it can be shown that the plasma successfully can be made ion-rich, collisional, or subject to some volume ionization. Viable variations can be devised by adjusting system geometry, work functions, ionization potentials and densities of the working gas, conductivity of parts, and groundings. For example, the hypothetical system would be successful if potassium were replaced with sodium and tungsten replaced with rhenium (adjusting the neutral density so as to achieve the same plasma density and potential). Or the need to reverse the piston stroke could be obviated by running it in a hollow toroidal chamber, or by replacing it with a radiometer-like system with S1 and S2 vane faces (as in Figure 7.1).

Cavity Length,L	0.5 m
Cavity Diameter,D_1	0.5 m
S2 Patch Diameter,D_2	10^{-2}m
Work Function S1 (Hf),Φ_{Hf}	3.5V
Surface Area S1,$(SA)_1$	1.6m^2
Work Function S2 (W),Φ_W	4.5V
Surface Area S2,$(SA)_2$	8×10^{-5}m^2
Ionization Potential (K),$I.P.$	4.3eV
Cavity Temperature, T	2200K
Richardson Constant, A	6×10^5A/m^{2}K^2
Statistical Weight,$(\frac{g_n}{g_i})_K$	2
Cavity Neutral Density,n_n	5.4×10^{15}m^{-3}
S1 Ionization Probability,$P_{i,1}$	7.3×10^{-3}
S2 Ionization Probability,$P_{i,2}$	0.59
Plasma Density,$(n_{e,p})$	4×10^{13}m^{-3}
Plasma Potential,V_{pl}	-2.1V
Debye Length,λ_D	5×10^{-4}m
Charge-Charge Collision Length,	$\sim$ 30m
Neutral-Neutral Collision Length,	$\sim$ 900m
Neutral Differential Pressure,ΔF_n	-4.8×10^{-5}Pa
Electronic Differential Pressure,ΔF_e	-6.3×10^{-7}Pa
Ionic Differential Pressure,ΔF_i	$+1.8 \times 10^{-4}$Pa
Net Differential Pressure, ΔF_{net}	$+1.3 \times 10^{-4}$Pa
Output Power ($v_{piston} = 0.05\overline{v_n}$)	6×10^{-7}W

Table 8.2: Summary of system parameters for hypothetical Plasma II system.

8.3.2 Experimental

8.3.2.1 Apparatus and Protocol

The primary operating criterion for Plasma II is this: that different materials simultaneously sustain different steady-state pressures over their surfaces (due to different ionic, electronic, and neutral emissions) in a single blackbody environment. Solid state and plasma theory and experiments strongly corroborate this hypothesis, but there does not appear to be any unambiguous experimental evidence for it.

Piston or radiometer experiments could be performed, but these are likely to be problematic. (Sheehan performed several radiometer-torsion balance experiments under high-temperature, high-vacuum conditions, but the results were deemed inconclusive due to confounding effects of outgassing, temperature, and thermal gradients.) A double-ended Q-machine with equipotential, equithermal hot plates of varied compositions could serve as a good two-dimensional model for the paradoxical system. Electron and ion velocity distributions could be measured with standard directional energy analysers (neutrals might be more difficult to diagnose) and pressure from each surface calculated quantitatively. Anecdotal accounts abound of spotty emission from Q-machine hot plates, suggesting thermionic emission can vary from surface to surface at a single temperature.

The experiments reported here did not measure differential pressures; differential thermionic currents were inferred from pairs of similar and dissimilar sub-

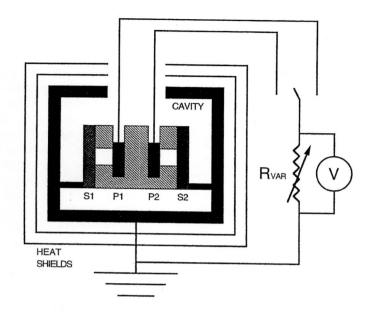

Figure 8.6: Experimental apparatus for tests of Plasma II.

stances under blackbody plasma conditions. In particular, tantalum, hafnium, and zirconia (ZrO_2) were compared.

Differential current measurements were made in an apparatus similar to that described in detail for Plasma I in §8.2. A schematic is given in Figure 8.6. The blackbody cavity and heat bath were approximated by a heated, hollow 5 cm diameter, 5 cm high molybdenum cylinder, with interior diameter and height of 3.8 and 4.7 cm, respectively. The cavity was lined with thin tantulum foil and heated by 16 tantalum heater wires buried at equidistant radial locations in the cavity wall. Cavity temperature were taken with a Type C thermocouple buried centrally in the cavity wall.

Cavity temperatures could be varied between 290 and 2060 K. Good thermal stability was achieved; e.g., at 1400 K temperature drift could be held to 12 K over 30 minutes for an average temperature variation of 7×10^{-3} K/sec. The cavity was insulated by a series of ten nested heat shields. The entire assembly was housed in a cylindrical vacuum vessel (L = 61 cm, diam = 28 cm) and operated at a base pressure of 4×10^{-6} Torr. Pressure measurements inside the cavity were not made. Cavity temperatures were probably too low for significant volume ionization, so it is assumed plasma was created primarily by thermionic emission.

Thermionic emissive materials (S1 and S2 in Figure 8.6) in the form of small square plates (1.6 x 1.6 cm) were supported at the cavity center, as shown in Figure 8.6 These were electrically grounded to the cavity walls, but electrically insulated from each other and from the probes (P1 and P2 in Figure 8.6) by thin layers (1.5 mm thick) of zirconia cloth. Small windows (4 mm x 4 mm) were cut in the center of the cloth to allow direct, well-controlled exposure of the emissive surface to its probe. The probes consisted of thin tantalum squares (1.2 x 1.2 cm) centered and sandwiched between S1 and S2. They, also, were electrically insulated from each other by 1.5 mm thick zirconia cloth. Probe leads were fine tantalum wires.

The diagnostic circuit was simple and passive. The primary measurements were of the potential difference between the probes and ground, called V_{pg}, and the electrical resistance between the probe and ground, called R_{pg}. Both varied with cavity temperature. Here V_{pg} and R_{pg} could be inferred from voltage drops across the variable resistor, R_{var}. (The value of R_{var} could be varied as $0 \leq R_{var} \leq 2M\Omega$.) From measurements of V_{pg} and R_{pg}, the current flowing from the probe, I_{pg}, could be inferred from Ohm's law. The value of I_{pg} should be dictated by its local plasma environment, therefore, strongly dictated by the emissive characteristics of its nearest surface, either S1 or S2.

Surfaces were tested in pairs in order to make direct comparisons between surface types under the most similar plasma blackbody conditions possible. Absolute measurements of plasma currents and properties could not be made due to the emissive nature of the probes – self-emissive probes are unsuitable for standard Langmuir probe measurements [6]. Furthermore, at elevated temperatures the electrical characteristics of S1 and S2 became increasingly coupled due to increased plasma conductivity and bulk conductivity of the zirconia insulation. Since absolute measurements of plasma currents were suspect, the salient experimental quantity was taken to be the difference in currents collected by P1 and P2: $\Delta I = I_{pg,2} - I_{pg,1}$. If S1 and S2 emit differently, non-zero values of ΔI should result; on the other hand, if surfaces 1 and 2 are identical, ΔI should be zero, by symmetry. Identical S1 and S2 serve as controls and for testing uniformity in cavity properties.

Care was taken to avoid the several possible solid state thermoelectric effects – the potentially most serious of which was considered to be the Seebeck effect – that could confound plasma-related voltage and current measurements. For example, diagnostic wires leading out of the cavity and vacuum vessel were made of tantalum to match the composition of P1 and P2. They were brought to a single isothermal reference block outside the vacuum vessel before continuing as copper wires to the diagnostic circuit.

8.3.2.2 Results and Interpretation

Several surface combinations were investigated for differences in thermionic emission in a blackbody plasma environment. In Figure 8.7 are plotted ΔI versus temperature data for the following S1/S2 combinations: Ta/Hf, Ta/ZrO$_2$, Ta/Ta, and Hf/Hf. The latter two homogeneous pairs were controls. Several points are noteworthy:

   1) Heterogeneous combinations (Ta/Hf and Ta/ZrO$_2$) displayed significant non-zero average ΔI consistent with differential thermionic surface emission. (Semilog plots of ΔI versus T confirmed the exponential current-temperature dependence.) Significant differential cur-

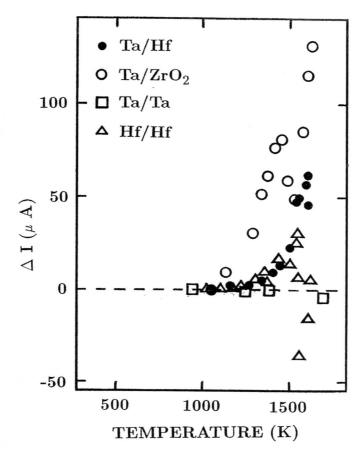

Figure 8.7: Differential current versus temperature for four S1/S2 pairs in Plasma II experiment. Heterogeneous pairs (Ta/Hf, Ta/ZrO$_2$) show greater average differential current than homogeneous pairs (Ta/Ta, Hf/Hf).

rents ($\sim 100\mu$A) could be maintained for several minutes at a single temperature.

2) Homogeneous combinations (Ta/Ta and Hf/Hf) displayed smaller ΔI values which on average were near zero. (Semilog ΔI vs T plots did not indicate exponential current-temperture dependencies.)

3) Differential current values on successive thermal cycles were smaller, probably due to outgassing from the cavity of volatile, ionizable species.

4) Both positive and negative values of ΔI were observed, possibly due to transitions between ion-rich and electron-rich plasmas.

These results strongly corroborate the assertion that different materials can simultaneously emit distinctly in a steady-state fashion in a single blackbody environment. However, these experiments also suffered from several non-idealities which included (i) spatial and temporal temperature variations within the cavity; (ii) influx and efflux of particles from the cavity; (iii) possibly unaccounted for or unwanted thermoelectric potentials; (iv) potentials possibly generated by high temperature chemical reactions; (v) increased electrical coupling between S1 and S2 at elevated temperatures; and (vi) possible differential outgassing of ionizable species from S1 or S2. One or more of these probably degraded the quality of the results and their quantitative effects cannot be assessed with certainty. It would be helpful if similar experiments could be conducted closer to room temperature, however, this would probably require either materials with unnaturally low work functions or plasma species with exceptionally low ionization potentials. Experiments involving radiometer torsion balances or particle energy analysers in Q-machines are possible.

8.4 Jones and Cruden Criticisms [8, 16]

Although both systems ostensibly utilize plasmas, they rely on different aspects to achieve their paradoxical effects. Plasma I relies on fundamental asymmetries between electrons and ions. Electrons are thermionically emitted from surfaces more easily than ions; they are also lighter and more mobile. As a result, thermally-sustained charge separations, boundary sheaths, and non-zero floating potentials are possible, by which work can be derived. In the Plasma II system, two surface ionize the working gas to different extents (via chemical potential differences) and thereby render different gas pressures. These augment and extend the sheath effects of Plasma I.

Plasma I has been questioned and criticised both on theoretical and experimental grounds, notably by Jones [8] and Cruden [16]. Jones considers possible flaws in the experimental set-up and Cruden considers the underlying theory. These criticisms are serious but inconclusive, suggesting additional experiments and theory are warranted.

Jones points out that the experimentally reported probe potentials might be due to: i) temperature gradients in the cavity; ii) cool atoms entering the black-body cavity from its upper opening; or iii) thermoelectric effects arising from dissimilar metals used between the probe and voltmeters. Surely, material and temperature gradients existed in the apparatus, however, in considering the magnitudes and temporal variations of the probe voltages, Sheehan argues against gradients as their explanation. The details of the discussion between Jones and Sheehan can be found elsewhere [8]. Given the relatively small values of voltage reported ($\leq 1V$) and the experimental intricacies of high temperature plasma experiments, Jones' suggested systematic errors of this magnitude cannot be ruled out, and more conclusive experiments should be conducted.

Cruden performs a thoughtful analysis of the plasma system based on thermodynamics and plasma probe theory [16] and concludes that the probe in Figure 8.1 should not be able to supply current at a voltage through the load. Sheehan made no formal response in the open literature. The researchers' thermodynamic

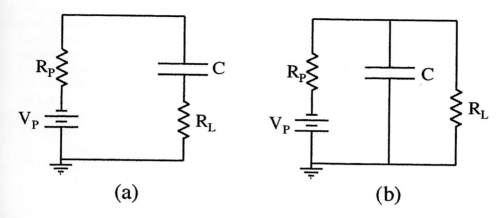

Figure 8.8: Equivalent circuits for Plasma I system: (a) Cruden [16] and (b) Sheehan [2, 3].

and probe analyses are in fundamental disagreement.

In examining the experimental set-up (Figure 8.1), Cruden argues that the first law should be written

$$\frac{dE}{dt} = \frac{dE_{hb}}{dt} + \frac{dE_{load}}{dt} + \frac{dE_p}{dt} = 0, \qquad (8.19)$$

where E subscripted hb, $load$, and p refers to the energy of the heat bath, load and plasma, respectively. He finds

$$\frac{dE_{load}}{dt} = \frac{V_{probe}^2}{R_L}; \qquad \frac{dE_p}{dt} = -\frac{V_{probe}^2}{R_L} \qquad (8.20)$$

and, thus, $\frac{dE_{hb}}{dt} = \frac{dQ_{hb}}{dt} = 0$. From this it follows that $\frac{dS_{hb}}{dt} = 0$ and $\frac{dS}{dt} = \frac{dQ_{load}}{dt} + \frac{dQ_p}{dt} \geq 0$. Thus, the second law is upheld.

This thermodynamic argument is unsatisfactory. Indeed, as Cruden asserts, the work in the load can be traced to the energy in the plasma sheath's electric field, but if this is exhausted by the load and if the sheath is to remain in some form of steady state, then, as argued previously, this energy must come ultimately from the heat bath. In Cruden's analysis, apparently the sheath is able to provide energy indefinitely without exhaustion, which is surely not the case unless the probe potential is zero, which begs the question. The necessary condition for second law compliance — zero probe voltage — however, conflicts both with theory and experiment.

The root of Cruden's and Sheehan's disagreement appears to lie in their differing views of the system's equivalent circuit. The plasma system can be modeled as an electrical RC circuit, with the plasma sheath modeled as an emf V_p with an internal resistance R_p; the load modeled as a resistor R_L; and the probe-wall combination modeled as capacitor C. It appears that, in Cruden's analysis (Figure 2 in [16]), the equivalent circuit is Figure 8.8a, which from Kirchhoff analysis shows that, indeed, no steady-state current flows in the circuit. Sheehan asserts that this circuit does not model the system correctly. The correct circuit is Figure 8.8b, where the load is in *parallel* with the probe-wall capacitor, rather than in *series* with it (Figure 8.8a). As a result, there can be a steady-state current through the load, whose power consumption is

$$P_L = \frac{V_p^2}{(\frac{R_p}{R_L} + 1)(R_L + R_p)} \tag{8.21}$$

Here, P_L is zero only for $R_p = \infty$, $R_L = 0$, or $R_L = \infty$.

Cruden's probe theory analysis [16] is too limited to draw any definitive conclusions, even in Cruden's estimation. He lists several conditions which would obviate zero probe work; to these others can be added, for instance, if the discharge of the probe capacitor is made intermittant with a switch (Figure 8.1).

Solid state and plasma physics are closely related. The capacitor interpretation of the solid state linear electrostatic motor (§9.3) is analogous to the Plasma I system; in fact, the former was discovered by direct analogic comparison to the latter. The plasma Debye sheath is the gaseous equivalent to the solid state depletion region and the probe-wall charging and capacitance are equivalent to the J-II gap in Figure 9.1. Theory, experiment, and numerical analysis strongly support the existence of the thermally-charged solid-state capacitor; in turn, this corroborates the plasma capacitor concept.

In summary, both theoretical [16] and experimental challenges [8] to the Plasma I paradox have been raised. Since the Plasma I experiments are admittedly corroborative, but not conclusive, Jones' objections could be correct. Cruden's theoretical analysis does not square with Sheehan's.

The Plasma II system has not attracted the critical response of Plasma I; proposed resolutions include: (1) all surfaces in a blackbody cavity identically Richardson emit, ionize and recombine the working gas (*i.e.*, all work functions in the cavity are rendered effectively equal under ideal conditions); (2) despite their differences in thermionic emission, by some unspecified means, materials achieve identical pressures over their surfaces (*e.g.*, particles do not leave in thermal equilibrium); or (3) system parameters conspire on distance scales long compared with microscopic surface processes to thwart macroscopic pressure differentials (*e.g.*, large-scale, steady-state density depletions are somehow established in regions where pressures would otherwise be high). None of these are compelling. Resolution (1) is not supported either by theory or experiment and (2) and (3) have no motivation aside from preservation of the second law, which begs the question.

Plasma systems I and II are untenable on earth because terrestrial temperatures ($T \sim 300$K) are too low to generate plasmas. Were ionization energies and work functions ten times smaller than they are — then perhaps they might. On the other hand, high-temperature, near blackbody conditions are ubiquitous in stellar atmospheres and interiors. A true second law violator might consist of a sealed tungsten cavity housing a probe, high-temperature motor, and a cesium plasma buried shallowly in a red dwarf star. A red dwarf star with surface temperature 2000 K would have a lifetime well approaching 10^{13} years – this is about 1000 times the current age of the universe. The convective zone of the such a star would closely approximate the steady-state, high-temperature, isothermal heat bath envisioned for these experiments. For practical systems, however, one must also consider real material limitations, for example, the evaporation of metal surfaces and high-temperature chemical reactions. For systems similar to the present laboratory experiments thermal or plasma degradation could be minor since the vapor pressures of the refractory metals (Ta, Mo) are small, since their chemical reactivity with alkali metals (K, Cs) are negligible, and since the electron and ion energies (0.2 eV) are too small to damage metal surfaces significantly.

Today the average temperature of the cosmos, as measured by the cosmic microwave background, is about 2.73K — far to low to support thermionic emission from known solids. About 10^{10} years ago, however, the universal blackbody temperature was about 2000 K — hot enough to sustain both plasmas and solids. This suggests that, at one time, work could have been extracted directly from the universal heat bath. Sadly, the universe has simply cooled below the temperature required for this type of second law challenge.

References

[1] Sheehan, D.P., Phys. Plasmas **2** 1893 (1995).

[2] Sheehan, D.P., Phys. Plasmas **3** 104 (1996).

[3] Sheehan, D.P. and Means, J.D., Phys. Plasmas **5** 2469 (1998).

[4] Čápek, V. and Sheehan, D.P., Physica A **304** 461 (2002).

[5] Motley, R.W. *Q-Machines* (Academic, New York, 1975).

[6] Hershkowitz, N., Nelson, B., Pew, J. and Gates, D., Rev. Sci. Instrum. **54** 29 (1983).

[7] Rynn, N. and D'Angelo, N., Rev. Sci. Instrum. **31**, 1326 (1960).

[8] Jones, R., Phys. Plasma **3** 705 (1996); Sheehan, D.P., Phys. Plasma **3** 706 (1996).

[9] Langmuir, I. and Kingdom, K.H., Proc. Roy. Soc. Ser. A **107** 61 (1925).

[10] Datz, S. and Taylor, E.H., J. Chem. Phys. **25** 389 (1956).

[11] Kaminsky, M. *Atomic and Ionic Impact Phenomena on Metal Surfaces* (Academic Press, New York, 1965).

[12] Motley, R.W. and Kawabe, T., Phys. Fluids **14** 1019 (1971).

[13] Buzzi, J.M., Doucet, H.J. and Gresillon, D., Phys. Fluids **13** 3041 (1970).

[14] Kaye, J. and Walsh, J.A., Editors, *Direct Conversion of Heat to Electricity* (Wiley, New York, 1960).

[15] Hatsopoulos, G.N. and Huffman, F.N., in *McGraw-Hill Encyclopedia of Science and Technology* **18** (McGraw-Hill, New York, 1987) pg. 273.

[16] Cruden, B., Phys. Plasma **8** 5323 (2001).

9

MEMS/NEMS Devices

Two challenges are discussed that involve room-temperature, micro- and nanoscopic semiconductor structures and utilize the thermally-sustained electric fields of p-n junctions. Analytic calculations and numerical simulations support the feasibility of these devices.

9.1 Introduction

This chapter is concerned with two experimentally-testable solid-state second law challenges that can operate at room temperature and which could, in principle, have commercial applications [1, 2, 3]. These are based on the cyclic electrome-chanical discharging and thermal recharging of the electrostatic potential energy inherent in the depletion region of a standard solid-state p-n junction. Essentially, the depletion region can be considered a thermally-rechargable capacitor which, in these incarnations, are used to power either a linear electrostatic motor (LEM) or a high-frequency, MEMS/NEMS[1], double-cantilever resonant oscillator. Numerical results from a commercial semiconductor device simulator (Silvaco International – Atlas) verify primary results from one dimensional analytic models. Present day micro- and nanofabrication techniques appear adequate for laboratory tests of principle. Experiments are currently being planned. The initial impetus to

---

[1] Micro-Electro-Mechanical Systems/ Nano-Electro-Mechanical Systems

explore such devices was given in 1995 by J. Bowles, who noted that solid state and plasma physics are kissing cousins; hence, there should be solid-state analogs to the previously proposed plasma paradoxes (Chapter 8) [4, 5, 6].

As detailed in this monograph, a number of concrete, experimentally-testable second law challenges have been proposed, some of which have been corroborated by laboratory experiments. No experiment has yet demonstrated actual violation, however, since in all cases the entropy generated by experimental apparatus (e.g., heaters/coolers, vacuum pumps) has always exceeded the theoretical maximum reduction in entropy that could be achieved by the proposed negentropic process itself. The present solid-state challenges appear different in this respect: whereas other challenges purport the *potential violability* of the second law, they offer no practical hope of *actual violation* under everyday terrestrial conditions. These solid state challenges, on the other hand, make positive claims on both, for, whereas previous challenges are viable only under extreme thermodynamic conditions (e.g., high temperatures ($T \geq 1000K$), low temperatures ($T \leq 100K$), or low pressure ($P \leq 1$ Torr)), the present systems should be viable at room temperature and pressure and they do not require ancillary entropy-generating apparatus.

This chapter is organized as follows. In §9.2 the physics of p-n junctions and thermally-charged capacitors — which undergird the solid-state challenges — is introduced and developed via one-dimensional analytical models and numerical simulations. In §9.3, a linear electrostatic motor (LEM) is discussed. It is substantiated three ways: via a 1-D analytical model, by analogy with an R-C network, and through 2-D numerical simulations. The device is shown to be viable within a broad range of realistic physical parameters. In §9.4, a resonant double-cantilever oscillator (*hammer-anvil*) is developed along similar lines as for the LEM. Finally, in §5, prospects for laboratory experiments are briefly considered.

9.2 Thermal Capacitors

9.2.1 Theory

The present challenges are based on the physics of the standard p-n junction diode [7, 8]. At equilibrium, the depletion region of a diode represents a minimum free energy state in which bulk electrostatic and diffusive forces are balanced. It follows that when individual n- and p- materials are joined, there is a transient current (due to rapid charge carrier diffusion) and energy release as a depletion region forms and equilibrium is attained. Space charge separation gives rise to a built-in potential (typical values, $V_{bi} \sim 0.5-1V$) and an internal electric field which arrests further charge diffusion. Typical depletion regions are narrow, ranging from $10\mu m$ for lightly-doped semiconductor to $0.01\mu m$ for heavily-doped ones. Although these distances are small, the broadest depletion regions have scale lengths visible to the naked eye and the narrowest are two orders of magnitude larger than atoms. They are large enough to interact with some present-day and many envisioned micro- and nano-scale devices [9, 10]. The thermally-generated electrostatic potential energy of the depletion region fuels this challenge. Practically speaking, a semiconductor

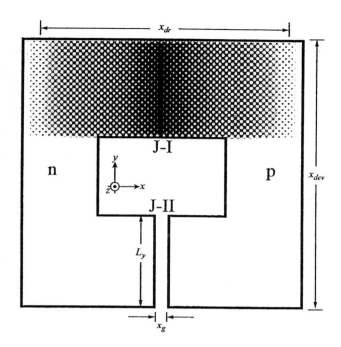

Figure 9.1: *Standard device* with Junctions I and II and physical dimensions and standard coordinates indicated. Depletion region at Junction I is shaded.

depletion region constitutes a thermally-charged capacitor. Whereas standard capacitors dissipate their electrostatic energy through internal parasitic resistance (R_i) on a timescale $\tau \sim R_i C$, thermal capacitors can remain energized indefinitely; they can also recharge thermally under appropriate circumstances.

Consider a p-n device (Figure 9.1) consisting of two symmetric horseshoe-shaped pieces of n- and p-semiconductor facing one another. At Junction I (J-I), the n- and p-regions are physically connected, while at Junction II (J-II) there is a vacuum gap whose width (x_g) is small compared to the scale lengths of either the depletion region (x_{dr}) or the overall device (x_{dev}); that is, $x_g \ll x_{dr} \sim x_{dev}$. Let the n- and p-regions be uniformly doped and let the doping be below that at which heavy-doping effects such as band gap narrowing are appreciable. The p-n junction is taken to be a step junction; diffusion of donor (D) and acceptor (A) impurities is negligible; the depletion approximation holds; impurities are completely ionized; the semiconductor dielectric is linear; and the system operates at room temperature. For a silicon device as in Figure 9.1, representative physical parameters meeting the above conditions are: $N_A = N_D = 10^{21}$ m^{-3}, $x_{dev} = 10^{-6}$ m on a side, $x_{dr} = 1.2 \times 10^{-6}$ m, and $x_g = 3 \times 10^{-8}$m. This dopant concentration results in a built-in potential of $V_{bi} \simeq 0.6$ V. For the discussion to follow, the p-n device (Figure 9.1) with these parameters will be called the *standard device*.

Standard one-dimensional formulae have been used to estimate V_{bi} and x_{dr} [7, 8]:

$$V_{bi} = \frac{kT}{q} \ln(\frac{N_A N_D}{n_i^2})$$ (9.1)

and

$$x_{dr} = \left[\frac{2\kappa\epsilon_o V_{bi}}{q} \frac{(N_A + N_D)}{N_A N_D} \right]^{\frac{1}{2}}$$ (9.2)

Here kT is the thermal energy; q is an electronic charge; n_i is the intrinsic carrier concentration of silicon ($n_i \simeq 1.2 \times 10^{16}$ m^{-3} at 300K); ϵ_o is the permitivity of free space; and $\kappa = 11.8$ is the dielectric constant for silicon.

That an electric field exists in the J-II gap at equilibrium can be established either via Kirchhoff's loop rule (conservation of energy) or via Faraday's law. Consider a vectorial loop threading the J-I depletion region, the bulk of the standard device, and the J-II gap. Since the electric field in the J-I depletion region is unidirectional, there must be a second electric field somewhere else along the loop to satisfy Faraday's law ($\oint \mathbf{E} \cdot dl = 0$). An electric field elsewhere in the semiconductor bulk (other than in the depletion region), however, would generate a current, which contradicts the assumption of equilibrium. Therefore, by exclusion, the other electric field must exist in the J-II gap. Kirchhoff's loop rule establishes the same result. Conservation of energy demands that a test charge conveyed around this closed path must undergo zero net potential drop; therefore, to balance V_{bi} in the depletion region, there must be a counter-potential somewhere else in the loop. Since, at equilibrium, away from the depletion region there cannot be a potential drop (electric field) in the bulk semiconductor – otherwise there would be a nonequilibrium current flow, contradicting the assumption of equilibrium – the potential drop must occur outside the semiconductor; thus, it must be expressed across the vacuum gap.

In Figure 9.2, the energy ($\mathcal{E}$), space charge density (ρ), and electric field ($\mathbf{E}$) are depicted versus horizontal position (x) through J-I and J-II. There are several important differences between the two junctions. The most noticable is that, while physical properties vary continuously with position across the J-I region, there are marked discontinuities for J-II. These are due to the inability of electrons to jump the vacuum gap (x_g). This restricts the diffusion of charge carriers that would otherwise spatially smooth the physical properties. As a result, Junction II suffers discontinuities in energies, voltages and space charge. Because the J-II gap is narrow and the built-in potential is discontinuous, there can be large electric fields there, more than an order of magnitude greater than in the J-I depletion region. Treating the gap one-dimensionally, the J-II electric field is uniform, with $|\mathbf{E_{J-II}}| \simeq \frac{V_{bi}}{x_g}$, while in the J-I bulk material it has a triangular profile, with average magnitude $|\mathbf{E_{J-I}}| \sim \frac{V_{bi}}{x_{dr}}$. The ratio of the electric field strength in the J-II gap to that in the middle of the J-I depletion region scales as $\frac{\mathbf{E_{J-II}}}{\mathbf{E_{J-I}}} \sim \frac{x_{dr}}{x_g} \gg 1$. For the *standard device*, the average value of the field strength is $|\mathbf{E_I}| \sim \frac{0.6V}{1.2 \times 10^{-6}m} \simeq 5 \times 10^5$ V/m and $|\mathbf{E_{II}}| \sim \frac{0.6V}{3 \times 10^{-8}m} \sim 2 \times 10^7$ V/m, rendering $\frac{\mathbf{E_{J-II}}}{\mathbf{E_{J-I}}} \sim 40$.

Let a switch bridge the J-II gap, physically connecting the entire facing surfaces of the n- and p-regions. For the present discussion, let the switching element be

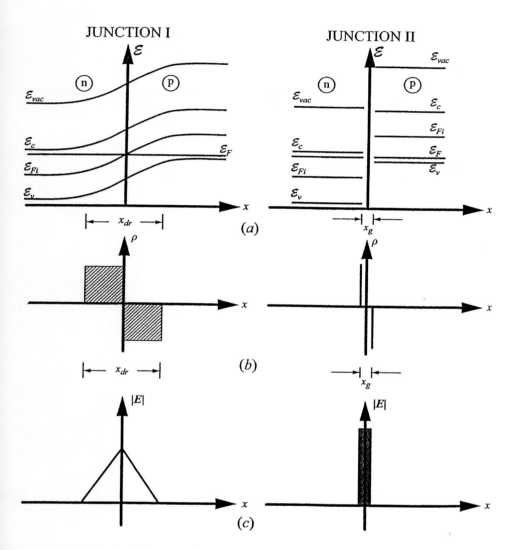

Figure 9.2: Physical characteristics versus position x through Junctions I and II. Left ($x < 0$) and right ($x > 0$) sides of each graph corresponds to n- and p-regions, respectively. (a) Energy levels for vacuum ($\mathcal{E}_{vac}$), conduction band edge ($\mathcal{E}_c$), intrinsic Fermi level ($\mathcal{E}_{Fi}$), Fermi level ($\mathcal{E}_F$), valence band edge ($\mathcal{E}_v$). (b) Charge density (ρ). (c) Electric field magnitude ($|\mathbf{E}|$) Note that vertical scales for $\mathbf{E}$ are different for J-I and J-II ($|\mathbf{E}_{J-II}| \gg |\mathbf{E}_{J-I}|$).

simply a slab of intrinsic semiconductor inserted into the J-II gap. If the current transmission through the slab is good (that is, its effective resistance and junction potentials are small), then when equilibrium is reached, the physical characteristics of J-II will be approximately those of J-I, as depicted in Figure 9.2.

Theoretical limits to the energy released from J-II during its transition from an open- to a closed-switch configuration can be estimated from the total electrostatic energy $\mathcal{E}_{es}$ inherent to the J-II junction. Let $\Delta\mathcal{E}_{es}(J - II) = [\mathcal{E}_{es}(J - II, open) - \mathcal{E}_{es}(J - II, closed)]$ be the difference in electrostatic energy in J-II between its closed- and opened-switch equilibrium configurations (Figure 9.2). Within the 1-D model constraints, this can be shown to be roughly:

$$\Delta\mathcal{E}_{es}(J - II) \simeq \frac{\epsilon_o}{2}[\frac{x_{dr}kT}{q}\ln(\frac{N_A N_D}{n_i^2})]^2 \cdot [\frac{1}{x_g} - \frac{1}{3}\frac{\kappa}{x_{dr}}], \qquad (9.3)$$

By eliminating V_{bi} and x_{dr} with (9.1) and (9.2) and using $N_A = N_D \equiv N$, (9.3) can be recast into:

$$\Delta\mathcal{E}_{es}(J - II) \simeq \frac{16\kappa\epsilon_o^2}{qN}\{\frac{kT}{q}\ln[\frac{N}{n_i}]\}^3\{\frac{1}{x_g} - \frac{2}{3}\kappa(\frac{2\kappa\epsilon_o}{Nq}(\frac{kT}{q})\ln[\frac{N}{n_i}])^{-1/2}\} \qquad (9.4)$$

It is evident from (9.4) that the device's energy varies strongly with temperature, scaling as $(T)^3$. This is not surprising since primary determinants of the energy are V_{bi} and x_{dr}, both of which originate from thermal processes.

Positive energy release ($\Delta\mathcal{E}_{es} > 0$) is subject to limits in x_g, N, and T. From (9.3), an energy crossover ($+\Delta\mathcal{E}_{es}$) to ($-\Delta\mathcal{E}_{es}$) occurs at $x_g = \frac{3}{\kappa}x_{dr}$; for silicon, this is $x_g \simeq \frac{x_{dr}}{4}$. That is, only for $x_g \leq \frac{x_{dr}}{4}$ will net energy be released in switching from open- to closed-gap configurations. Since x_{dr} is normally restricted to $x_{dr} \leq 10^{-5}$m, this implies $x_g \leq 2 \times 10^{-6}$m, thus, thermal capacitors must intrinsically be microscopic in the gap dimension; and, at least for the vacuum case, mechanical considerations will probably also similarly limit the other two dimensions. Equation (9.4) indicates that energy crossover for N occurs for the *standard device* at $N \sim 10^{22}$m^{-3}. Finally, $\Delta\mathcal{E} \longrightarrow 0$ when T falls below the freeze-out temperature for charge carriers; for silicon, $T_{freeze} \leq 100$K.

For the *standard device*, (9.3) predicts the J-II region contains roughly three times the electrostatic potential energy of the J-I region. Equivalently, the whole p-n device contains twice the energy in its open-gap configuration as it does in its closed-gap configuration and the majority of this excess energy resides in the electric field of the open J-II vacuum gap.

The energy release in closing the J-II gap is equivalent to the discharge of a capacitor. For the *standard device*, (9.3) gives the net energy release as $\Delta\mathcal{E}_{es}(J - II) \sim 5.2 \times 10^{-17}$J ~ 320 eV. When J-II is open, there are about 330 free electronic charges on each gap face (calculable from Gauss' law); when it is switched closed, most of these disperse through and recombine in the J-II bulk. This net flow of charges is due to particle diffusion powered by concentration gradients and to particle drift powered by the large capacitive electric field energy of the open J-II vacuum gap. Thermodynamically, this energy release may be viewed as simply the relaxation of the system from a higher to a lower energy equilibrium state.

This thermal capacitor can remain charged indefinitely (until discharge) since the open-gap configuration is an equilibrium state of the system.

The device output power P_{dev} scales as: $P_{dev} \sim \frac{\Delta\mathcal{E}_{es}(J-II)}{\tau_{dis}}$, where τ_{dis} is the characteristic discharging time for the charged open-gap J-II region as it is closed. If τ_{dis} is short, say $\tau_{dis} \simeq 10^{-7} - 10^{-8}$ sec – a value consistent with the size of micron-sized p-n junctions or typical inverse slew rates of micron-sized transistors – then the instantaneous power for a single, switched *standard device* should be roughly $P_{dev} \simeq 0.5 - 5 \times 10^{-9}$ W. Instantaneous power densities can be large; for the *standard device* it is on the order of $\mathcal{P}_{dev} = \frac{P_{dev}}{(10^{-6}m)^3} \sim 0.5 - 5 \times 10^9$ Wm^{-3}.

9.2.2 Numerical Simulations

Two-dimensional numerical simulations of this system were performed using Silvaco International's semiconductor Device Simulation Software [Atlas (S-Pisces, Giga)]. Junctions were modeled as abrupt and the physical parameters for charge carriers were generic. Output from the simulations were the two-dimensional, steady-state, simultaneous solutions to the Poisson, continuity, and force equations, using the Shockley-Read-Hall recombination model. There is good agreement between the results of the 2-D simulator and those of the 1-D analytic model.

Devices identical to and similar to the *standard device* were studied. Over a wide range of experimental parameters ($10^{17} \leq N_{A,D} \leq 10^{26}$m^{-3}; $10^{-8} \leq x_g \leq 3 \times 10^{-7}$m), the two-dimensional numerical simulations showed good agreement with the primary findings of the 1-D analytic model, most significantly that much larger electric fields reside in the J-II vacuum gap than in the J-I junction, and that significant electrostatic energy is both stored in the J-II region and is released upon switching. Their differences can be traced primarily to the unrealistic discontinuities in physical parameters in the 1-D model, which were smoothed by the more realistic 2-D simulator.

In Color Plate IV, the electric field magnitude is shown for three related variations of the *standard device*. Color Plate IVa (hereafter, Case 1) depicts the electric field for the *standard device*, with the J-II gap closed. As expected, the electric fields are modest ($|\mathbf{E}| \leq 10^6$ V/m) and are centered on the depletion regions, which, as predicted in the 1-D model, extend over the length of the device. The field structure demonstrates perfect symmetry with respect to its horizontal mirror plane and rough mirror symmetry with respect to its vertical mirror plane. The imperfect vertical mirror symmetry is due to the differences in the physical properties of the charge carriers.

Color Plate IVb (Case 2) depicts the electric field magnitude for the *standard device*. While the electric fields in the J-I depletion regions of Cases 1 and 2 are similar, in the J-II regions they are significantly different. The J-II electric field in Case 2 is $E \sim 7 \times 10^6$ V/m versus an average of $E \sim 5 \times 10^5$ V/m for Case 1. Numerical integration of the electrostatic field energy over the entire region (vacuum and bulk) indicates the total electrostatic energy of Case 2 is roughly 1.5 times that of Case 1. Considering only the J-II region of each device, Case 2 stores roughly twice the electrostatic energy of Case 1. These are within 50% of the the energy estimates of the 1-D analytic model.

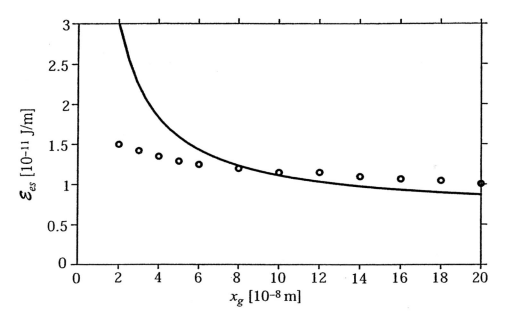

Figure 9.3: Z-normalized electrostatic potential energy $\mathcal{E}_{es}$ versus gap width (x_g) for *standard device* in 1-D model (solid line) and 2-D model (open circles).

Color Plate IVc depicts Case 3, a configuration intermediate between Cases 1 and 2, and one in which the J-II gap of the *standard device* is 20% bridged at its center by a slab of undoped silicon ($l_x = 300\mathring{A}$, $l_y = 600\mathring{A}$). As expected, the bridge allows electron-hole transport between the n and p regions, thereby reducing the large fields of Case 1 closer to values of Case 2. The field is attenuated most across the bridge, but in fact, attenuation extends over the entire length of the channel (L_y). The electrostatic energy of Case 3 is intermediate between Cases 1 and 2. This can be viewed as partial shorting out of the thermal capacitor. The electric fields for all cases are primarily in the x-direction, and especially so for Case 2.

Electrostatic potential energy is stored in the J-I and J-II regions of the device in both the open- and closed-gap configurations. In Figure 9.3, $\mathcal{E}_{es}$ is plotted versus x_g for the *standard device*, comparing the 1-D and 2-D models. (Note that the energy is normalized here with respect to the z-direction (J/m) so as to conform with the output of the 2-D model.) The total electrostatic energy is the sum of the contributions from the vacuum energy density ($\frac{\epsilon_o E^2}{2}$) and n-p bulk energy density ($\frac{\kappa \epsilon_o E^2}{2}$), integrated over their respective regions. In the 1-D model, we take the electric field in the J-II gap to be constant, while in the J-I region it is taken to have a triangular profile as in Figure 9.2, with maximum electric field strength of $E_{max} = \frac{2V_{bi}}{x_{dr}}$. For both models, the   device energy decreases

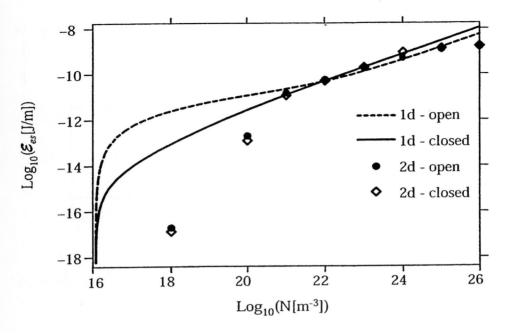

Figure 9.4: Z-normalized electrostatic energy $\mathcal{E}_{es}$ versus dopant concentration (N) for *standard device* for open- and closed-gap configurations. Comparison of 1-D and 2-D models.

monotonically with increasing gap width, however their magnitudes and slopes differ due to the differing model assumptions. At small gap widths ($x_g \leq 10^{-7}$m), the 1-D model predicts greater energy than the 2-D model, owing principally to its vacuum energy, however, at larger gap widths ($x_g \geq 10^{-7}$m) the energy in the 2-D model's n-p bulk dominates, as will be shown later. The 1-D model explicitly ignores contributions of energy to the open-gap configuration arising from the p-n bulk semiconductor on either side of the gap. (See (9.3).) In the density vicinity of the *standard device* the two models agree to within about 50%.

The stored electrostatic potential energy of the device strongly depends upon the dopant concentration. In Figure 9.4, $\mathcal{E}_{es}$ is plotted for the *standard device* versus dopant concentration N, for both open- and closed-gap configurations, comparing 1-D and 2-D models. Above $N = 10^{18}$m^{-3} the 1-D model shows roughly constant logarithmic increase in $\mathcal{E}_{es}$ with increasing N, while the 2-D model shows a roughly constant logarithmic increase up to about $N \simeq 10^{21}$m^{-3}, at which point $\mathcal{E}_{es}$ begins to flatten out and saturate for both open- and closed-gap configurations.

Both models display a crossover in energy between the open-gap and closed-gap configuration (See Color Plate IV) above the dopant concentration of the *standard device* ($N = 10^{21}$m^{-3}). The crossover density N_{cross} is where $\Delta\mathcal{E}_{es}$ reverses sign. In the 2-D model the energy crossover occurs at $N_{cross} \simeq 7 \times 10^{22}$m^{-3}, while in the 1-D model it occurs at $N_{cross} = 8 \times 10^{21}$m^{-3}. Above N_{cross} the closed-gap

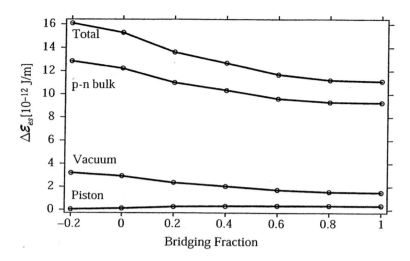

Figure 9.5: Z-normalized electrostatic energies $\mathcal{E}_{es}$ versus gap bridging fraction by undoped silicon slab.

configuration is more energetic than the open-gap one. As a result, above N_{cross} one cannot expect to extract energy by closing the J-II gap. The *standard device* operates at $N = 10^{21} \text{m}^{-3}$, which is a factor of 8 below the 1-D crossover and a factor of 70 below the 2-D crossover density.

Energy release due to gap closing can be made continuous. Let a tightly fitting rectangular slab of silicon be inserted into the gap, thereby allowing the transport of charge between the separated n- and p-regions and the relaxation of the J-II region into an equilibrium state like J-I (Figure 9.2). Figure 9.5 displays the electrostatic energy of the *standard device* ($\mathcal{E}_{es}$) versus bridging fraction by a slab of undoped silicon ($x_g = 300\mathring{A} \times 3000\mathring{A} = L_y$). Here, 0% bridging corresponds to a completely open configuration and 100% bridging corresponds to a completely closed configuration.

As expected, the total system, vacuum, and bulk energies decrease as the silicon is inserted; the silicon bridge energy increases slightly with its insertion. At full insertion, the system's energy is partitioned between bulk, vacuum and piston energy in a ratio of roughly 6 : 1 : 0.25. These data suggest that for minimal investment in piston energy, roughly 10 times more energy is released in the device as a whole. The 2-D simulations indicate a faster-than-linear decrease in system energy with bridging fraction. This can be explained by diffusion of charge into the bulk, ahead of the silicon slab.

9.3 Linear Electrostatic Motor (LEM)

9.3.1 Theory

The 1-D analytic and the 2-D numerical models verify that significant electrostatic potential energy resides in the J-II region of the *standard device* and that it can be released when the device is switched from an open to a closed configuration. Both configurations (Cases 1 and 2 in Color Plate IV) represent equilibrium states; that is, these are states to which the device relaxes when left alone in a heat bath. Their energies are different because of their differing boundary conditions, specifically in the J-II gap, which frustrates the diffusive transport of electrons and holes between the n- and p-regions. Since each configuration is a state to which the system naturally thermally relaxes, the device may be made to cycle between Cases 1 and 2 simply by opening and closing (bridging) the J-II gap with a piston (as done in Figure 9.5). Many energy extraction schemes can be imagined; here we consider one that can be rigorously analysed: a linear electrostatic motor (LEM).

The motor consists of a dielectric piston in the J-II gap which is propelled by a self-generated, electrostatic potential energy minimum (pulse). This electrostatic pulse propagates back and forth through the channel, carrying the piston with it. The piston itself creates the potential energy minimum in which it rides by electrically bridging the J-II gap locally. The free energy that drives the piston resides in the gap electric field; its thermal origin was discussed earlier (See (9.4).). In essence, the piston perpetually 'surfs' an electrostatic wave that it itself creates. As will be shown, the piston can surf under load (thus performing work) in the presence of realistic levels of friction and ohmic dissipation. In accord with the first law of thermodynamics, the net work performed must come from the surrounding heat bath; however, if the first law is satisfied, then the second law is compromised.

Consider a dielectric slab piston situated outside a charged parallel plate capacitor, as in Figure 9.6a. Let its motion be frictionless. It is well known that the dielectric slab will experience a force drawing it between the capacitor plates; this is indicated by the accompanying force diagram, which gives the force density experienced by the dielectric at a given horizontal position. The force can be calculated either by integration of the $(\mathbf{p} \cdot \nabla)\mathbf{E}$ force over the piston volume, or equivalently, by invoking the principle of virtual work since the total energy of the the piston-capacitor system is reduced as the slab enters the stronger field region between the plates. As the force diagram indicates (Figure 9.6a), the piston experiences a force only so long as it in the inhomogeneous field near the end of the capacitor. Specifically, the y-force (F_y) requires gradients in the y-component of the electric field; i.e., $[(\mathbf{p} \cdot \nabla)\mathbf{E}]_y = F_y = (p_x \frac{\partial}{\partial x} + p_y \frac{\partial}{\partial y})E_y$.

Now let the stationary dielectric piston be situated symmetrically between two identical capacitors (Figure 9.6b). Here the net force on the piston is zero and it rests at equilibrium. However, as the accompanying force diagram indicates, this equilibrium is unstable since any infinitesimal y-displacement increases the net force on the piston in the direction of its displacement, while simultaneously reducing the net force in the opposite direction. As a result, the piston will accelerate in the direction of its initial displacement.

Next, consider Figure 9.6c, which depicts a semiconducting dielectric piston at

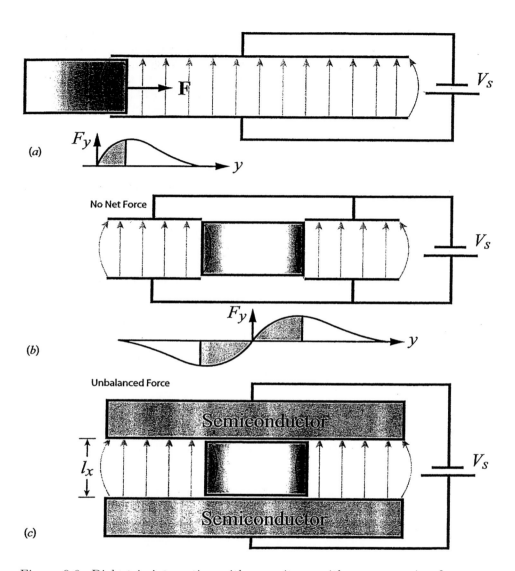

Figure 9.6: Dielectric interacting with capacitors, with accompanying force versus displacement graphs. a) Dielectric piston is drawn into charged capacitor via $[(\mathbf{p} \cdot \nabla)\mathbf{E}]_y$ force. b) Dielectric piston situated equidistantly between two equivalent capacitors in an unstable equilibrium; unbalanced force in direction of displacement. c) Linear electrostatic motor (rail gun): semiconducting dielectric piston in unstable equilibrium between semiconducting capacitor plates.

rest, situated between two semiconductor capacitor plates. (Compare this to Case 3 in Color Plate IV.) The semiconducting dielectric piston allows charge transport between the plates, and so it locally reduces the electric field in and around the piston; thus, the piston sees more intense fields to either side. Essentially, it is in the same unstable equilibrium depicted in Figure 9.6b. If displaced, it will accelerate in the direction of its displacement.

From the principle of virtual work, one can write the frictionless electrostatic acceleration (a_{es}) of the piston (mass density ρ; physical dimensions l_x, l_y, l_z; dielectric constant ϵ) inside a long parallel plate capacitor as

$$a_{es} \simeq \frac{\epsilon - \epsilon_o}{2l_y\rho}(E_1^2 - E_2^2) = \frac{\epsilon - \epsilon_o}{2\rho l_y l_x^2}V_s^2(\alpha_1^2 - \alpha_2^2), \qquad (9.5)$$

where E_1 and E_2 are the electric field strengths at the ends of the piston and $\alpha_{1,2} = \frac{E_{1,2}}{E_o}$, where $E_o = \frac{V_s}{l_x}$ is the strength of the undisturbed electric field far from the piston.

If the piston is at rest, then by symmetry $E_1 = E_2$, and there is no acceleration, but if the piston is displaced, then $E_1 \neq E_2$ and the piston accelerates in the direction of motion. In the frictionless case, the piston is unstable to any displacement. In essence, this motor is an electrostatic rail gun, the electrostatic analog of the well-known magnetic rail gun.

We note that $a_{es} \neq 0$ *only* for the case of both a semiconductor capacitor and a semiconductor piston; if either the piston or the capacitor plates are perfectly conducting or perfectly insulating, then $a_{es} = 0$. If the capacitor plates are perfect conductors (approximated by metallic plates), then the plate surfaces must be equipotentials, in which case there cannot be a net electric field difference between the front and back ends of the piston ($E_1 - E_2 = 0$), therefore $a_{es} = 0$. On the other hand, if the plates are perfect insulators, then their surface charges are immobile and the electric field remains the same throughout the capacitor despite any displacement of the piston and again $E_1 - E_2 = 0$. Conversely, if the piston is a perfect conductor, its surfaces must be equipotentials so the electric field at the front and back must be the same ($E_1 - E_2 = 0$), or alternatively, one can say that, as a conductor, electric fields cannot penetrate into the piston interior so as to apply the $[(\mathbf{p} \cdot \nabla)\mathbf{E}]_y$ force, and again there can be no net force exerted on it. Finally, if the piston is an insulator, then charge residing on the capacitor plates cannot flow through it so as to diminish the electric field; again, $E_1 - E_2 = 0$. Thus, it is only when both the piston and the plates have finite, non-zero conductivities that they can act as an electrostatic motor.

Assuming the piston to be a semiconducting ($0 < \sigma < \infty$) dielectric (ϵ), then using Ohm's law ($\mathbf{J} = \sigma\mathbf{E}$), the continuity equation ($\nabla \cdot \mathbf{J} = -\frac{\partial\rho}{\partial t}$), and Gauss' law ($\nabla \cdot \mathbf{E} = \frac{\rho}{\epsilon_o}$), one can describe the acceleration of the piston a_{es} in terms of its electromechanical properties as it locally shorts out the electric field in the channel through which it passes:

$$a_{es} = \frac{\epsilon - \epsilon_o}{2\rho l_y}[\frac{V_s}{l_x}]^2 \exp[-\beta\eta]\{1 - \exp[-\eta]\} = \frac{\epsilon - \epsilon_o}{2\rho l_y}E_o^2 \exp[-\beta\eta]\{1 - \exp[-\eta]\}$$

$$(9.6)$$

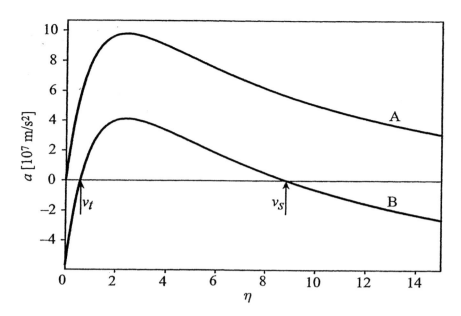

Figure 9.7: Acceleration of piston a_{es} versus η for *standard device*. Curve A: No friction or load. Curve B: Non-zero friction or load ($a = 5.6 \times 10^7 \mathrm{m/sec}^2$); minimum starting velocity and terminal velocity indicated.

Here $\eta = \frac{2\sigma l_y}{\epsilon v_y}$, v_y is the velocity of the piston, and β is a phenomenological constant that is a measure of how far ahead of the moving piston the electric field is affected. β must be positive to avoid unphysical delta function charge densities. Small β values are evidenced in later 2-D simulations (Plate V); here we take $\beta = 0.1$.

Consider a rectangular slab of silicon ($l_x = 300\mathring{A}; l_y = 600\mathring{A}$; $l_z = 10^4\mathring{A}$, $\sigma = 4 \times 10^{-3}(\Omega m)^{-1}$, $\kappa = 11.8$), hereafter called the *standard piston*. In Figure 9.7, the *standard piston*'s acceleration is plotted versus η for the *standard device*. Curve A represents the frictionless case. In the limits of $v_y \to 0$ ($\eta \to \infty$) and $v \to \infty$ ($\eta \to 0$), one has $a_{es} \to 0$, as expected. The former case ($v_y = 0$) has been treated previously. For $v_y \to \infty$, the piston moves too quickly for the capacitor's charge to cross the piston and short out the field, so $E_1 - E_2 = 0$ and $a_{es} = 0$. Since σ and v_y are reciprocals in η, this model also predicts, as before, that $a_{es} = 0$ if the piston is perfectly conducting ($\sigma = \infty$) or perfectly insulating ($\sigma = 0$) and, therefore, accelerates only for the semiconductor case.

The form of a_{es} in (9.6) is handy for introducing friction on, and loading of, the piston. This model considers loading to be constant over the range of velocities of the piston, with the result that its acceleration curves are simply shifted down by an amount equal to the magnitude of the loading. Thus a non-zero start-up velocity and a bounded terminal velocity are imposed on the piston dynamics. We point out that the negative portion of Curve B to the right of v_s does not signify negative acceleration, but simply indicates values of η for which motion is forbidden.

Numerical integration of (9.6), incorporating friction and load resistances, allows investigation of the piston's complex nonlinear dynamics. For example, in Figure 9.8a, the *standard piston*'s velocity is plotted versus time for three values of friction/loading for the *standard device*. Curve (i) corresponds to the unloaded, frictionless piston case; it has no classically-defined terminal velocity. Curve (ii) corresponds to the piston subject to a constant frictional/load acceleration of 5.6×10^7 m/sec². In this case, the piston has a terminal velocity of roughly 8 m/sec. Finally, for Curve (iii) ($a = 9.4 \times 10^7$ m/sec² friction/load), the piston has only a narrow range of velocities for which it has positive acceleration; for greater friction or loading the piston does not begin to move.

Figure 9.8c plots piston power versus v_y for the previous three cases. In the frictionless case (Curve (i)), power increases monotonically, but is bounded. Cases (ii) and (iii) display local maxima. The power maximum for case (ii) occurs below its terminal velocity, indicating that the most efficient power extraction schemes should use velocity-governed loads, rather than constant loads. Also, notice that case (ii) and (iii) show initially negative excursions, evidence that energy must be supplied to kick-start the piston's motion.

There are three characteristic times scales pertinent to the operation of the *standard device*: (i) the plate discharge time along the piston ($\tau_{dis} \simeq \frac{l_y}{v_y}$); (ii) the recharging time for the plates (τ_{rec}); and (iii) the period of oscillation of the piston in the channel ($\tau_{osc} \simeq \frac{2L_y}{v_y}$), where v_y is the average velocity of the piston. The discharge time (τ_{dis}) must allow a sufficient difference in electric field to be maintained between the leading and trailing edges of the piston so that it is pulled through the channel.

Circuit theory shows that the recharge time (τ_{rec}) will be longer than the discharge time and should not present an operational problem. Typically, τ_{rec} for p-n diodes of physical dimensions comparable to the *standard device* are $\tau_{rec} \simeq 10^{-7} - 10^{-8}$ sec. However, in order for the electric field in the gap to thermally regenerate enough to maintain force on the piston, the period of oscillation of the piston in the channel (τ_{osc}) must be longer than τ_{rec}, and ideally, much longer. Therefore, for the smooth operation of the motor, the ordering for characteristic time scales should be $\frac{l_y}{v_y} \simeq \tau_{dis} < \tau_{rec} \ll \tau_{osc} \simeq \frac{2L_y}{v_y}$.

The electrostatic motor (Figure 9.9a) can be modeled as a network of discrete resistors and capacitors (Figure 9.9b). The semiconductor capacitor plates are modeled as a distributed network of resistors (R) and their interior surfaces as a sequence of aligned parallel plate capacitors (C). The network is powered by a battery (V_s).

The piston is represented by a resistor and by an accompanying switch. The piston's motion is modeled by the sequential closing and opening of the local switches. As the piston leaves a capacitor$_n$ region, a closed switch$_n$ opens up, while the next switch$_{n+1}$ in line closes, signaling the arrival of the piston. The trailing capacitor recharges while the leading capacitor discharges.

It can be shown from basic circuit theory – and has been confirmed by parametric studies of this system using PSpice network simulations – that the time constant

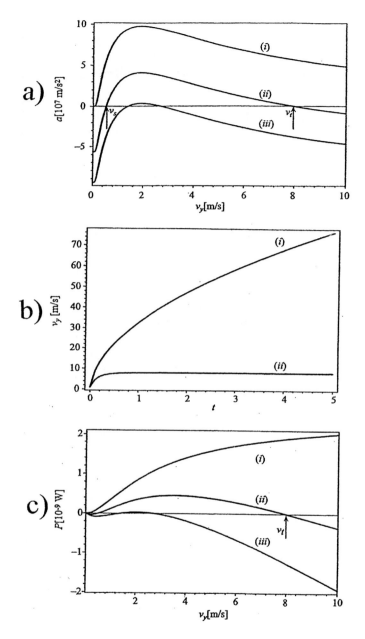

Figure 9.8: Piston dynamics for *standard device*. a) Acceleration versus v_y for three cases: (i) frictionless; (ii) friction/load acceleration $a = 5.6 \times 10^7 \text{m/sec}^2$; and (iii) friction/load acceleration $a = 9.4 \times 10^7 \text{m/sec}^2$. b) Velocity versus time for cases (i) and (ii) above; case (iii) absent for lack of sufficient start-up velocity. c) Piston power versus v_y for cases (i) - (iii) above.

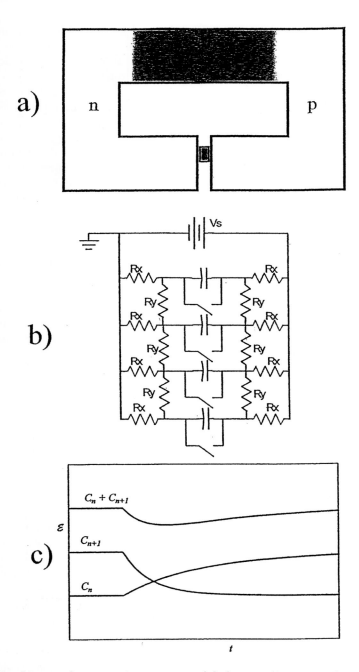

Figure 9.9: Linear electrostatic motor modeled as a discrete resistor-capacitor network. a) Piston in *standard device*. b) Analog resistor-capacitor network model. c) Electrostatic energy versus time for sequential firing of two capacitors; traveling negative potential energy pulse evident in $(C_n + C_{n+1})$ curve.

for the discharging capacitor is less than the time constant for the recharging capacitor. As a result, the moving piston always finds itself moving in the direction of more intense electric fields and field gradients. In other words, it perpetually moves forward toward a lower local energy state, riding in a self-induced potential energy trough. The traveling piston can also be viewed as the material equivalent of an electrical pulse propagating through a resistive-capacitive transmission line. This semiconducting piston acts analogously to the conducting piston in a magnetic rail gun which, by completing the circuit between the gun's two electrified rails, establishes a current and magnetic field by which the resultant Lorentz force on the piston's current drives the piston along the rails. In the present electrostatic case, the piston is propelled forward by the greater $(\mathbf{p} \cdot \nabla)\mathbf{E}$ force on its leading edge.

When the piston reaches the end of the R-C network, where the field ahead has dropped off, but where field behind has regenerated, the piston reverses its motion. As a result, it will move cyclically through the network. It is remarkable that this motion does not require any electronic timing circuitry; instead, the timing is set by the piston itself. As long as it overcomes friction, the piston will run perpetually for the life of the battery.

Via the substitution $V_s \rightarrow V_{bi}$, the piston in Figures 9.9 and 9.10 may now be identified as the semiconductor piston in Case 3 (Plate IVc). The same physics applies, except that, whereas the free energy for the linear electrostatic motor (rail gun) above is supplied by a battery, now it is supplied by the free energy of the thermally-powered p-n depletion region.

9.3.2 Numerical Simulations

Essentials of the above 1-D dynamical nonequilibrium model of the linear electrostatic motor are corroborated by the equilibrium solutions of the 2-D model. Color Plate V presents a sequence of 2-D equilibrium solutions simulating aspects of the motion of the piston through the J-II region of the *standard device*. It is strongly emphasized that this is *not* a dynamical simulation in which the piston is modeled as moving; rather, these are quasi-static equilibrium configurations of the system simulated by the Atlas program in which the piston is held at rest at different locations in the J-II region, despite implicit force imbalances. Nevertheless, much physics can be inferred by stepping the piston through the channel in this fashion.

In Plate Va, the leading edge of the piston is visible above the J-II channel. The electric field is fairly uniform in the gap interior ($E \simeq 7 \times 10^6$ V/m), decreasing in strength at its ends, as expected. As the piston enters the gap, thereby initiating the bridging of the separated n- and p-regions, the electric field strength falls throughout the J-II vacuum and p-n bulk regions, but most strongly near the piston. This substantiates the β term in (9.6). The field and field gradients are stronger below the piston (in the direction of implied motion) than above it (outside the channel); as a result, should the piston be free to move, it would be drawn further into the channel. In Plate Vc, with the piston now squarely within the channel, the electric fields in and near the piston have been reduced by a factor of 3 below pre-insertion values, but they remain larger in the channel ahead of the

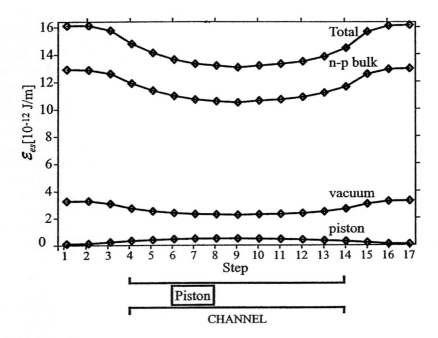

Figure 9.10: Electrostatic potential energy of *standard device* versus piston position in J-II gap: Total energy, p-n bulk energy, vacuum energy, and piston energy indicated. Piston located at Step 7, corresponding to Plate Vd.

piston and, therefore, continue to draw it in.

In Plate Vd, as the piston approaches mid-channel, the field ahead of the piston continues to be more intense than the one behind. At mid-channel, (Plate Ve), the field is roughly balanced on both sides of the piston. Here, a resting piston would experience roughly no net force, but it would be in the unstable equilibrium position depicted in Figure 9.8b. Were it in motion, then it should continue to see stronger fields and field gradients ahead of it than behind it and, thus, would continue to move in the direction of motion. Furthermore, since presumably it has already accelerated to mid-channel from the gap ends, its inertia should carry it past this mid-channel equilibrium point.

Now compare the upper channel in Plates Vd and e. Notice the field has been partially restored between Vd and Ve after the 'passage' of the piston. Finally, in Plate Vf, the piston has reached the bottom of the channel. As before, the field is locally reduced, but it has regenerated behind. Since the field is now stronger behind the piston, it should exert a net force upward so as to reverse its motion. It is instructive to view this 'motion sequence' in reverse, proceeding from Ve→Va so as to appreciate how the piston's motion can be cyclic. This is most evident perhaps in the inversion symmetry seen between Plates Vc and Vf.

Figure 9.10 displays the equilibrium electrostatic potential energies of the *standard device* and *piston* for a sequence of steps through the channel, calculated with the Atlas 2-D simulator. Frames a, b, c, d, e, and f in Plate V correspond to Steps 1, 3, 5, 7, 9 and 13, respectively, in Figure 9.10. The total, vacuum, and p-n bulk energies of the *standard device* decrease significantly and symmetrically as the piston enters the channel from either direction and reaches the mid-channel (Step 9). The fractional change in field energy in the vacuum is greater than for the p-n bulk, but the greatest absolute change occurs in the bulk. The electrostatic energy invested in the piston itself is small compared with the bulk and vacuum contributions.

The data in Figure 9.10 are equilibrium solutions and assume the piston to be at rest. The energy depression seen in Figure 9.10 would occur only locally around the piston and would be spatially asymmetric, with its greatest strength and gradient in the direction of the piston's motion, both as suggested in individual frames of Plate V, in Figure 9.9c, and in the 1-D analytic model. In summary, the sequential 2-D numerical simulations (Plate V and Figure 9.10) support the 1-D nonequilibrium analyses preceding it.

9.3.3 Practicalities and Scaling

The steady-state operation of this solid-state electrostatic motor constitutes a *perpetuum mobile* of the second type. It pits the first law of thermodynamics against the second. If the piston cycles perpetually while under load, performing work, then this energy must come from somewhere. Assuming the first law is absolute, the only possible source of this unlimited energy must be the [infinite] heat bath surrounding the device. Since the device operates in a thermodynamic cycle, heat is transformed solely into work, in violation of the second law.

This section addresses the practical details of this device, paying especial attention to the operational limits imposed by physically realistic parametric values: mass, physical dimensions, electric field, friction, electrical conductivity, characteristic time scales (e.g., τ_{dis}, τ_{rec}, τ_{osc}), and statistical fluctuations. It is found that there exists a broad parameter space at and below the micron-size scale for which a semiconducting piston should be able to overcome realistic levels of friction and load so as to perform work indefinitely, while being driven solely by the thermally-generated electric fields of a p-n junction. It is found that these devices should be able to convert heat energy into work with high instantaneous power densities, perhaps greater than 10^8 W/m^3.

Consider the *standard piston* situated in the J-II gap of the *standard device*. From Figure 9.10, the *standard piston* should reside in a potential well approximately 3×10^{-18} J deep. From Figures 9.8 and 9.9, in the frictionless case the piston should experience a maximum acceleration of 10^8 m/sec^2 and be capable of instantaneous power outputs of 2×10^{-9} W. We will now consider a realistic model for friction.

Let the J-II channel walls be tiled with a thin, low-friction surface such as graphite. Let the outer surfaces of the piston be only partially tiled with a matching low-friction surface such that the contact fraction between the piston and the channel walls (f_c) is small ($0 < f_c \ll 1$). On the other hand, let f_c be sufficiently large that: (i) there are sufficient numbers of atoms projecting out from the piston surfaces in contact with the channel walls to hold and guide the piston; and (ii) there is sufficiently good electrical conduction between the piston and the channel walls that one can use standard Ohmic current rather than quantum mechanical tunneling current to describe the system's electrical behavior.

It is well known that, at micron and sub-micron size scales, atomic, ionic, and electrostatic forces (e.g., van der Waals' interactions, induced surface charge, molecular and hydrogen bonding, surface tension) can play dominant roles in system dynamics. In order to minimize friction between the piston and channel walls, f_c should be as small as possible. The smallest non-zero coefficients of static and kinetic friction yet measured experimentally are found in nested multi-walled carbon nanotubes (MWNT) [11, 12]. Upper-limit values of coefficients of static (s) and kinetic (k) friction have been experimentally measured to be: $\mathcal{F}_s < 2.3 \times 10^{14}$ N/atom $= 6.6 \times 10^5$ N/m^2, and $\mathcal{F}_k < 1.5 \times 10^{-14}$ N/atom $= 4.3 \times 10^5$ N/m^2. Theoretical arguments suggest true values could be much lower than these. This friction is presumed to arise purely from van der Waals' interactions between the sliding carbon contact surfaces. The friction can be reduced by reducing the contact fraction f_c. Experimental observations suggest that MWNT operate as totally wear-free bearings [13].

The static or kinetic friction $F_{f(s,k)}$ between two surfaces of area A, where normal forces are not imposed and asperities are absent, should scale as: $F_{f(s,k)} = f_c A \mathcal{F}_{(s,k)}$. For the piston in the J-II channel, the acceleration is:

$$a_{f(s,k)} = \frac{F_{f(s,k)}}{m} = \frac{f_c A \mathcal{F}_{(s,k)}}{\rho_{Si} l_x l_y l_z} = \frac{2 f_c \mathcal{F}_{(s,k)}}{\rho_{Si} l_x} \tag{9.7}$$

For the piston to begin moving in the channel the electrostatic acceleration must exceed the static friction:

$$\frac{a_{es}}{a_{f,s}} = \frac{(\epsilon - \epsilon_o) V_{bi}^2 (\alpha_1^2 - \alpha_2^2)}{4 l_x l_y f_c \mathcal{F}_s} > 1 \tag{9.8}$$

This inequality is the starting point for delimiting a viability regime for the operation of this device. For the *standard piston* in the *standard device* (letting $\mathcal{F}_s$ be the upper-limit value for MWNT and taking $((\alpha_1^2 - \alpha_2^2) \simeq 0.5)$, (9.8) reduces to:

$$\frac{a_{es}}{a_{f,s}} \simeq (3.2 \times a^{-18}) \frac{1}{l_x^2 f_c} \tag{9.9}$$

In Figure 9.11 is plotted $\text{Log}_{10}(\frac{a_{es}}{a_{f,s}})$ versus l_x for various contours of constant f_c. For $\text{Log}_{10}(\frac{a_{es}}{a_{f,s}}) < 0$, the frictional acceleration exeeds the electrostatic acceleration, so the piston cannot move. This places a lower bound on the viability regime of the *standard device*. Above this bound, the piston can experience sizable accelerations, on the order of $10^7 - 10^8$ m/sec^2, but these accelerations are still within mechanical strength limits for small structures.

A left-most viability bound for the *standard device* is found by requiring that l_x significantly exceed the size of individual atoms and, preferably, be large enough that the system can be treated by classical, rather than quantum, theory. If the piston thickness l_x is greater than about 50-100 atoms, or about 10^{-8} m, this

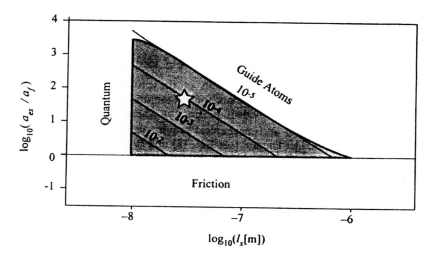

Figure 9.11: Ratio of piston acceleration to frictional acceleration ($\mathrm{Log}_{10}(\frac{a_{es}}{a_f})$) versus gap width ($x_g$) with contours of constant f_c indicated. Viability regime delimited by labeled boundary lines. Star indicates location of *standard device*.

system should be essentially classical. This criterion sets the left-most bound of the viability regime in Figure 9.11. The last bound is set by restricting f_c such that some reasonable minimum number of atoms act as guide surfaces between the piston and the channel walls. Choosing 10 atoms/piston face as sufficient, the sigmoidal right-top viability bound is determined. This bound can later be modified to satisfy electrical conductivity constraints.

The viability regime has been delimited using realistic, but conservative, choices for the system parameters. More liberal choices (e.g., letting $\mathcal{F}_s \to \mathcal{F}_k$ or $(\alpha_1^2 - \alpha_2^2) = 0.8$) would expand the regime somewhat. Even as it stands, however, the viability regime for the electrostatic motor spans two orders of magnitude in size (10^{-8}m $\leq l_x \leq 10^{-6}$m) and over three orders of magnitude in $\frac{a_{es}}{a_{f,s}}$.

Several observations can be made from Figure 9.11:

a) The spontaneous acceleration of the piston by self-generated fields appears possible only for micron and sub-micron pistons. This is especially evident in (9.6) where $a_{es} \sim \frac{1}{l_x^3}$. Given the severe physical and mechanical requirements for positive acceleration against friction (See (9.8) and (9.9), it is not surprising that this phenomenon has not been discovered accidentally.

b) a_{es} can exceed a_f by more than 3 orders of magnitude, thus allowing significant loading of the piston with which to perform work.

c) More frictional contact surfaces appear feasible (up to 10^3 times more frictional than MWNT), without precluding piston motion or loading.

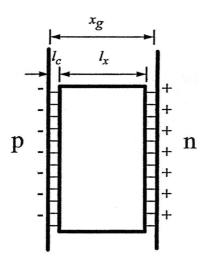

Figure 9.12: Physical dimensions of *standard piston*.

The magnitude of the piston's acceleration ($a_p = a_{es} - a_f$) can be calculated, including friction, using (9.5-9.7). For the *standard device*, using graphitic surfaces and assuming $f_c = 10^{-4}$ (corresponding to 1.7×10^6 atoms on each piston face), one finds $a_f = 1.9 \times 10^6$ m/sec^2, $a_{es} = 6.7 \times 10^7$ m/sec^2, and $a_p = 6.5 \times 10^7$ m/sec^2. The average velocity during a piston stroke is roughly $v_y \simeq \sqrt{2 L_y a_p} \simeq 2$m/s. The oscillation period of the piston in the channel is $\tau_{osc} = (\frac{2 L_y}{v_y}) \simeq 2 \times 10^{-7}$sec; the oscillation frequency is $f_{osc} = \tau_{osc} \simeq 5 \times 10^6$Hz. τ_{osc} is significantly longer than the typical inverse slew rates for p-n transistors of comparable physical dimensions ($\tau_{dis} \lesssim \tau_{rec} \sim \tau_{trans} \sim 10^{-8} - 10^{-7}sec<< \tau_{osc} \simeq 2 \times 10^{-7}$sec); therefore, the electric field in the wake of the piston traversing the channel can recharge before the piston's return. On the other hand, given a typical piston velocity and length ($v_y \simeq 2$m/sec, $l_y \simeq 6 \times 10^{-8}$m), these field decay rates are sufficiently high for the electric field in the channel walls to decay along the length of the piston ($\tau_{dis} \simeq \frac{l_y}{v_y} \sim 3 \times 10^{-8}$sec) so as to admit significant difference in the magnitude of the electric field between the leading and trailing edge of the piston; therefore, the *a priori* estimate of (($\alpha_1^2 - \alpha_2^2$) = 0.5) is plausible. A similar conclusion is supported by evaluation of the exponential decay model (in (9.6)). Overall, the time scale ordering developed earlier ($\tau_{dis} \lesssim \tau_{rec} \ll \tau_{osc}$) is reasonably well satisfied.

The viability regime depicted in Figure 9.11 is favorable to ohmic treatment of the piston and channel. The piston acts as a sliding electrical resistor – essentially a motor brush – between the positive polarity n-region and the negative polarity p-region, as depicted in Figure 9.12. The piston's electrical resistance can be written as

$$R_{piston} = R_b + 2R_c = \frac{1}{l_y l_z}[\frac{x_g - 2l_c}{\sigma_b} + \frac{2l_c}{f_c \sigma_c}] \simeq \frac{1}{l_y l_z}[\frac{l_x}{\sigma_b} + \frac{2l_c}{f_c \sigma_c}] \qquad (9.10)$$

where $R_{b(c)}[\Omega]$ is the electrical resistance of the piston bulk (contacts); $\sigma_{b(c)}[(\Omega m)^{-1}]$

is the electrical conductivity of the bulk (contact) material; and l_c is the x-length of the contacts. It is assumed that $l_c \ll l_x \simeq x_g$. The values of the l_c and f_c are both small and offset one another, while σ_c can in principle be varied over many orders of magnitude such that R_c can be made negligible compared with R_b. Consider, for example, the *standard device* with a silicon piston lined with graphite, operating with the following parameters: $l_x = 300\mathring{A}$, $l_y = 600\mathring{A}$, $l_z = 10^4\mathring{A}$, $l_c = 5\mathring{A}$, $\sigma_b = 10\sigma_{Si} = 4 \times 10^{-3}(\Omega m)^{-1}$, $\sigma_c = \sigma_{graphite} = 7.1 \times 10^4(\Omega m)^{-1}$, and $f_c = 10^{-4}$. With these parameters, one has from (9.6-9.10): $R_b \simeq 10^8\Omega \gg R_c = 2.5 \times 10^3\Omega$ and $(a_{es} = 6.7 \times 10^7 \mathrm{m/sec^2}) > (a_p = 6.5 \times 10^7 \mathrm{m/sec^2}) \gg (a_{f,s} = 1.9 \times 10^6 \mathrm{m/sec^2}) > (a_{f,k} = 1.2 \times 10^6 \mathrm{m/sec^2})$. Ohmic losses for this system, concentrated in the piston region, can be engineered to be insignificant.

For objects in this size range, the effect of statistical fluctuations should be considered, especially since they have been the foil of many past challenges. Earlier analysis indicates the *standard device* can be modeled as an R-C network, so it is appropriate to consider fluctuations in electronic charge. Charge is also naturally salient since it is through charge-induced electric fields that the system is powered. Spectral analysis in the spirit of the Nyquist and Wiener-Khintchine theorems [14] allows one to write the rms charge fluctuation for a resistor capacitor system as $\sqrt{<\Delta Q^2>} = \Delta Q_{rms} \sim \sqrt{4RC^2kT\Delta f}$, where C is capacitance of the J-II region and Δf is the spectral width of the fluctuations measured. Taking characteristic values for the *standard device* ($R = 10^9\Omega$, $C = \frac{\epsilon_o L_y l_z}{l_x} = 10^{-16}$F, $\Delta f \sim f_{osc} \simeq 5 \times 10^6$Hz, $T = 300$K), one obtains $\Delta Q_{rms} \simeq 2$ electronic charges. Since the total charge in the *standard device*'s J-II region is found to be $Q_{total} \simeq 330q$, one expects less than one percent statistical fluctuation in electronic charge over the entire J-II channel capacitor during a piston's oscillation period. Since the fractional statistical fluctuation is much less than the fractional change in charge due to electrical operation of the piston itself ($0.01 \simeq \frac{\Delta Q_{rms}}{\Delta Q_{total}} \ll \frac{\Delta Q_{op}}{\Delta Q_{total}} \simeq 0.4$), by this measure, statistical fluctuations should not play a primary role in the operation of the *standard device*. A similar conclusion can be reached by equating the thermal energy to the piston's kinetic energy.

Assuming that a_{es} is constant in magnitude and that $a_{es} \gg a_f$, the average power per cycle can be shown to be $< P_{sd} >= m_{piston}(2a_p^3 L_y)^{1/2} \simeq 2 \times 10^{-9}$W, where $m_{piston} = \rho_{Si}l_xl_yl_z$ is the mass of the piston. The average power densities for the *standard device* are, therefore, $\mathcal{P}_{sd} \sim 2 \times 10^9 \mathrm{Wm^{-3}}$. The *standard device* appears capable of producing significant output power and power densities in the presence of realistic levels of friction, while satisfying the conditions for classical electrical conductivity, providing substantial numbers of guide/contact atoms, and overcoming statistical fluctuations.

In Color Plates VIa,b power (W) and power density (Wm^{-3}) are explored for a range of device sizes, scaled in direct physical proportion to the standard device (i.e., $l_y = 2l_x$, $L_y = 5l_y$, $l_z = 33.3l_x$, etc.). The other physical specifications of the *standard device* are retained (i.e., silicon matrix, $N_A = N_D = 10^{21} \mathrm{m^{-3}}$, etc.). The previously discussed viability bounds (Figure 9.11) are still enforced. In Plate VIa, the maxima of the power curves (Figure 9.8c) are calculated over an extended viability regime (as in Figure 9.11) and plotted versus l_x and f_c. Power contours extend linearly in value from a maximum of 1.2×10^{-8} W/device (yellow, center) down to 1×10^{-9} W/device (red). The star indicates the location of the *standard device*.

Perhaps a more meaningful figure of merit than maximum power per device (Plate VIa) is maximum power density (Wm^{-3}). It is a better indicator of how rapidly thermal energy can be transformed into work by a given volume of working substance; thus, it is a better measure of how significantly this device challenges the second law. In Plate VIb, maximum power density (Wm^{-3}) is presented for a range of devices versus f_c and x_g, scaled as before in direct physical proportion to the *standard device*. Whereas in Plate VI the contour values vary linearly with adjacent contours, in Plate VIb they vary logarithmically in value from 10^{10}Wm^{-3} (left-most, yellow) to 10 Wm^{-3} (right-most, red). Again, the *standard device* is located by the star. The greatest power density obtains for small devices, while the greatest unit power obtains for larger devices.

The parameter space available for this device (spanned by x_g, x_{dev}, N_A, N_D, T, etc.) is far greater than can be explored here, and only modest attempts have been made to optimize the performance of the *standard device*. Nonetheless, it appears the theoretical instantaneous power densities achievable by it are sizable. To put this in perspective, one cubic meter of *standard devices* (amounting to 10^{18} in number) could, in principle, convert thermal energy into work with instantaneous power output on par with the output of a modern-day nuclear power plant; or, in 1 second, produce the work equivalent of the explosive yield of 500 kg of high-explosive. This, of course, is only *instantaneous* power density since, were the device to convert thermal energy into work at this rate without compensatory heat influx from the surroundings, the device would cool at an unsustainably fast rate of about 100 K/sec.

More advanced designs for the motor can be envisioned. For example, the linear *standard device* could be circularized. This rotary motor would consist of concentric cylinders of n- and p-regions (the stator) joined at their base (to create a depletion region) and having a gap between them in which a multi-piston rotor runs. Multiple rotor pistons could be yoked together so as to balance radial forces and torques. In the limit of large radius, the rotor pistons would move in what is essentially a linear track, so the above discussion for linear motors should apply. The rotor pistons would be driven by the local electric field energy in the cylindrical gap. If they are spaced sufficiently far apart azimuthally, then the field in the wake of a given piston could thermally regenerate in time to power the advancing piston.

9.4 Hammer and Anvil Model

9.4.1 Theory

A more immediate laboratory test of the thermal capacitor concept appears feasible, one sidestepping the high-tolerance micromachining required of the LEM. This will be called the *hammer-anvil*. It is a thermally-charged semiconductor parallel-plate capacitor, with one plate fixed and the other mounted on a flexible double cantilever spring. For mechanical $Q \geq 10^3$, and for matched electrical and mechanical time constants ($\tau_e \sim \tau_m$), the system can execute steady-state, resonant oscillation by which thermal energy is converted into mechanical energy. An example based on Sandia National Laboratories' SUMMiT^{TM} process is examined, however, more advanced designs are possible. As for the *standard device*, the *hammer-anvil* relies on the depletion region of a n-p junction to establish a potential difference and electric field in the active, open-gap region at the middle of the device. This device can be constructed within present-day NEMS and MEMS fabrication art and so represents a more immediate and cogent challenge than the LEM.

NEMS and MEMS cantilever oscillators have many proven and potential applications, including as accelerometers, motors, clocks, sensors (e.g., temperature, pressure, electronic charge, magnetic fields, environmental contaminants, microbes), beam steerers, choppers, and modulators, computing elements, and switches [15, 16, 17]. These are usually driven by AC electrical signals whose frequencies are commensurate with their mechanical oscillation frequencies, but under suitable circumstances, DC signals can also effectively drive them. Owing to their utility, the art of NEMS-MEMS cantilevers is relatively advanced. DC-driven, resonant micro-cantilevers have been explored [18].

Consider the macroscopic electromechanical device pictured in Figure 9.13a, consisting of a battery (V_o), resistor (R), and a variable capacitor in which the bottom plate (the *anvil*) is fixed, while the top plate (the *hammer*, mass m) is supended from a conducting spring with spring constant k. This will be called the *hammer-anvil*. It is a hybrid of well-known mechanical and LRC oscillators. The hammer is free to move with respect to the anvil and when they contact any accumulated charge on the plates is assumed to flow between them without resistance. The electrical capacitance of the device varies with the dynamic separation of the plates according to

$$C(y) = \frac{\epsilon_o A}{y_{gap} - y},\tag{9.11}$$

where A is the area of the plates, $y = 0$ is the static mechanical *equilibrium position* of the hammer, y_{gap} is the *equilibrium separation* of the plates, and $y = y(t)$ is the *instantaneous position* of the hammer, with the positive direction downward. For convenience, we denote by $C_o = \frac{\epsilon_o A}{y_{gap}}$ the capacitance when the spring is in its undeflected equilibrium state.

Two independent time constants characterize this system: one electrical ($\tau_e \sim RC_o$) and one mechanical ($\tau_m = 2\pi\sqrt{\frac{m}{k}}$). The electromechanics of the hammer is described by the coupled pair of equations:

$$F = F_{diss} + F_s + F_{es} = m\ddot{y} = -\frac{1}{Q}\dot{y} - ky - \frac{q^2}{2\epsilon_o A},\tag{9.12}$$

where the instantaneous charge on the plates $q(t)$ satisfies:

$$\dot{q} = (V_o - \frac{q(y_{gap} - y)}{\epsilon_o A})\frac{1}{R}; \quad q < q_{sat},\tag{9.13}$$

and $\dot{q} = 0$ for $q \geq q_{sat}$. Here the rhs of (9.12) gives of the dissipative, spring, and

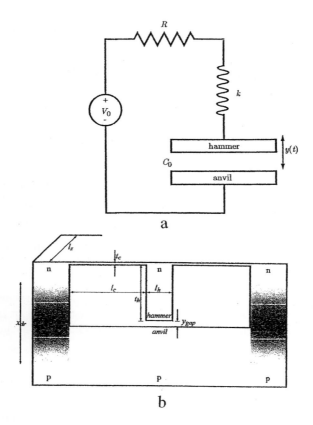

Figure 9.13: *Hammer-anvil* electromechanical oscillator. (a) macroscopic device schematic; (b) schematic of NEMS-MEMS device, with engineering dimensions.

electrostatic forces, respectively. q_{sat} is the maximum (saturated) charge on the plates, set by geometry and composition of the plates.

This system is electromechanically unstable: if the charged capacitor plates electrostatically draw together and electrically discharge, the attractive electric field collapses, the spring retracts the plates, the plates recharge on time scale τ_e, and the cycle can repeat. If the hammer's mechanical oscillation time constant $(\tau_m \sim 2\pi\sqrt{m/k})$ is comparable to the circuit's electrical time constant $(\tau_e \sim RC)$, and if the mechanical quality factor, Q_m, is sufficiently large, then the system can execute resonant, sustained electromechanical oscillation, converting electrical into mechanical energy.

A macroscopic laboratory model similar to Figure 9.13a was built and tested (scale length ~ 50 cm); it validated the operating principles of this device. The model consisted of a 60 cm long tungsten spring (spring constant k = 0.8 N/m) attached to a mobile, circular capacitor plate (hammer, dia = 10 cm, m = 4 gm,

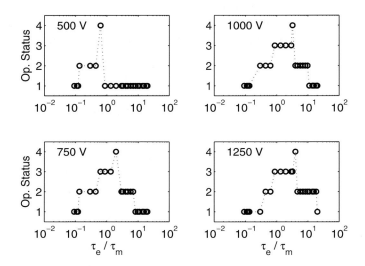

Figure 9.14: Operational status (OS) of macroscopic laboratory model *hammer-anvil* versus ratio of electrical and mechanical time constants (τ_e/τ_m). Legend: (1) no oscillation; (2) sub-harmonic or super-harmonic oscillation; (3) nearly harmonic oscillation; (4) harmonic oscillation. [19]

metallic), suspended above a fixed metallic plate, in series with a variable resistor ($5 \times 10^4 \Omega \leq R \leq 2 \times 10^7 \Omega$) and power supply ($500V \leq V_0 \leq 2000V$). (A booster capacitor ($C_{boost} = 2\mu$ F) was placed in parallel with the *hammer-anvil* capacitor to allow $\tau_m \sim \tau_e$.) As the resistance R was varied and the resonance condition was met ($\tau_e \sim RC \simeq 2\pi\sqrt{\frac{m}{k}} \sim \tau_m$), the hammer-capacitor fell into steady-state oscillation, while outside this regime, the oscillation either could not be started or, if it was jump-started, the oscillation quickly died out.

Figure 9.14 depicts the operational status (OS) of the laboratory model at four bias voltages (500V, 750V, 1000V, 1250V). Operational status levels 1-4 on the ordinate correspond to: (1) no oscillation; (2) sub-harmonic or super-harmonic; (3) nearly harmonic; (4) harmonic oscillation. The abscissa gives the ratio of electrical to mechanical time constants (τ_e/τ_m), adjusted via the variable resistor R (Figure 9.13a). These are response curves comparable to those of typical forced resonant oscillators. As expected, Figure 9.14 indicates harmonic response (OS-3,4) at $\tau_e/\tau_m \sim 1$; and non-harmonic response elsewhere. The best resonance shifts to higher τ_e/τ_m values with increasing bias voltage. The OS-2 and OS-3 plateaus broaden with increasing bias voltage; this is consistent with the oscillator being driven harder and, thus, requiring less stringent (τ_e/τ_m) criterion to achieve gap closure. Similar peak broadening is predicted for the MEMS and NEMS *hammer-anvil*.

9.4.2 Operational Criteria

This macroscopic oscillator should scale down to the micro- and nanoscopic realms. Force equation (9.12) still applies, but with the addition of a van der Waals force term:

$$F_{vdW} = \frac{HA}{6\pi(y_{gap} - y)^3},\qquad(9.14)$$

where H is the Hamaker constant ($H = 0.4 - 4 \times 10^{-19}$J for most non-polar materials; for silicon $H_{Si} = 10^{-19}$J).

Consider the p-n device Figure 9.13b, a microscopic version of the *hammer-anvil*, consisting of two p-n diodes (columns) on either side connected on top to a block of n-type material suspended by two flexible cantilever springs over the central p-type base. Comparing Figures 9.13a and 9.13b, the top-center n-semiconductor mass in Fig 13b acts as the *hammer* in Figure 9.13a; likewise, the lower stationary p-semiconductor in Figure 9.13b acts as the lower, fixed *anvil*. The spring is replaced by a double cantilever. For long, thin cantilevers ($t_c \ll l_c$) and for small vertical displacements ($y_{gap} \ll l_c$), a linear spring constant can be defined: $k = \frac{Yl_z}{2}[\frac{t_c}{l_c}]^3$ where l_c, l_z, and t_c are length, width, and thickness, and Y is Young's modulus ($Y_{silicon} = 1.1\times10^{11}$N/m^2). The entire device can be regarded as a distributed network of resistors and capacitors; the long, thin cantilevers can dominate device resistance. The column depletion regions impose the built-in voltage across the central gap, similarly as for the p-n diode in Figure 9.1.

The electric field across its central gap provides negative electrostatic pressure which drives and sustains the mechanical oscillations. For sustained oscillation, three fundamental criteria must be met:

(i) The electrical and mechanical time constants must be comparable ($\tau_e \sim \tau_m$) to achieve electromechanical resonance.
(ii) The hammer's mechanical energy gain per cycle ($\Delta\mathcal{E}_{es}$) must equal or exceed its mechanical dissipation ($\Delta\mathcal{E}_{diss}$), otherwise the oscillation will damp out; and
(iii) The cantilever spring force retracting the hammer after contact with the anvil must exceed the maximum attractive forces (van der Waals + electrostatic), otherwise the hammer will stick to the anvil.

We will address each criterion separately, then combine (ii) and (iii) into a more general, combined criterion.

Criterion (i) ($\tau_e \sim \tau_m$): The electrical time constant τ_e for the p-n *hammer-anvil* junction (Figure 9.12b) should be on the order of the inverse-slew rate of a comparably-sized p-n diode. This is typically $10^{-6} - 10^{-8}$s for micron-size silicon diodes, corresponding to frequencies of $f \sim 1-100$MHz. The approximate resonant mechanical frequency of a double-cantilever is given by:

$$f_m \sim B_n\sqrt{\frac{Y}{\rho}\frac{t_c}{l_c^2}},\qquad(9.15)$$

where B_n is a constant of order unity, and ρ is the mass density of the cantilever ($\rho_{Si} = 2.3 \times 10^3$ kg/m^3). For comparison, a silicon cantilever of dimensions $t_c = 10^{-6}$m and $l_c = 10^{-5}$m should have a resonant frequency of approximately $f_m \sim 10^8$Hz. (The mass of the hammer could lower this frequency.)

Since f_m can be made comparable to f_e, the first criterion appears capable of being met by NEMS or MEMS systems. Mechanical resonant frequencies for cantilevers in excess of 10^9 Hz have been achieved, however, the quality factors (Q) of these are significantly reduced, possibly due to dissipative surface states which can dominate the physics at short distance scales [20, 21]. (For initial tests, f_e should be minimized so as to minimize f_m since this would imply physically larger devices, which are generally easier to fabricate and diagnose; also, larger devices imply larger Q_m, which should reduce power requirements.)

Criterion (ii) (Work versus dissipation): Criterion (ii) requires that the energy gained through electrostatic work on the hammer per mechanical cycle ($\Delta\mathcal{E}_{es}$) exceed the mechanical dissipation per cycle ($\Delta\mathcal{E}_{diss}$). The relatively small sizes of the electrostatic and dissipative forces compared with the mechanical spring forces allow use of the harmonic approximation, which yields a closed-form integral solution to the coupled equations (9.12) and (9.13). Thus, we assume the hammer executes lightly-damped ($Q \gg 1$) simple harmonic motion: $y(t) \simeq y_{gap}cos(\omega_o t)$, with $\omega_o = \sqrt{\frac{k}{m}}$. Substituting this into (9.13) yields an uncoupled equation for the evolution of the electric charge on the capacitor plates:

$$\dot{q} + q\frac{(1 - \cos(\omega_o t))}{RC_o} - \frac{V_o}{R} = 0 \tag{9.16}$$

whose solution for homogeneous initial conditions is found to be

$$q(t) =$$

$$\frac{V_o}{R} \exp[\kappa \sin(\frac{2\pi t}{\tau_m}) - \frac{t}{\tau_e}] \int_0^t \exp[\frac{\tau'}{\tau_e} - \kappa \sin(\frac{2\pi\tau'}{\tau_m})]d\tau' = \frac{V_o}{R} \exp[g(t)] \int_0^t \exp[-g(\tau')]d\tau', \tag{9.17}$$

where $\tau_m = 2\pi/\omega_o$, and $\kappa \equiv \frac{\tau_m}{2\pi\tau_e}$, and $g(t) = [\kappa \sin(\frac{2\pi t}{\tau_m}) - \frac{t}{\tau_e}]$. Here $\tau_e = RC_o$ need not be an RC time constant as for the macroscopic oscillator (Figure 13a); rather it will likely be set by microscopic thermal processes, for instance, charge carrier diffusion, generation and recombination rates. The type, doping, and temperature of the semiconductor should have a strong influence on τ_e; for example, GaAs should have significantly shorter time constants than Si.

The Gaussian parallel plate approximation $E = \frac{q}{\epsilon_o A}$ allows the electrostatic energy gain over one period of mechanical oscillation τ_m to be written as

$$\Delta\mathcal{E}_{es} = \oint F_{es} \cdot dy(t) = \int_0^{\tau_m} \frac{q^2(t)}{2\epsilon_o}[-\frac{1}{\tau_m}y_{gap} \sin(\frac{t}{\tau_m})]dt \tag{9.18}$$

Meanwhile, for lightly-damped oscillators ($Q_m \gg 1$), the dissipation can be expressed in terms of Q_m as:

$$\Delta E_{diss} \simeq \frac{\pi k y_{gap}^2}{Q_m} \tag{9.19}$$

Combining (9.18) and (9.19), the second criterion can be written:

$$\frac{\Delta\mathcal{E}_{es}}{\Delta\mathcal{E}_{diss}} = \frac{-Q}{2\pi\epsilon_o\tau_m k y_{gap}} \int_0^{\tau_m} q^2(t)\sin(\frac{t}{\tau_m})dt. \qquad (9.20)$$

Resonant oscillation develops only for $\tau_e \sim \tau_m$. Away from this condition (either $\tau_m \gg \tau_e$ or $\tau_m \ll \tau_e$), it can be shown (taking $\tau_e \to 0$ or $\tau_e \to \infty$ in (9.17)) that the electric field becomes essentially static, so there is no net energy gain per cycle ($\Delta\mathcal{E}_{es} \longrightarrow 0$), while ΔE_{diss} remains constant (See (9.19).); thus the oscillation damps out. In the regime $\tau_e \sim \tau_m$, however, the asymmetry critical to resonance is realized: more work is performed *on* the spring by the field during gap closure than is work performed *by* the spring *against* the field on gap opening. It is also required that the Q of the oscillator is sufficiently large that the energy gain per cycle exceeds the energy loss per cycle. For lightly damped oscillators, oscillation can be sustained by minimal energy input.

Criterion (iii) (Non-stick hammer): The third criterion arises from the disparity in magnitude and spatial variation of the strengths of the forces acting on the hammer. For systems of interest, dissipative and electrostatic forces are subordinate to spring and van der Waals forces over the critical distances near where the hammer makes contact with the anvil. For the hammer not to stick to the anvil, the spring force at $y = y_{gap}$ must exceed the sum of the electrostatic and van der Waals (vdW) forces at the latter's cut-off (saturation) distance, typically $y_{cut-off} \sim 1.6 \times 10^{-10}$m, roughly an atomic radius. In this model, the dissipative and electrostatic forces act mechanically non-conservatively and can be of the same order of magnitude. The spring and vdW forces, on the other hand, are conservative and vary spatially with significantly different power laws and intrinsic strengths; while the spring force varies as $F_s \sim y$, the vdW force varies as $F_{vdW} \sim [y_{gap} - y]^{-3}$. Because of the latter's stronger spatial dependence, it can exceed the spring force at small gap distances – leading to stiction – unless steps are taken. Varying surface composition, one can alter the vdW force magnitude roughly over an order of magnitude via the Hamaker constant, but it can be most directly and easily reduced by reducing the contact area between the surfaces.

Since the parameter space for viable *hammer-anvil* oscillators is quite broad, for the sake of clarity and because experimental prototypes will most likely be pursued first in the MEMS regime, we will restrict much of the following discussion to parameters closely aligned with a well-known and specified MEMS production standard: the SUMMiT^{TM} process as developed and supported by Sandia National Laboratories.

In Color Plate VII the principal forces exerted on the hammer (excluding dissipation) are plotted versus gap opening for three typical oscillators, as specified by cantilever length (10-90μm). The electrostatic force is given for $V_o = V_{bi} = 0.6$V, with an electric field saturating at a maximum strength of 2×10^7V/m, similarly as for the p-n *standard device*. (This follows the conservative assumption that the vacuum gap electric field strength will remain below the dielectric strength of silicon (3×10^7V/m). It should also render a conservative (under-) estimate of actual device performance.) The vdW force is presented for five values of surface contact fraction ($10^{-4} \le \eta \le 1$). (For optimal designs of the *hammer-anvil*, η can be up to about $\eta \simeq 10^{-2}$, indicating 1% direct physical contact between hammer and anvil surfaces.)

The third criterion can be written $F_s > F_{vdW} + F_{es}$, or

$$ky_{gap} > \frac{H\eta A}{6\pi y_{cut-off}^3} + \frac{\epsilon_o E_{max}^2 A}{2} \qquad (9.21)$$

Since $F_{es} \ll F_{vdW}$ in the contact region, the third criterion can be reduced to:

$$\frac{6\pi ky_{gap}y_{cut-off}^3}{H\eta A} > 1 \qquad (9.22)$$

If one sets the third criterion ((9.22)) to equality and combines it with the second criterion ((9.20)), one obtains a combined criterion

$$\frac{Q}{4\pi\epsilon_o^2\tau_m ky_{gap}} \int_0^{\tau_m} q^2(t)\sin(\frac{t}{\tau_m})dt > 1 \qquad (9.23)$$

If one extracts the dimensional term q^2 from the integrand, assumes the integral resolves to order one, and re-expresses slightly, then a general dimensionalized condition for steady-state operation of the oscillator can be written:

$$\frac{6\pi y_{cut-off}^3\epsilon_o E_{max}^2 Q}{\eta H} \simeq \frac{Q}{\eta} \cdot \frac{\text{Electrostatic Pressure}}{\text{van der Waals Pressure}} > 1 \qquad (9.24)$$

This general condition is the product of two simple ratios. The pressure ratio incorporates the pressure driving the oscillation (electrostatic) and the 'stiction pressure' (van der Waals), while the other ratio indicates the importance of minimum dissipation (Q) and minimum surface contact (η). Interestingly, the spring force, which figures prominently in both criteria (ii) and (iii), drops out of this combined criterion entirely.

9.4.3 Numerical Simulations

Numerical simulations using MatLab and commercial semiconductor device simulators verified the principal results of the 1-D model of the dc-driven resonant oscillator. Two-dimensional numerical simulations of the *hammer-anvil*, performed using Silvaco International's Semiconductor Device Simulation Software [Atlas (S-Pisces, Giga)], verified the equilibrium aspects of the system's electric field. Output from the simulations were the steady-state, simultaneous solutions to the Poisson, continuity, and force equations, using the Shockley-Read-Hall recombination model. Simulations verified that the magnitude of the open-gap electric field can exceed that of the local depletion region by almost an order of magnitude, topping out in excess of 2×10^7V/m, similarly as for Color Plate IVb. Although the gap volume is significantly less than the depletion region volume ($y_{gap} \ll y_{dr}$), since electrostatic energy density is proportional to E^2, the electrostatic potential energy of the open gap can significantly exceed that of the depletion region. Numerical simulations also verified that the electric field and electrostatic energy in the gap is lost upon gap closure as a new depletion region forms. Like the *standard device* earlier, the *hammer-anvil* constitutes a microcapacitor that can be discharged by gap closure.

The electrostatic pressure P_e for the open-gap *hammer-anvil* with even modest biasing (*e.g.*, $V_{bias} = 0.6$V) will be at least $P \sim 10^3$Pa. In principle, this can be supplied by the built-in potential, ($V_{bias} \equiv V_{bi}$). Although the absolute electrostatic force exerted on the *hammer* is small, under Criteria (i-iii) it is sufficient to resonantly drive and maintain a high-Q oscillation. NEMS-MEMS cantilevers have documented Qs as high as $Q \sim 10^5$ in vacuum [20]. This implies that a small energy gain per cycle ($\sim 10^{-5}$ total mechanical energy) should be sufficient to sustain oscillation. In Plate VIII is plotted a range of viability for *hammer-anvil* devices constructed with physical dimensions achievable with the SUMMiT^{TM} process, and identical with the $l_c = 30\mu m$ case from Plate VII. Plate VIII presents minimum bias voltage required for sustained mechanical oscillation consistent with Criteria (i-iii) and realistic physical parameters for silicon based devices. Voltages are plotted as a function of quality factor Q and the ratio of electrical to mechanical time constants ($\frac{\tau_e}{\tau_m}$). Equipotentials (0.6V - 90V) are overlayed for comparison. Simulations are bounded above by the condition: $Q < 10^6$. Other areas not colored represent unviable regions of parameter space wherein the device requires a net input of reactive energy to oscillate, above and beyond the work required to offset presumed dissipation.

Plate VIII, as expected, indicates that the *hammer-anvil* performs most efficiently – *i.e.*, at the lowest dc-voltage – at the resonance condition ($\frac{\tau_e}{\tau_m} \sim 1$) and at large Q values. Away from these, either large driving voltages are required (*e.g.*, $V_o = 50$V for $\frac{\tau_e}{\tau_m} \sim 10^{-1}$, $Q \sim 10^3$) or else the device fails entirely (*e.g.*, $\frac{\tau_e}{\tau_m} \sim 20$, $Q \sim 10^4$). In the sweet spot of Plate VIII ($0.25 \leq \frac{\tau_e}{\tau_m} \leq 4$, $2 \times 10^3 \leq Q \leq 10^5$), the device can be driven at relatively low voltages ($1V \leq V_o \leq 5V$) and should have a resonant electromechanical frequency of about $f \sim 1$MHz. Note it should be viable using $V_o = V_{bi} \simeq 0.6$V. This device is almost macroscopic in size (maximum dimension ~ 0.1mm) and can be fabricated within the current art of MEMS technology. Analysis shows that this device should scale down well into the sub-micron regime and operate well at biases comparable to standard built-in voltages.

The most sensitive device dimensions and tolerances occur in the *hammer-anvil* gap. Optimal gap width will probably be less than 0.1 μm. The contacting surfaces must be highly parallel and their morphology must be tightly controlled so as to meet the condition of low contact fraction ($\eta \ll 1$). Contact wear is inevitable and may place limits on the total number of oscillations the device can execute [22, 23].

The *hammer-anvil* envisioned here will almost certainly require a kick start since the maximum achievable electrostatic pressure, although sufficient to sustain oscillation, appears insufficient to intitiate it. The kick start might be delivered in a number ways, including: a) a large, transient dc voltage spike across the gap; b) a small, short-lived, resonant, ac tickler voltage; or c) piezoelectric ac mechanical drive of the entire device. Device operation might be monitored either by laser interferometery of the hammer's motion, or by coupling its vibrational energy to piezolectric sensors. The latter would be propitious since, in principle, a piezo could be used both to jump start the oscillation and also to detect it.

9.5 Experimental Prospects

Prospects are good for laboratory construction and testing of these solid-state electromechanical devices in the near future[2]. Present-day micro- and nano-manufacturing techniques are adequate to construct the necessary structures, however, the art of surface finishing, which is crucial to reducing friction and stiction, may not yet be adequate, particularly for the LEM. State-of-the-art molecular beam epitaxy can reliably deposit layers to monolayer precision, but control of surface states is still problematic. Self-assembly of the requisite surfaces is plausible. Large scale biotic systems (e.g., DNA, microtubules) are well-known to self-assemble with atomic precision, as are abiotic ones (e.g., carbon nanotubes [11, 12]). Molecularly catalysed construction (e.g., RNA to protein transcription inside ribosomes) is accomplished with atomic precision. Scanning tunneling microscopes have also been used to assemble complex systems atom by atom. In light of these accomplishments, it seems plausible that experimental tests of these solid-state challenges may be on the horizon. We predict laboratory tests for the LEM will become feasible within 5 years; tests of the *hammer-anvil* concept are feasible today. Such tests are currently being pursued by the USD group.

---

[2] *Capacitive Chemical Reactor*: One can also conceive of non-mechanical challenges emerging from the thermal capacitor concept [24]. Consider gas molecules, having ionization energies less than the work function of the semiconductor, infusing the J-II gap (Figure 9.1). The neutral gas molecules positively ionize at the positive gap surface (n-side), then desorb, at which time the gap electric field accelerates them across the gap up to superthermal energies ($qV_{bi} \gg kT$). (Likewise, gas molecules with large electron affinities could form negative ions at the p-side and accelerate in the opposite direction. Together, positive and negative ion fluxes would constitute a diffusion current that is otherwise forbidden by the vacuum gap.) The positive ion current is unidirectional since, with $qV_{bi} \gg kT$, once an ion crosses the gap it cannot return until it is neutralized. The ion kinetic energy is sufficient to drive chemical reactions (or at least catalyse them). In principle, low-energy chemical reactants can enter the capacitor gap and emerge as high-energy products. In this way, chemical energy can be created solely from heat (via the gap electrostatic energy) — this in conflict with the Kelvin-Planck form of the second law.

References

[1] Sheehan, D.P., Wright, J.H., and Putnam, A.R., Found. Phys. **32** 1557 (2002).

[2] Wright, J.H., *First International Conference on Quantum Limits to the Second Law*, AIP Conference Proceedings, Vol. 643, Sheehan, D.P., Editor (AIP Press, Melville, NY 2002) pg. 308; Putnam, A.R., *ibid.*, (2002) pg. 314.

[3] Wright, J.H., Sheehan, D.P., and Putnam, A.R., J. Nanosci. Nanotech. **3** 329 (2003).

[4] Sheehan, D.P., Phys. Plasmas **2**, 1893 (1995).

[5] Sheehan, D.P., Phys. Plasmas **3**, 104 (1996).

[6] Sheehan, D.P. and Means, J.D., Phys. Plasmas **5**, 2469 (1998).

[7] Yuan, J.S. and Liou, J.J. *Semiconductor Device Physics* (Plenum Press, New York, 1998).

[8] Neudeck, G.W. *Vol. II: The pn Junction Diode*, 2nd Ed., in *Modular Series on Solid State Devices*, Pierret, R.F. and Neudeck, G.W., Editors (Addison-Wesley, Reading, 1989).

[9] Trimmer, W., Editor, *Micromechanics and MEMS: Classic and Seminal Papers to 1990* (IEEE Press, New York, 1997).

[10] Meyer, E. and Overney, R.M., Dransfeld, K. and Gyalog, T. *Nanoscience* (World Scientific, Singapore, 1998) Chapters 3,5.

[11] Dresselhaus, M.S., Dresselhaus, G., and Avouris, P., Editors, *Carbon Nanotubes – Synthesis, Structure, Properties and Applications* (Springer-Verlag, Berlin, 2001).

[12] Harris, P.J.F., *Carbon Nanotubes and Related Structures* (Cambridge University Press, Cambridge, 1999).

[13] Cumings, J. and Zettl, A., Science **289** 602 (2000).

[14] Pathria, R.K., *Statistical Mechanics* (Pergamon, Oxford 1972) pp. 467-74.

[15] MacDonald N.C. in *Nanotechnology*, Timp, G., Editor (AIP-Springer Verlag, New York, 1999).

[16] Pelesko, J.A. and Bernstein, D.H., *Modeling MEMS and NEMS* (Chapman and Hall/CRC, Boca Raton, 2003).

[17] Roukes, M.L. in *Technical Digest of the 2000 Solid-State Sensor and Actuator Workshop*, Hilton Head Island, SC. (2000).

[18] Bienstman, J., Vandewalle, J., and Puers, R., Sensors and Actuators **A 66** 40 (1998).

[19] Perttu, E.K. and Sheehan, D.P., unpublished data (2003).

[20] Mohanty, P., Harrington, D.A., Ekinci, K.L., Yang, Y.T., Murphy, M.J., and Roukes, M.L., Phys. Rev. B **66**, 085416 (2002).

[21] Lifshitz, R. and Roukes, M.L., Phys. Rev. B **61**, 5600 (2000).

[22] Patton, S.T. and Zabinski, J.S., Tribology **35**, 373 (2001).

[23] Komvopoulos, K., Wear **2**, 305 (1996).

[24] Seideman, T. and Sheehan, D.P., private communication, *Frontiers of Quantum and Mesoscopic Thermodynamics*, Prague, Czech Republic (2004).

10

Special Topics

Four special topics are discussed: (i) two proposals for a common rubric for classical second law challenges; (ii) speculation on a third type of life — *thermosynthetic* — which would sustain itself by converting heat into biochemical energy; (iii) speculations on the far future (physical eschatology); and finally, (iv) a collection of quotations from the sciences and humanities pertaining to the second law.

10.1 Rubrics for Classical Challenges

Of the several second law challenges that have been advanced over the last 20-25 years, the majority have been forwarded as independent counterexamples, narrow in their thermodynamic regimes of validity and specific in their physical constructions. No attempt has been made to place all under a single theoretical rubric. Perhaps such an enterprise is misguided. Just as the many attempts over the last 150 years to produce a general proof of the second law have failed — it has been proven only for a few highly idealized systems — so too, perhaps, a general theory of second law exceptions might prove equally elusive. Surely, one valid counterexample suffices to refute absolute validity, but finding the common thread among the many would be both satisfying and useful to future investigations.

In this section, we explore the similarities among the several ostensibly disparate challenges investigated at the University of San Diego (USD) since the early 1990's [1-9]. Several common threads bind them: geometric and thermody-

System	Temp (K)	Pressure (Torr)	Power Density (W/m³)
Plasma I (electrostatic)	> 1500	$< 10^{-3}$	10^{-2}
Plasma II (chemical/electrostatic)	> 1500	$< 10^{-3}$	10^{-2}
Chemical (chemical)	> 1000	< 1	10^{-3}
Gravitational (gravitational)	> 10	$< 10^{-10}$	10^{-9}
Solid State (chemical/electrostatic)	100<T<1000	—	$> 10^{8}$

Table 10.1: USD challenges, comparing viable temperature and pressure ranges and estimated power densities. MPG listed below the system name.

namic asymmetries, macroscopic potential gradients, and asymmetric momentum fluxes. A notable theorem by Zhang and Zhang [10] asserts that the latter of these threads should bind *all* second law challenges to date.

10.1.1 Macroscopic Potential Gradients (MPG)

Nearly all natural and technological processes are nonequilibrium in character and can be understood in terms of a working fluid moving under the influence of a *macroscopic* field expressible as the *gradient* of a *potential*. Examples are endless: water falling from the clouds under gravity; molecular hydrogen and oxygen combining in a fuel cell to form water; current in an electrical circuit.

For this discussion, *potential gradient* refers to any potential whose spatial derivative is capable of directing a fluid in a preferred spatial direction (*i.e.*, $\nabla\Phi = -\mathbf{F}$) and can transform equilibrium particle velocity distributions into nonequilibrium ones. Directional, nonequilibrium particle fluxes are the hallmarks of standard work-producing processes. *Macroscopic* refers to length scales long compared to atomic dimensions and to those of statistical fluctuations. In order to extract wholesale work, a system's potential gradients should be *macroscopic*.

Working fluids can be transformed from one MPG to another, descending a ladder, starting from high-grade directed energy to lower-grade undirected energy (heat), ending in a maximum entropy state (equilibrium). For practice, it is instructive to trace the million year journey of energy via various working fluids and MPGs, starting from hydrogen fusion in the Sun's core, to its release as light from

the solar photosphere, its absorption by chloroplasts and conversion into carbo-
hydrates in Iowan corn fields, to its ultimate exhaustion as frictional heat as a
child pedals down a country road in rural Indiana after a breakfast of corn flakes.
Here one can identify a dozen different fluids and MPGs. It is fair to say that
nonequilibrium macroscopic potential gradients make the world go round.

So far, only *nonequilibrium* MPGs have been considered, but *equilibrium* MPGs
are also common. Whereas nonequilibrium MPG derive their energy from exhaustible free energy sources (*e.g.*, nuclear reactions, sunlight, chemical reactions),
equilibrium MPG derive their energy from purely thermal processes (or none at
all). As a result, their exploitation to perform work can imply the use of heat from
a heat bath, and thereby a challenge to the second law.

Each USD challenge consists of (i) a blackbody cavity surrounded by a heat
bath; (ii) a working fluid (*e.g.*, gas atoms, electrons, ions, holes); (iii) a work
extraction mechanism (*e.g.*, electrical generator, piston); and (iv) an equilibrium
MPG (*e.g.*, gravitational field, electric field of a Debye sheath or depletion region).
Work is extracted as the working fluid cycles through the potential gradient. On
one leg of its cycle the working fluid 'falls' through the MPG and is transformed
into a spatially-directed nonequilibrium flux, by which work is performed. On the
return leg, the fluid and system returns to its original thermodynamic macrostate
via thermal processes (*e.g.*, diffusion, evaporation). The work performed by ex-
ploiting the MPG (either intermittently or in steady state) represents a *large, local*
excursion from equilibrium, but only a *small* excursion for the system *globally*. The
USD challenges range in size from nanoscopic to planetary ($10^{-7} - 10^7$ m), operate
over more than an order of magnitude in temperature (100 - 2000 K), and over
more than 8 orders of magnitudes in pressure ($\sim 10^3 - 10^{-6}$ Torr). They span
chemical, plasma, gravitational and solid state physics.

A summary of the USD challenges can be found in Table 10.1, comparing their
viable temperature and pressure regimes, and theoretical achievable power densities. The following is a summary of the *equilibrium* MPG that are discussed in
Chapters (6-9).

Plasma (Chapter 8; [1, 2, 3]) Electrons and ions at a single temperature have
different average thermal speeds $(\frac{kT}{m})^{1/2}$, owing to their different masses. In a
sealed blackbody cavity, in order to balance thermal flux densities in and out of
the plasma, the plasma resides at an electrostatic potential (the so-called *plasma
potential*, V_{pl}) with respect to the confining walls. This potential drop occurs
across a thin layer between the plasma and the blackbody walls, called the Debye
sheath (thickness λ_D). Typical plasma parameters render plasma potentials up to
several times $\frac{kT}{q}$ and gradients of order $\nabla V \sim \frac{V_{pl}}{\lambda_D}$. Sheath electrostatic gradients
of the order of 10^3V/m are common.

Chemical (Chapter 7; [4, 5, 6]) In a sealed blackbody cavity, housing a low density gas (*e.g.*, A_2) and two surfaces (S1 and S2) which are distinctly chemically
reactive with respect to the gas-surface reaction ($2A \rightleftharpoons A_2$), a chemical potential gradient can be supported, expressed as steady-state differential atomic and
molecular fluxes between the surfaces.

Gravitational (Chapter 6; [7, 8]) All finite masses exhibit gravitational potential gradients (gravitational fields) that can direct working fluids (gases) preferentially along field lines. No thermodynamic processes are required to sustain this MPG.

Solid State (Chapter 9; [9]) When n- and p-doped semiconductors are joined (forming a standard p-n diode) the requirement for uniform chemical potential (Fermi level) throughout the diode, and the thermal diffusion of electrons and holes down concentration gradients between the two regions, gives rise to an electrostatic potential difference (built-in potential, V_{bi}) between the two regions, across the so-called *depletion region* (thickness x_{dr}). One can say that, in the depletion region, a balance is struck between electrostatic and chemical potentials. The equilibrium electrostatic potential gradient scales as $\nabla V \sim \frac{V_{bi}}{x_{dr}}$, which for typical p-n diodes is on the order of $\frac{1V}{10^{-6}m} = 10^6 V/m$. (The similarities between the plasma and solid state systems is not coincidental.)

The above *equilibrium* MPGs and their working fluids possess *all* of the required physical characteristics by which standard *nonequilibrium*, free-energy-driven MPGs are known to perform work in traditional thermodynamic cycles. The equilibrium MPGs are well understood and experimentally verified; all are macroscopic structures. They possess potential gradients of sufficient magnitude and directionality to overcome thermal fluctuations and to perform macroscopic work. They differ from their nonequilibrium counterparts only in that they are generated and maintained under equilibrium conditions.

If MPGs are so common, it is natural to ask why their aptitudes for challenging the second law were not discovered earlier. First, the thermodynamic regimes in which they thrive are extreme (*e.g.*, high-temperatures and low-pressure for chemical and plasma systems) or else their operational scale lengths are either too large (planetary for gravitator) or too small (sub-micron for p-n diode) to be easily studied. Furthermore, the paradoxical effects are usually secondary in magnitude to other system effects and must be carefully isolated.

Several of the MPG systems are thermodynamically non-extensive in the sense described by D.H.E. Gross [11, 12]; that is, they are "finite systems of size comparable to the range of the forces between their constituents," and they are "thermodynamically unstable." Their energies and entropies do not scale linearly with size — the hallmark of non-extensivity. This is most apparent in the plasma, solid state, and gravitational systems, whose electric and gravitational fields energies scale quadratically with field strength. (Recall, electrostatic energy density scales as $\rho \propto E^2$.)

Boundaries are critical to these systems; without them they would either not possess usable MPGs or could not utilize them, in which case they would not pose second law challenges. And, because of boundaries, these systems should properly be evaluated outside the usual thermodynamic limit assumptions of infinite particle number and volume. Microcanonical approaches like those championed by Gross might be more fruitful [11, 12].

10.1.2 Zhang-Zhang Flows

In 1992, Kechen Zhang and Kezhao Zhang (hereafter Zhang-Zhang) [10] demonstrated a number of new aspects to Maxwell demons within the framework of classical mechanics. These were developed in the context of what they termed *spontaneous momentum flows* (SMF). A SMF is *a sustaining and robust momentum flow inside an isolated mechanical system*. All qualifiers must be simultaneously satisfied. The momentum flow must arise spontaneously within the system without resort to internal exhaustible free energy sources or to external sources of free energy. It must be *robust* in the sense that minor perturbations do not destroy it and it must be able to restore itself if it is interrupted. It must be *sustaining* in the sense that the long-term average of the momentum flow is non-vanishing.

Zhang-Zhang show the second law can be formulated in terms of the nonexistence of SMF. They begin by arguing that [10]

> (i) the existence of a perpetual motion machine of the second kind implies the existence of SMF; and the existence of SMF implies the existence of a perpetual motion machine.

They demonstrate that "the Kelvin-Planck statement of the second law is equivalent to the statement *in any isolated system, no spontaneous momentum flow exists.*"

From here, Zhang-Zhang prove a nonexistence theorem for SMF for classical systems of N interacting point particles, defined through the force equation:

$$m_i \ddot{\mathbf{r}}_i = \mathbf{F}_i(\mathbf{r}_1, ...\mathbf{r}_N; \dot{\mathbf{r}}_1, ...\dot{\mathbf{r}}_N), \tag{10.1}$$

where $\mathbf{r}_i$ and $\dot{\mathbf{r}}_i$ are the position and velocity of the ith mass m_i. The Zhang-Zhang theorem shows that a system of N interacting point particles cannot harbor a SMF under two conditions:

> (i) Its energy function E is symmetric under momentum reversal, *i.e.*, $E(q_i, p_i) = E(q_i, -p_i)$, where q_i and p_i are generalized coordinates and momenta; and
> (ii) Its phase space volume $d\Omega = dq_1...dq_N dp_1...dp_N$ is temporally invariant.

The theorem does not require system ergodicity, although it applies equally well to the ergodic case. It is valid for classical systems; its demonstration in the quantum realm has not been pursued.

Using the Zhang-Zhang theorem, one can immediately discount challenges which demonstrate symmetric and invariant measure on their phase space energy surfaces. Zhang-Zhang develop an example of a purported Maxwell demon governed by the Lorentz force

$$m\ddot{\mathbf{r}} = q(\dot{\mathbf{r}} \times \mathbf{B}(\mathbf{r})) - q\nabla\phi(\mathbf{r}) \tag{10.2}$$

and show that, since the phase space volume of this system is preserved by its dynamics, it cannot support a SMF — therefore, cannot violate the second law — regardless of its complexity and intricacy of construction.

The Zhang-Zhang theorem demonstrates the deep relationship between the second law and phase space volume invariance of classical systems. They state:

> ... the validity of the second law relies on the dynamics in the underlying mechanical system, [but] this validity cannot be justified by the laws of classical mechanics alone. Invariance of phase volume appears as an additional factor which is responsible for this validity. ... Looking from another angle, if the second law is taken as a fundamental assumption, then the invariance of phase volume may be considered as a constraint imposed by the second law on the allowable dynamics of the mechanical systems.

Independently of Zhang-Zhang, in 1998 Sheehan [13] explained the necessary and sufficient conditions for the several USD challenges in terms of *asymmetric momentum fluxes*. He noted that each challenge relies on two broken symmetries — one thermodynamic and one geometric — in the physical design of the system. The broken thermodynamic symmetry creates an internal, steady-state asymmetric momentum flux and the geometric asymmetry converts this flux into work at the expense of the heat bath. Sheehan's asymmetric momentum fluxes are equivalent to the Zhang-Zhang spontaneous momentum flows. More recently, it has been recognized that the asymmetric momentum fluxes common to the USD systems arise due to macroscopic potential gradients (MPG), as discussed above (§10.1.1).

In their article, Zhang-Zhang assert "the concept of SMF is an imaginary one since no known physical systems exhibit it." The several challenges in this volume stand as counterexamples to this claim. As derived, the Zhang-Zhang theorem is descriptive of all classical challenges to date, but rather than refuting them, it gives insight into their common basis. We suspect that deeper physics will be eventually uncovered linking them further. Attempts to fathom this *deeper* physics behind the second law have begun [14].

10.2 Thermosynthetic Life

10.2.1 Introduction

In this section the hypothesis is explored that life might exploit second law violating processes. All evidence will be theoretical and circumstantial since there is no experimental evidence that life violates the second law in the least, either macroscopically or microscopically. Quite the contrary, life is often considered to be a strong ally of the second law since biotic chemicals typically generate far more entropy in the world than they would otherwise create in an abiotic state.

It is estimated that since the emergence of life on Earth roughly 3.8 - 4 billion years ago, on the order of a billion species have existed and that, of these, roughly

10 million currently exist. Of this extant 1% of total species, perhaps only about 20% have been identified by name. Of those identified, few have been studied well enough to make general claims about their thermodynamics; however, those that have been studied carefully have shown compliance with the second law. In highly studied species — for instance, E. coli, C. elegans, mice, monkey, men — only a small fraction of proteins and biochemical pathways have been thoroughly studied. (Human are estimated to contain roughly 32,000 genes and about 10^6 proteins of which less than 10% have been positively identified.) Despite this apparent dearth of biochemical knowledge, there is significant biochemical commonality among species such that close study of a few should give an overview of the many. Nonetheless, current understanding of biology and biochemistry across all species is insufficient to rule conclusively that *all* biological processes, structures, and systems comply with the second law.

Several theoretical proposals for second law challenges have been inspired by biological systems. Modern work in this area dates back 25 years to the seminal proposals of L.G.M. Gordon (§5.2). The Crosignani-Di Porto mesoscopic adiabatic piston (§5.4) suggests that the second law might fail at scale lengths characteristic of living cells. There are suspicions by many that if the second law is violable, life would have exploited it.

The exigency of natural selection — *whatever survives to reproduce survives* — suggests that if subverting the second law confers a reproductive evolutionary advantage on organisms and if the second law can, in fact, be subverted, then the soaring cleverness, diversity and resourcefulness of Nature would, with high probability, achieve this end. Since there are several imaginable scenarios in which *thermosynthetic life* (TL) [15] would compete well with or even outcompete ordinary *free-energy life* (FEL) by subverting the second law, it is reasonable to seek biological violation. Standard FEL is divided into *chemosynthetic* and *photosynthetic* forms; the former derives its primary energy from chemical processes, while the latter relies on electromagnetic radiation (photosynthesis). Thermosynthetic life, which would derive its energy from heat, would constitute a third distinct type of life.

Basic characteristics of TL are suggested by thermodynamic considerations. Multicellular life is less likely to exploit thermal energy than unicellular life because it has an intrinsically lower surface-to-volume ratio and, thus, has access to less thermal energy per cell than unicellular life. Therefore, thermosynthetic life, if it exists, is most likely to be small and unicellular. The lower limits to the size of organisms (viruses, viroids, and prions aside) is a subject of debate. Recent studies suggest that it may be as small as 20 nm [16].

If TL must compete against FEL for material resources, then it will likely be outcompeted under everyday circumstances where rich free-energy sources are abundant, e.g., sunlight, plant and animal tissue, or raw energetic chemicals spewing from hydrothermal vents. Pure TL would eschew these and so would be at a severe energetic and evolutionary disadvantage. Thermosynthetic life might best compete against FEL in free-energy poor environments; thus, it is also likely to be anaerobic and isolated geologically and hydrologically from chemosynthetic and photosynthetic life. This suggests TL might be best suited to life deep inside the earth. (In recent years it has become apparent that deep-rock microbes enjoy significant diversity and might actually represent greater biomass than surface dwelling life [17].) These environments can also be quite stable so as to allow TL to evolve and thrive without direct competition from FEL for long periods of time.

Thermosynthetic life might have an edge in high-temperature environments. First, higher temperatures imply higher thermal power densities to drive biochemical reactions. Second, for reasons to be discussed shortly, high temperatures favor some second law violating mechanisms. Again, deep subsurface environments fit the bill. Temperatures rise with increasing depth in the Earth at a rate of roughly $1.5 - 3 \times 10^{-2}$K/m. At depths of 5 km rock temperatures approach 400 K. If it exists in deep rock, TL is likely superthermophilic and hyperbarophilic. In deep-sea hydrothermal vent environments, the high pressures augment thermophilic tendencies by raising the boiling point of water and by compressing molecular structures that might otherwise thermally disintegrate. In the laboratory, microbes have survived exposure to pressures of 1.6 GPa (1.9×10^4atm). Microbes have been discovered in continental rocks down to depths of 5.2 km [18]. The limit to the depth of life is likely set not only by temperature and pressure, but also by the pore sizes between rock grains, which are reduced at high pressures. It is expected that pressure, temperature and pore size would disallow carbon-based life below about 10 km.

In summary, if thermosynthetic life exists, it is likely to be small, unicellular, anaerobic, hyperbarophilic, superthermophiles confined to free-energy poor environments, well isolated from FEL in long-term stable locations. Deep rock microbes (Archaea) seem to be the best candidates, situated at and beyond the fringes of where FEL is known to survive.

Archaea are among the most ancient life forms; current molecular evidence based on RNA analysis places them near the base of the tree of life. (At present, three domains are generally recognized: eukaryotes (cells with nuclei), bacteria and archaea (cells without nuclei); the latter two display superthermophilicity.) Among the most ancient archaea are hyper- and superthermophiles, suggesting — but certainly not proving — that life may have originated in high temperature environments, like deep-sea marine vents or in the earth's crust.

The current high temperature record for culturable microbes is $T = 394$K, set by *Strain 121*, an Fe(III)-reducing archaea recovered from an active "black smoker" hydrothermal vent in the Northeast Pacific Ocean [19]. It grows between $85°$ and $121°$C — temperatures typically used in autoclaves to sterilize laboratory equipment and samples. The upper temperature limit for microbes has been estimated to be roughly 425K ($150°$C) [20-26].

Archaea and bacteria have many special adaptations for survival at high temperatures. Unlike bacteria and eukaryotes which have lipid bilayer membranes, archaea have monolayer lipid membranes (ester cross-linked lipids) that resist thermal separation. Thermophiles also employ stiff, long-chain carotenoids that span their membranes, thereby reinforcing them against thermal separation. Carotenoids are highly conjugated organics also known for good electrical conductivity via long-range, delocalized molecular orbitals. They can mediate direct charge transport across membranes.

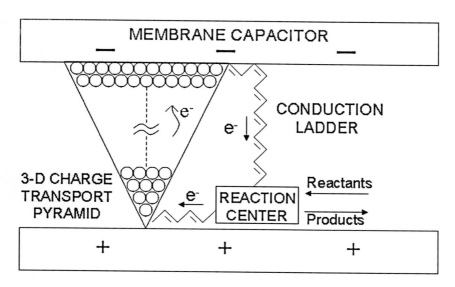

Figure 10.1: Schematic of proposed biochemical machinery for thermosynthetic life. Charge cycles clockwise: diffusively up through the pyramid and ballistically down the conduction ladder through the reaction center, where high-energy chemical products are formed.

Another possibility is that standard FEL life may harvest thermal energy as a supplement to standard free energy sources, or might resort to it under dire circumstances when its traditional free energy sources are cut off. This suggests that long-entombed and dormant microbes might be good candidates for TL. For example, bacteria have been found to remain viable for millions of years trapped in ancient salt crystals essentially absent of free energy sources and nutrients [27].

Thermosynthetic organisms would not necessarily be expected to arise *ex nihilo* from abotic chemicals, but one can envision strong evolutionary forces by which they might evolve from standard FEL. Free energy life, wherever it first evolved — deep-ocean vents, surface, or deep rock — would naturally spread to all possible habitable regions. Where conditions were not initially favorable, in time, evolutionary forces would reshape the organism as far as possible. FEL would extend deeply into subsurface environments — as has been discovered [17, 18, 25] — down to the biochemical limits of heat and pressure, and to the lower limits of material and free energy resources. One can imagine conditions at the limits of free energy resources where FEL would face an evolutionary imperative to convert some (or all) of its cellular machinery over to the reclamation of heat — if this is possible — such as to push into regions uninhabitable by its competitors. As will be discussed below, variants of standard cellular machinery (membranes, carotenoids, enzymes) might be conducive to this enterprise.

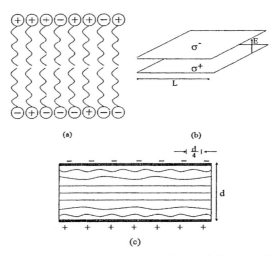

Figure 10.2: Charged biomembrane as a capacitor: a) Schematic of biomembrane lipid bilayer; b) Lipid bilayer as parallel plate capacitor; c) Equipotentials in biomembrane with finite point charges separated by $\frac{d}{4}$.

10.2.2 Theory

Here we consider the possibility that *macroscopic potential gradients* (§10.1.1, [15]) might be exploited by life to circumvent the second law. Consider a biochemical system consisting of five parts, depicted in Figure 10.1: 1) membrane capacitor; 2) 3-D pyramidal array of charge transport molecules; 3) electrically conducting molecular ladder spanning the membrane from pyramidal base to vertex; 4) chemical reaction center for utilization of superthermal charge; 5) small number of mobile charges (electrons or protons) that can circulate through the system.

One of life's most basic biological structures is capable of supporting a MPG and is also quite exposed to the thermal field: the cellular membrane. Particulars of cell membrane vary considerably across life forms so we will consider archetypical membranes consisting of ambipolar lipid layers. Simple polar lipids have a charged functional group on one end of a long organic skeleton. These can self-assemble electrostatically end-to-end to form a bilayer as shown in Figure 10.2a. Polar molecules (*e.g.*, water) outside the hydrophilic ends can stabilize the membrane, trapping the hydrophobic, non-polar organic skeleton within. In principle, opposite sides of the membrane can support permanent opposite surface charge densities.

Consider a planar section of membrane (Figure 10.2b) with fixed surface charge density σ. (These charges might be anions or cations fixed on the polar ends of the phospholipids.) In the infinite plane approximation ($l \gg d$) the membrane acts like a charged capacitor (Figure 10.2b). For biomembranes, typical scale lengths are $d \sim 10^{-8}$m and $l \sim 10^{-6}$m. For our model membrane, let $l = 2 \times 10^{-6}$m and $d = 10^{-7}$m. This structure (or multiple copies) could be sequestered within the cell rather than be exposed on its outer surface.

The maximum charge density and steady-state voltage drop supportable by a biomembrane can be estimated from the dielectric strengths and compressive strengths of typical organics. If the electrostatic pressure due to σ exceeds the material compressive strength, the membrane will collapse; if the the electric field exceeds its dielectric strength, it will arc through. Taking the dielectric strength of the membrane to be $E_{max} = 10^7$ V/m, dielectric constant $\kappa = 3$, and compressive strength to be 10^7N/m^2, one finds that the maximum charge density supportable on the membrane surface to be roughly $\sigma_{max} = \kappa\epsilon_o E_{max} = 2.5 \times 10^4$ C/m^2 $= 1.5 \times 10^3$ e$^-$/μm^2. (Estimates from compressive strength yields larger σ_{max}.) This σ implies that the distance between excess charges is roughly $\frac{1}{\sqrt{\sigma}} \simeq 2.5 \times 10^{-8}m\simeq \frac{1}{4}d$; thus, the majority of the molecules comprising the membrane need not contribute to the net surface charge. The electrostatic equipotentials are fairly parallel near the midplane of the membrane and are undulatory near the surfaces; likewise the electric field vectors are parallel in the interior and less so near the surfaces (Figure 10.2c). (Note that archaean cross-linked monolayer membranes would be relatively good at retaining capacitive charge separation since membrane molecules would be less likely to invert than standard bilayer lipids in bacteria and eukaryotes. Cross-linking also adds structural strength.)

The maximum potential drop across the model membrane surfaces will be on the order of $V_m \simeq E_{max}d = \frac{\sigma_{max}d}{\kappa\epsilon_o} = 1$V; for this model, let $V_m = 0.8$V. (This agrees quantitatively with common membrane potentials, scaled to membrane thickness $d = 10^{-7}$m.) An electronic charge falling through this potential would have roughly the energy required to drive typical chemical reactions: $qV_m = \Delta\mathcal{E} \sim 0.5 - 4$eV. Electrons and protons should be the most convenient mobile charges for this system; they are standard currency in biochemical reactions. Low-energy chemical reactions have been exploited by life. For instance, forms of bacterial chlorophyll have spectral absorption in the near IR, corresponding to energies of roughly 1eV. The hydrolysis of ATP (adenosine triphosphate) into ADP (adenosine diphosphate) releases roughly 0.56eV of free energy. ATP is the primary energy releasing molecule in most cells.

In principle, multiple membranes might be stacked in electrical series and self-triggered sequentially — the high energy charge from a previous step triggering the subsequent one — so as to create a series-capacitive discharge with resultant energy in multiples of a single membrane energy[1], $\Delta\mathcal{E}$. Or, each discharge could create low-energy chemical intermediates that could be brought together to drive a single, more energetic chemical reaction, similarly to how ATP is utilized in cells. In this way, the energy necessary to drive even high-energy chemical reactions might obtain.

It is biologically requisite that $\Delta\mathcal{E} \gg kT$ for at least some biochemical reactions because if they could all be thermally driven, none would be irreversible and the organism would find itself essentially at thermal equilibrium — and dead. (Also, unless $\Delta\mathcal{E} \gg kT$, molecules tend to thermally disintegrate.) A 0.8eV potential energy drop across the membrane far exceeds the typical thermal energy associated with life (kT(300K)$\sim \frac{1}{40}$eV$\sim \frac{1}{30}V_m$). At first glance, it seems improbable that

---

[1]Many species are known to utilize series-capacitive discharge, *e.g.*, electric eels, rays, and catfish, achieving up to hundreds of volts in total potential.

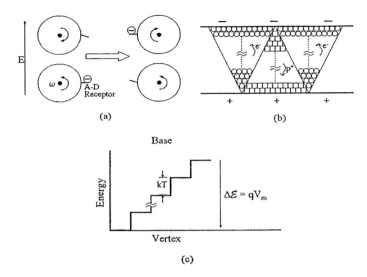

Figure 10.3: Proposed charge transport in a biomembrane: a) Rotary molecules transferring charge; b) Diffusive transport of electrons and protons between membrane faces by A-D molecules, up electrostatic potential gradients; c) Energy diagram for charge thermal transport and quantum transition in biomembrane.

thermal energy alone could drive charges up such a large electrostatic potential at room temperature, but if accomplished in steps it is less daunting.

Let the model membrane be embedded with charge transporting molecules arranged in a 3-D pyramidal structure, as depicted in Figure 10.3a. The actual molecular structure is unspecified — it could be simply a conductive-diffusive molecular matrix — but for this model consider it to be composed of rotary molecules, each with charge acceptor-donor (A-D) sites. The transport molecules spin freely, driven by thermal energy. When two A-D sites meet a charge can be transferred between them, with minimal energy of activation. This is taken to be a random process. Once charged, the electric force within the membrane will tend to constrain the rotor against charge movement up the potential gradient, but if the step sizes are small ($kT \simeq q\Delta V$) and if the transfer probability favors diffusion up the potential gradient either by favorable multiplicity of states, or in this case, by a favorable multiplicity of transfer molecules in the direction up the gradient (Figure 10.3b), then one can expect appreciable transport by diffusion alone.

Charge transport via diffusion can be placed on more realistic footing. Let the membrane have area $l^2 \sim 4 \times 10^{-12} \text{m}^2$ and let the area of an individual A-D molecule be $\delta^2 = (2 \times 10^{-9}\text{m})^2$. In the membrane depicted in Figure 10.2c, the base of the molecular pyramid accommodates roughly $\frac{l^2}{\delta^2} \sim 10^6$ molecules, the vertex accommodates one; the pyramid consists of roughly 50 molecular layers. At $T = 400\text{K}$, the electrostatic potential energy increase per level in the pyramid is roughly $0.6kT$, thus allowing charges to rise comfortably upward against the gradient by thermal diffusion.

Applying Boltzmann statistics, the relative charge occupation probability (p_{rel}) for the base versus the vertex should be on the order of

$$p_{rel} \simeq \frac{g_{\text{base}}}{g_{\text{vertex}}} \exp\left[-\frac{\Delta \mathcal{E}}{kT}\right] \tag{10.3}$$

Here $\frac{g_{\text{base}}}{g_{\text{vertex}}}$, the ratio of multiplicity of states, is assumed to be the simple ratio of A-D molecules at each site: $\frac{g_{\text{base}}}{g_{\text{vertex}}} = \frac{l^2}{\delta^2} \simeq 10^6$. The energy difference $\Delta\mathcal{E} = qV_m \simeq$ 0.8eV. Let the temperature be that survivable by the superthermophile *Strain 121*: 400K. These parameters render $p_{rel} \sim 10^{-3}$. (p_{rel} can be increased by further grading the quantum multiplicity of states in molecules toward the pyramid's base, for instance, by having more A-D sites per molecule at the base than at the vertex, or by simply packing more A-D molecules in the array.) For comparision, dropping the temperature by just 25% (T = 300K) reduces p_{rel} by over three orders of magnitude: $p_{rel} \simeq 4 \times 10^{-8}$. In all, despite a sizable potential difference ($V_m \gg \frac{kT}{q}$), charges have reasonable probability of traversing the membrane by purely thermal diffusive processes. The molecular pyramid creates a natural mechanism for charge transport and also offsets the deleterious Boltzmann exponential.

Once a charge has climbed the electrostatic potential via multiple small, diffusive sub-kT steps (Figure 10.3c), it can fall through the entire potential (base to vertex) in a single, nearly-lossless quantum transition ($qV_m = \Delta\mathcal{E}$) which, in principle, could drive a chemical reaction at the reaction center (*e.g.*, formation of ATP). Quantum transitions that require large and specific energies cannot be triggered prematurely (*i.e.*, by thermal energy steps); furthermore, the reactions can be triggered and mediated by structure dependencies, for instance, those that might occur only at the membrane boundaries. Or, they could require the participation of the reactants in the reaction center to complete the circuit. Thus, the multiple low-energy thermal transitions up the pyramid can be insulated from the single high-energy transition down the ladder.

Charge transfer from the pyramid's base back to its vertex can be plausibly executed across the thickness of the membrane ($d = 10^{-7}$m) along electrically conducting polymers. Aromatic and highly conjugated organics that span the entire membrane are the most promising candidates. Carotenoids are highly conjugated linear organics that are known to be electrically conducting and are also found in the membranes of thermophilic archaea, presumably to give structural strength against thermal disruption. Individual conducting organic molecules have demonstrated electron transport rates of roughly 10^{11}/sec and current densities far in excess of the best metallic conductors [28]. The directionality of the current might also be promoted by fashioning the conduction ladder as a molecular electronic diode [28, 29, 30]. These have been engineered [28] by attaching electron-donating and electron-withdrawing subgroups to a conducting carbon skeleton. Biological counterparts are conceivable.

Electrons falling down the conducting ladder to the reaction center in Figure 10.1 could drive useful biochemical reactions since their energy is far greater than kT; in this model, $\Delta\mathcal{E} = 0.8\text{eV} \simeq 23kT$. Reactants, diffusing in from other parts of the cell, could adsorb preferentially to specific locations (enzymes) at the reaction center, undergo reaction, and their products could preferentially desorb and diffuse back into the cell. Like pieces of metal laid together on an anvil and artfully struck to join them, reactants could be enzymatically forged using the energy of superthermal electrons delivered down the conducting ladder, perhaps in ways akin to ATP synthase. ATP synthase is a protein complex that resides in the membranes of chloroplasts and mitochondria. It catalyses the production of ATP using proton current through the membrane [31]. A proton gradient is established across the membrane by catabolic processes and the leakage current back across the membrane, down the gradient, through a channel in the ATP synthase catalyses the formation of ATP via phosphorylation of ADP. It is conceivable that thermosynthetic life could utilize similar intermolecular protein complexes as their reaction centers.

Specificity in the reaction direction could be further enhanced if the binding sites in the reaction center were chemically tuned to strongly adsorb reactants and quickly desorb products. In other words, reactants would be tightly bound on the reaction template until a high energy electron falls through the membrane potential and drives the reaction forward, at which time the product is loosely bound and, therefore, quickly desorbs and diffuses away, thereby supressing the reverse reaction. This type of specificity in adsorption and desorption are hallmarks of enzymes.

The reactants, coupled with reactant-induced conformational changes in the binding enzyme, could act as an electronic switch in the reaction center; that is, electrons in the base of the pyramid would discharge through the reaction center when (and only when) the reactants are present in the reaction center. This description closely resembles the plasma (Chapter 8) and solid state (Chapter 9) thermal capacitive discharges. An electrical circuit is similar to that of the plasma paradox (Figure 8.8b, §8.4) is easily envisioned.

Thus, between the large potential energy rise upward through the carbon conduction ladder, the possibly diodic nature of the ladder, the reactant-product binding specificity of the reaction center, and its possible switching action, it appears that a unidirectional current is favored for this circuit. At first glance, this seems to violate the principle of detailed balance — which would demand no net current flow in the circuit — but since this system involves an inherently nonequilibrium process (a 0.8eV electrons driving a chemical reaction), detailed balance need not — and in this case, does not — apply. This chemical activity bears strong resemblance to those proposed by Gordon (§5.2.2) and Čápek (§3.6.4).

The power $\mathcal{P}_m$ delivered by the thermosynthetic membrane should scale as $\mathcal{P}_m \sim \frac{p_{rel} N_q \Delta\mathcal{E}}{\tau_{vb}}$, where N_q is the number of charges in play in the membrane and τ_{vb} is the average transit time for charges from the vertex to the base. N_q must be substantially less than σl^2 in order to not significantly distort the membrane's electric fields; for this model, let $N_q = 10 \ll l^2\sigma$. τ_{vb} depends critically on the mechanism of charge transport; as a measure we take it to be the diffusion time of a simple molecule (H_2) through water the thickness of the membrane; that is, $\tau_{vb} \sim \tau_{\text{diff}} \sim \frac{x_{rms}^2}{2D} \simeq 10^{-6}\text{sec}$. For the model membrane, one has $\mathcal{P}_m \sim \frac{p_{rel} N_q \Delta\mathcal{E}}{\tau_{\text{diff}}} \simeq$

2×10^{-16}W. This is orders of magnitude less than what would be allowed for radiative or conductive heat transfer into a cell, in principle. It is also on the order of 0.1% of the resting power requirement for mammalian cells; thus, from the viewpoint of these energetics, this TL would probably not compete well against FEL. On the other hand, many FEL organisms have far lower power requirements than mammalian cells. For instance, some deep rock microbes are believed to have infinitesimal metabolisms; by some estimates they grow and reproduce on time scales approaching centuries, whereas surface dwelling bacteria can run through a generation in minutes. Based on the availability of free energy sources and nutrients and on the amount of carbon dioxide produced from the oxidation of organic matter, it is estimated that many deep rock microbes have metabolisms that are more than a billion times slower than surface free-energy microbes, thereby possibly putting them below TL in energy utilization. If so, then whereas its low metabolic rates makes it unlikely to compete successfully with FEL on the surface, thermosynthetic life might compete quite well with free energy life in the deep subsurface.

The above diodic biochemical circuit should, of course, be considered critically in light of the failure of solid state diodes to rectify their own currents, thereby failing as second law challenges. Superficially, this biochemical diode bears resemblance to the solid state diode considered by Brillouin [32], Van Kampen [33], McFee [34] and others. This one, however, relies on quantum molecular processes, rather than on more classical solid state processes, and so cannot be directly compared. More detailed physical chemical analysis is currently being undertaken.

Thermosynthetic life would seem to enjoy several evolutionary advantages over free energy life. Most obvious is freedom from reliance on free energy sources since it simply harvests energy from its own thermal field. It is unclear whether this would allow it to economize on its metabolic machinery, but insofar as it does not require large bursts of energy, one can imagine that TL would not be so dependent on energy storage mechanisms (*e.g.*, fats) as FEL. Once grown, TL could, in principle, operate as a closed thermodynamic cycle, neither taking in nutrients nor expelling wastes as FEL must. Its material needs would be satisfied by those required for reproduction. This economy would confer evolutionary advantage both in reducing the necessary conditions for survival but also in reducing local bio-pollution which is known to be deleterious or even fatal to many species in closed environments. Third, supposing TL competed with FEL at the spatial margins of energy resources, TL could more easily strike off, explore, and inhabit niches unavailable to FEL. For instance, were superthermophiles to push more deeply into the subsurface, according to the discussion above, the temperature rise could further favor TL's heat-driven biochemistry.

In summary, it is conceivable that thermosynthetic life could utilize macroscopic electrostatic potential gradients in biomembranes and other standard biochemical machinery to rectify thermal energy into chemical reactions so as to compete well with free energy life in some environments.

10.2.3 Experimental Search

An experimental search for thermosynthetic life would probably prove problematic at present. First, in most terrestrial environments FEL would dominate over TL, rendering TL a trace population, at best. In more extreme environments where FEL is crippled, perhaps in hot, deep, subsurface deposits, a search might prove fruitful. It has been estimated that the depth-integrated biomass of subsurface microbes might be equivalent to a meter or more thick of surface biomass over the entire planetary surface [17, 25]. Most of these organisms have not been cataloged or studied. Second, the vast majority of *all* microbes — surface or subsurface dwelling — are either difficult or impossible to culture in the laboratory; it is estimated that perhaps less than 10% of microbes from *any* random natural setting can be cultured with present techniques. Since, in principle, full-fledged TL might ingest little or no chemical fuel (aside from that required for reproduction) and might expel little or no chemical waste, many standard tests for life should fail.

The above profile for TL — small, unicellular, anaerobic, hyperbarophilic superthermophiles, geologically and hydrologically isolated from FEL in long-term, stable, free-energy poor environments — suggests how and where TL might be discovered. For pure TL, a search might be conducted in deep, geologically stable rock that is nearly biochemically barren both of material and energy resources; that is, look where there is little hope of finding anything alive. Rock samples with organisms, but lacking appropriate levels of biological wastes would be signposts for TL. Thermosynthetic life might be physically subsumed into the cells of standard free energy life, perhaps in a role similar to that of mitochondria which are believed to have once been free living cells before being captured and incorporated as the energy-producing cellular machinery of parent cells. Hybrid FEL-TL might display the advantages of both types of life.

Recovery of deep subsurface microbes has been carried out to depths of 5.2km in igneous rock aquifers. Boreholes (without microbe recovery) have been conducted to depths of 9.1 km, so searches for proposed thermosynthetic life appear feasible. Culturing would probably be more problematic. One might suppose that TL can be discriminated experimentally from FEL by isolating a mixed culture from all free energy sources for sufficiently long to starve FEL, thereby isolating only TL. Unfortunately, *sufficiently long* can be *very long*. Microbes can place themselves in low-energy, sporulating, static modes for years, decades, and perhaps centuries or millenia awaiting favorable conditions. With microbes, sometimes there seems to be no clear distinction between *fully alive* and *truly dead*.

Ancient samples might offer good candidates for TL. Recently, viable cultures have been made of a halotolerant bacterium believed to have been entombed in salt crystals for 2.5×10^8 years [27]. It is unclear how such organisms remain viable for millenia against the ravages of natural radioactivity and thermal degradation. There are a number of biochemical and genetic repair processes that contribute but, presumably, all require free energy. A reliable energy source like thermosynthesis would be convenient for such ongoing repair.

DNA/RNA analysis might eventually help identify TL. Currently, it can be used to identify species, but it cannot yet be used to predict the function, structure, or viability of an organism. Presumably, genomics will eventually advance to the point where cellular structure and biochemical function can be predicted from DNA/RNA analysis alone such that thermosynthetic life forms could be identified solely from their genetic maps. This type of search, however, is still years or decades away.

In summary, given the exigencies of natural selection and the plausibility of second law violation by biologically-inspired systems, we predict that thermosynthetic life will be discovered in the deep subsurface realm, and that perhaps some forms of free energy life will be found to resort to it under suitable circumstances in less extreme environments.

10.3 Physical Eschatology

10.3.1 Introduction

The history and future of the universe are inextricably tied to the second law. That the cosmos began in a low gravitational entropy state destined its subsequent thermodynamic evolution into the present. And, to a fundamental degree, its fate is defined by its inexorable march toward equilibrium. Although *heat death* in the Victorian sense has been commuted into merely *heat coma* by modern cosmology, the prognosis is still rather bleak.

Study of the far future and the end of the universe has become a popular scientific pastime, in many ways replacing religious myths with scientific ones [35-47]. The number of books and articles on the subject are almost beyond reckoning. In this section, we explore the *physical eschatology* for a universe in which the second law is violable.

Future projections are almost certainly seriously off the mark (if not entirely wrong) — just as are ancient myths, considered in the light of modern science — since even the most basic escatological questions have no definitive answers. Is the universe open or closed? Will it continue to expand or will it eventually collapse? Is space flat? Is the proton stable? What is the nature of dark matter and dark energy? What is the composition of 99% of the universe? What is the nature and energy of the vacuum state? Is there a theory of everything and if so, what is it? What is time and *when* did it begin? What is the role of intelligent life in the universe?

For gaining a true understanding about the future universe, eschatological studies are probably futile because they assume that our knowledge of physical laws is both *accurate* and *complete*. Historically, this assumption has always failed; therefore, there is little reason to believe it will not fail again [48, 49]. Nonetheless, as pointed out by Dyson [35], the value of eschatology may not lie so much in predicting the future, but in raising germane questions that might eventually lead to better understanding.

We stand with Dyson in asserting that scientific inquiry need not be divorced from issues of human purpose and values. Science finds fruition as and where it touches the human condition. We are bound by the second law and to it are tied the deepest human aspirations, fears, sufferings and joys. We would be remiss not to acknowledge this central truth. We hope one of the legacies of this volume will be to help illumine our future both in light of the second law and in its probable violation.

Most eschatologies ignore intelligent action as a factor in the development of the universe. This is partially attributable to the scientific habit of 'ignoring the observer,' but it is also largely in recognition of mankind's relative impotence within the cosmic order. The latter might not always be the case. Consider, for example, that humans are now the primary movers of soil on Earth, outstripping natural processes; we have significantly altered between 1/4 and 1/3 of the Earth's surface. Over the last several thousand years we have modified the composition of the atmosphere sufficiently to change the climate measurably. We now extract fish from the world's oceans at rates equaling or exceeding their rates of replenishment. As a species, we are both "the prince of creation" [50] and a principal instrument of mass extinction. If the last 100 years are any indication of the dominance humans can wield over their home planet, then extrapolating this last 10^{-7} fraction of Earth's history out to cosmological time scales, one cannot rule out the possibility that intelligent action — though almost certainly not *genetically human* and, hopefully, more than *humanly wise* — could have a significant effect on the cosmic order. Dyson alludes to this [35]:

> Supposing that we discover the universe to be naturally closed and doomed to collapse, is it conceivable that by intelligent intervention, converting matter to radiation and causing energy to flow purposefully on a cosmic scale, we could break open a closed universe and change the topology of space-time so that only a part of it would collapse and another part of it would expand forever?

(Actually, the reverse is now considered more likely.)

We wish to explore another scenario — that the second law is violable — and speculate how this might affect humanity and the universe in the distant future. This discussion will be relatively brief and general, in keeping with its speculative nature. We adopt the logarithmic unit of time η, introduced by Adams and Laughlin [36], defined by $\eta \equiv \mathrm{Log}_{10}(\tau)$, where τ is years since the Big Bang. In this notation, the present is $\eta \simeq 10$.

The fate of the universe for ($\eta \leq 100$) depends on many unknowns, some of which are listed above. Chief among these are whether the universe will expand forever and whether the proton is stable. Under the assumptions of continuing expansion and proton decay, the various cosmic ages proposed by Adams and Laughlin [36] are summarized as follows:

> *Radiation-Dominated Era* ($-\infty < \eta < 4$): Big Bang, expansion, particles and light elements created. Energy primarily in radiation form.
> *Stelliferous Era* ($6 < \eta < 14$): Energy and entropy production dominated by stellar nuclear processes. Energy primarily in matter form.
> *Degenerate Era* ($15 < \eta < 37$): Most baryonic matter found in stellar endpoints: brown dwarfs, white dwarfs, neutron stars. Proton decay and particle annihilation.

Black Hole Era ($38 < \eta < 100$): Black holes, the dominate stellar objects, evaporate via Hawking process.

Dark Era ($\eta > 100$): Baryonic matter, black holes largely evaporated. Detritus remains: photons, neutrinos, positrons, electrons.

This grim analysis ignores issues of dark matter and dark energy.

Classical heat death, discovered shortly after the discovery of the second law [51, 52, 53] is a process by which the universe moves toward thermodynamic equilibrium as it exhausts its available free energy sources. At its terminus, the universe becomes uniform in temperature, chemical potential, and entropy. It is lifeless. In the latter half of the 19th and through the early 20th centuries, before cosmological expansion was discovered, the general sentiment surrounding heat death was perhaps most famously expressed by Bertrand Russell [54]:

> All the labours of the ages, all the devotion, all the inspiration, all the noonday brightness of human genius are destined to extinction in the vast death of the solar system, and ... the whole temple of man's achievement must inevitably be buried beneath the debris of a universe in ruins.

Hope for temporary reprieve is offered by Bernal [55]:

> The second law of thermodynamics which ... will bring this universe to an inglorious close, may perhaps always remain the final factor. But by intelligent organizations the life of the Universe could probably be prolonged to many millions of millions of times what it would be without organization.

Even after the modifications wrought by cosmological expansion, the fateful effects of the second law remain, as pointed out more recently by Atkins [56]:

> We have looked through the window on to the world provided by the Second Law, and have seen the naked purposelessnes of nature. The deep structure of change is decay; the spring of change in all its forms is the corruption of the quality of energy as it spreads chaotically, irreversibly and purposelessly in time. All change and time's arrow, point in the direction of corruption. The experience of time is the gearing of the electrochemical processes in our brains to this purposeless drift into chaos as we sink into equilibrium and the grave.

This depressing vision — as inescapable as the second law itself — has penetrated so deeply into Western culture and consciousness as to be almost invisible. Arguably, the second law is no longer simply a scientific principle; it has become a sociological neurosis of Western civilization.

10.3.2 Cosmic Entropy Production

For this discussion we assume an ever-expanding universe since, in the last decade, strong evidence for it has accumulated. This includes observations of distant supernovae, cosmic microwave background (CMB) anisotropies, and inadequate gravitating matter inventories to cause collapse [57]. There are, of course, many eschatologies that assume collapse [44, 58, 59]. Since the apparent discovery of dark energy with its negative pressure [60, 61], the traditional topological terms *open* ($\Omega > 1$) and *closed* ($\Omega < 1$) are no longer adequate to describe cosmological evolution. Even a topologically closed universe can undergo continuous expansion given a positive cosmological constant (negative pressure), as is associated with dark energy.

In the standard cosmological model, the universe begins with exceptionally low gravitational entropy. That it does not immediately achieve something akin to thermal equilibrium — for instance, during the recombination era ($z \sim 1500$) when it moved from a fairly spatially uniform plasma state to a neutral atom state — is due primarily to the gravitational clumping and, to a lesser degree, to the cosmic expansion itself which continually creates more configuration states for matter. As large and small scale structures emerged (stars, galaxies, super-clusters), thermodynamic entropy production gave way to gravitational entropy production. This, in turn, allowed further thermodynamic entropy production, *e.g.*, in the form of stellar nucleosynthesis, planetary formation, geochemistry, and life.

Systems are driven forward thermodynamically — kept out of equilibrium — not by entropy *per se*, but by *entropy gradients*. This was recognized as far back as Poincaré. The Earth's biosphere, for instance, is driven by the solar-terrestrial entropy gradient more surely than it is driven by solar energy. This gradient has been conveniently expressed by Frautschi through photon counting [62]. For every fusion nucleon at the core of the Sun ($T \sim 1.5 \times 10^7$K), at the Sun's surface (photosphere, $T = 5800$K) roughly 5×10^6 photons are emitted. Each solar photon reaching Earth is converted, via photosynthesis, biochemical reactions, and thermalization, into roughly 20 photons ($\sim \frac{5800K}{300K} \simeq 20$). These are then mostly radiated back into space where they eventually equilibrate, each producing roughly another 100 CMB photons. Thus, a single, well-localized — and apparently entropy-reducing — fusion reaction in the Sun's core gives rise to $\sim 10^{10}$ photons and substantial entropy. Thus, it is fair to say that it is not the Sun's energy that drives the Earth's biosphere; rather, it is the solar-terrestrial entropy gradient. Were there not an entropy gradient between the two, then by definition they would be at mutual equilibrium and the nonequilibrium processes necessary for life would not be possible.

In the far future, the march toward equilibrium will step to a different beat: rather than being set by nuclear fusion, it will be set mainly by proton decay, black hole and positronium formation and evaporation. If the cosmos continues to expand, however, it can never reach constant temperature or equilibrium. Thus, while it may dodge classical heat death, it still dies in the sense that the entropy of comoving volumes asymptotically approach a constant value. This limiting value should be far below that set by the Bekenstein limit [62, 63]. Actually, in most

Entropy Source	$\Delta\eta$	Log_{10} [Relative Entropy Increase Over CMB]
Stellar Nucleosynthesis	11	-2
Black Hole Formation		
-Stellar	11	≤ 10
-Galactic	≤ 20	≤ 21
-Galactic Supercluster	≤ 20	≤ 24
Proton Decay	≥ 32	3
Black Hole Evaporation	≤ 106	≤ 24
Positron Formation		
and Decay	≥ 116	≥ 13

Table 10.2: Primary entropy production into the far-future ($\eta < 120$), after Frautschi [62]. Duration is expressed in $\Delta\eta$ and total entropy produced is compared to entropy of CMB.

cosmological scenarios, the entropy of a comoving frame falls further and further behind its theoretical maximum.

Entropy production is proposed to extend into the far-future ($\eta > 50$) [36, 62]. Several of the dominant sources are listed in Table 10.2. The predominant future entropy source should be black hole evaporation, whose total integrated entropy production should outpace stellar nucleosynthesis by roughly 26 orders of magnitude. Note, too, the entropy associated with Hubble expansion is negligible compared with it. This underscores the tremendously small initial entropy of the cosmos relative to its theoretical maximum.

The pinnacle of gravitational entropy is the black hole. The classical black hole is a thermodynamically simple object, specified entirely by three parameters: mass, angular momentum and electric charge. Particles falling into a black hole lose their individual identities and many of their normally conserved properties (*e.g.*, baryon number, lepton number). Since a hole's interior quantum states are not measurable, they are equally probable; thus, a black hole represents a real information sink and entropy source. The Bekenstein-Hawking entropy of a black hole of mass M_{BH} [64]

$$S_{BH} = \frac{4\pi k G M_{BH}^2}{\hbar c} \qquad (10.4)$$

is enormous compared to standard thermodynamic sources. For instance, the gravitational entropy for a one solar mass black hole is $\frac{S_{BH}}{k} \simeq 10^{77}$, while the thermodynamic entropy of the Sun is $\frac{S_{Sun}}{k} \simeq 10^{58}$. As for other gravitational entropies, black hole entropy is nonextensive, scaling as M_{BH}^2. Since M_{BH} is proportional to radius ($M_{BH} = \frac{R_{BH}c^2}{2G}$), a hole's entropy is proportional to its surface area A:

$$S_{\mathrm{BH}} = \frac{kc^3}{4\hbar G} A \qquad (10.5)$$

Since traditional black holes only accrete mass, thereby increasing their entropies and areas, a gravitational analog to the thermodynamic second law can be formulated. If S_m is defined as the entropy of mass-energy outside a hole, then a *generalized entropy* can be defined, $S_g \equiv S_m + S_{\mathrm{BH}}$. The *generalized second law* states that S_g never decreases. (Each law of classical thermodynamics has a black hole equivalent.)

In 1974, Hawking proposed black hole evaporation via emission of thermal quantum particles [65, 66]. To it an effective temperature can be ascribed:

$$T_{\mathrm{H}} = \frac{\hbar c^3}{8\pi G M_{\mathrm{BH}} k} \simeq 6.2 \times 10^{-8} \left[\frac{M_{\odot}}{M_{\mathrm{BH}}} \right] K. \qquad (10.6)$$

The evaporation timescale for a hole ($\tau_{\mathrm{H}}(\mathrm{yr})$) should be roughly

$$\tau_{\mathrm{H}} \simeq \frac{16\pi^2 G M_{\mathrm{BH}}^3}{\hbar c^3} \simeq 10^{62} \left[\frac{M_{\mathrm{BH}}}{M_{\odot}} \right]^3 \;(\mathrm{yr}) \qquad (10.7)$$

Clearly, for even the smallest black hole generated by standard astrophysical processes, the evaporation time is tens of orders of magnitude longer than the age of the universe: the far future. Evaporation is in the form of random thermal particles with the maximum entropy per unit mass.

The maximum value of entropy for a system of radius R and energy E is given by the Bekenstein limit [63]:

$$S \leq \frac{2\pi R E}{\hbar c}, \qquad (10.8)$$

This limit can be attained by black holes.

The action of a black hole can be likened a bit to a home trash compactor: relatively low-entropy items like fruits and vegetables are thrown in, ground up and mixed into a high entropy state, compacted and then, as they rot, slowly evaporate back out as a diffuse, non-descript, high-entropy gas.

10.3.3 Life in the Far Future

Whether one subscribes to traditional thermodynamic heat death or to merely heat coma mediated by proton decay and black hole evaporation, the second law ($\frac{dS}{dt} \geq 0$) is a prime mover and the grim reaper. If the second law can be violated and applied on astrophysical scale lengths, however, then it would seem our eschatological fate might be altered. The basic requirement for survival is this: that the natural rate of entropy production be less than or equal to the entropy reduction rate achieved by second law violation. Too little is known about either process to claim a theoretical victor at this point. Nonetheless, we briefly introduce three scenarios by which life might at least be extended in a dying universe.

Matrix: Within our cosmological horizon ($R \sim 10^{26}$m) it is estimated there are on the order of 10^{11} galaxies of average mass $10^{11} M_\odot$, giving for the entire universe a mass on the order of 10^{52}kg. The baryon number is about $N_b \sim 10^{79}$; there are roughly 10^9 photons per baryon, predominately in the CMB. (It is estimated the cosmic mass-energy is fractionated into roughly 73% dark energy (nature unknown), 23% dark matter (nature unknown), and 4% baryonic matter (only about 25% of which can be seen or inferred directly); thus, roughly 99% of the mass-energy of the universe remains mysterious.)

Let the mass and scale length of the universe be M and R, respectively. Let this mass be rolled into large sheets whose atomic scale length is r and whose thickness in number of atomic layers is n. The number of sheets of scale size R that can be constructed from M should be roughly $N_s \sim \frac{N_b r^2}{nR^2}$. For $n = 10$, $m = 10$amu and $M = 10^{52}$kg one has $N_s = 10^6$. In other words, the intrinsic mass of the universe could, in principle, be fashioned into a matrix of 10^6 partitions spanning the universe. Arranged uniformly, they would be spaced roughly every 10^4ly. If these partitions were antennae and second law violators for CMB or other far-future particle fields (*e.g.*, neutrinos, e^+/e^-), then the universe's energy might be recycled many times and heat coma forestalled. This, of course, presupposes the harvested energy would be sufficient to repair the ravages of the natural entropic process like proton decay [62]. Since the energy intercept time for this matrix should be on the order of 10^4yr, while primary decay processes like proton decay and black hole evaporation exceed 10^{32}yr, if the matrix were even only mildly efficient at entropy reduction, it should be able to keep pace. Easy calculations show that a matrix could, in principle, reverse the current entropy production by stellar nucleosynthesis [67]. Additionally, if the future civilization were able to *rearrange the furniture* on a cosmic level — see Dyson above — then perhaps spacetime topology could be modified either to avert continued expansion, so as to create a thermodynamically steady-state universe or sub-universe.

Outpost: On a more limited scale, the second law might be held at bay through construction of a finite outpost, perhaps the size of a solar system, surrounded by a well-insulated thermal shell. Like the matrix, it would harvest heat and particle from the exterior thermal background fields and concentrate it internally. Its success depends on it being well-insulated to heat and particle emission. Presumably, it would recycle its own interior heat into work. As the surrounding universal temperature falls and this is communicated to its interior, however, the outpost would have to rely on lower and lower temperature second law violation processes and lower power densities. Since negentropic process have been proposed down to superconducting temperatures already (Chapter 4), it is plausible that lower temperature processes will be discovered. The primary requirement for sustainability is that the heat harvested from the exterior space at least match the heat (and mass-energy due to particle decay) that leaves through the insulating shell[2].

---

[2]For thermal insulation, one might guess that the optimum would be a hollow black hole (*e.g.*, a thin spherical shell: $R = 200$ AU, $M = 10^{10} M_\odot$, $T_H = 10^{-17}$K, $\tau_H \sim 10^{92}$yrs). Unfortunately, general relativity forbids this as a stable geometry, although naive calculations indicate it might be possible.

Thermal Life There has been considerable debate as to whether life, intelligence and consciousness (LIC) can survive indefinitely in a dying universe [35, 62]. We will assume that if life is possible, so too are intelligence and consciousness. A general condition assumed by Dyson invokes entropy: $\int \frac{dS}{dt} dt = \infty$; that is, since LIC rely on inherently nonequilibrium processes, the entropy growth in a comoving volume must continuously increase. If the second law is violable, however, this limitation is suspect.

The very definitions of and conditions for LIC are themselves a matter of longstanding debate and we will not be drawn into the fray. We note, however, that since thermosynthetic life appears possible in principle (§10.2), there may be hope that life could survive in the far future by directly tapping the thermal fields alone, even if an entropy gradient is absent. Hopefully, such life would be more subtle than Hoyle's black cloud [68] or Čapek's sentient computer [69], but as before, its energy budget would be limited to what it could harvest from its surroundings, which, if the universe becomes increasingly diffuse, would constitute a long, slow road into heat coma.

Life is commonly considered antithetical to and distinguishable from thermodynamic equilibrium in at least two fundamental ways:

1) Biotic matter is physically arranged in a *non-random*, far from equilibrium configuration.
2) Biochemical reactions are superthermal and cannot be maintained at equilibrium.

Both assumptions should be re-examined in light of second law challenges. With regard to (1), there is no absolute, agreed upon definition of order, so conversely, there is no absolute definition of disorder. Furthermore, large-scale organized dynamic structures are not forbidden at equilibrium; Debye sheaths, for example, are dynamically-maintained large-scale structures that arise by thermal processes at the edges of plasmas within which highly non-equilibrium processes occur. Thus, it is not clear *a priori* that equilibrium could not support the sort of large scale energy-producing structures necessary for life.

Assumption (2) is undercut by the spectrum of second law challenges in this volume. In MPG systems (§10.1.1), superthermal nonequilbrium particle currents are maintained in equilibrium settings. Nikulov's persistent supercurrents arise out of purely quantum effects at low temperatures (§4.4). One cannot rule out that other contra-entropic effects will be discovered at even lower temperatures.

It is unclear what form life might take in the far future. Must it be solid-liquid like terrestrial life? Could it exist, for example, as complex, long-range, correlations in the electron-positron detritus predicted for the far future? It is not clear what the minimum requirements for life are, especially given that its definition is still a matter of debate.

Thermosynthetic life is predicted to be survivable in an equilibrium environment, but must it be at most an island in an equilibrium sea? Could it co-exist with thermal equilibrium; that is, might life assimilate itself directly into the thermal field? This is a modest extrapolation from J.A. Wheeler's mystic vision of the universe as "a self-synthesizing system of existences, built on observer-participancy" [70]. (Čápek's phonon continuum model (§3.6.6) might also be mined for ideas on how this might play out.) Presumably, this *thermal life* would wind down with the rest of the universe. Since this would not be a struggle against nature, but rather an acquiescence to it, this could be considered an *if you can't beat 'em, join 'em* strategy for the far future.

Although violation the second law appears useful for survival into the far future, one should be careful of what one wishes for. Entropy reduction, though useful, is an impoverishment of the universe's phase space complexity. Moreover, if carried out on cosmic scales, it would amount to a 'turning back' of the thermodynamic clock. This reversal of thermodynamic time would not imply time reversal in the palindromic sense that physical processes would precisely reverse — that the universe would retrace its exact path in phase space — but it would entail returning the universe to a lower entropy state. The clock would be turned back to a new and different clock each time. It would be akin to erasing a chalkboard in preparation for writing something new — and getting the chalk stick back, to boot[3]. Presumably, the cosmic horizon would not be affected since the mass-energy would not be changed, but simply rearranged. However, if carried out to the extreme, converting and removing all heat, the universe could freeze and die just as surely as by standard heat death[4]. A little bit of disorder (heat) is a good thing.

In conclusion, we reiterate that the continuing uncertainties surrounding cosmological issues, basic physical laws, and definitions of life, intelligence and consciousness make eschatological studies largely moot. The second law, however, appears integral to most, such that its possible violation should be considered in future studies.

10.4 The Second Law Mystique

Perhaps more has been written about the second law across the breadth of human knowledge and endeavor than about any other physical law. Direct and indirect references to it can be found in all branches of science, engineering, economics, arts, literature, psychology, philosophy, and popular culture.

Quite aside from its profound physical, technological, social, philosophical and humanistic implications, the second law is *famous for being famous*. It has become the epitome of scientific truth, notorious for its absolute status, virtually unquestioned by the scientific community. Much of its mystique can be traced to the imprimaturs of scientists like Einstein, Planck, Eddington, Fermi, Poincaré, Clausius, and Maxwell. Their reputations are reciprocally burnished by association

---

[3]Earlier thinking on time and change can be found in the Rubáiyát of Omar Khayyám (ca. 1100 C.E.): "The moving finger writes and having writ moves on. And all your poetry and wit cannot erase half a line, nor all your tears one word of it."

[4] "Some say the world will end in fire, some say ... that for destruction ice is also great, and would suffice." Robert Frost, *Fire and Ice* (1920).

with this fundamental law; thus, a cycle of mystique is established.

The following are representative quotes from the sciences [71] and humanities dealing with the second law and its absolute status. Outside of theatre, rarely does one find such melodrama.

Einstein [72]

> [Classical thermodynamics] is the only theory of universal content concerning which I am convinced that, within the framework of applicability of its basic concepts, it will never be overthrown.

Eddington [73]

> The law that entropy always increases — the second law of thermodynamics — holds, I think, the supreme position among the laws of Nature. If someone points out to you that your pet theory of the universe is in disagreement with Maxwell's equations — then so much the worse for Maxwell's equations. If it is found to be contradicted by observation — well, these experimentalists bungle things sometimes. But if your theory is found to be against the second law of thermodynamics I can give you no hope; there is nothing for it but to collapse in deepest humiliation.

Fermi [74]

> The second law of thermodynamics rules out the possibility of constructing a *perpetuum mobile* of the second kind ... The experimental evidence in support of this law consists mainly in the failure of all efforts that have been made to construct a *perpetuum mobile* of the second kind.

Clausius [75]

> Everything we know concerning the interchange of heat between two bodies of different temperatures confirms this, for heat everywhere manifests a tendency to equalize existing differences of temperature ... Without further explanation, therefore, the truth of the principle [second law] will be granted.

Maxwell [76]

> The second law of thermodynamics has the same degree of truth as the statement that if you throw a thimbleful of water into the sea, you cannot get the same thimbleful of water out again.

Çengel and Boles [77]

> To date, no experiment has been conducted that contradicts the second law, and this should be taken as sufficient evidence of its validity.

Horgan [78]

To Gell-Mann, science forms a hierarchy. At the top are those theories that apply everywhere in the known univese, such as the second law of thermodynamics and his own quark theory.

Maddox [79]

The issue is not whether the second law of thermodynamics is valid in the ordinary world; nobody doubts that.

Park [80]

Each failure, each fraud exposed, established the laws of thermodynamics more firmly. ... Extending mistrust of scientific claims to include mistrust of the underlying laws of physics, however, is a reckless game.

Maddox [81]

Maxwell's demon is therefore no longer regarded as a limitation of the second law.

Lieb and Yngvason [82]

No exception to the second law of thermodynamics has ever been found — not even a tiny one.

Brillouin [83]

Nobody can doubt the validity of the second principle, no more than he can the validity of the fundamental laws of mechanics.

In the humanities, explicit, implicit and interpretable references to the second law are uncountable. Many dwell on the futility of existence within its sphere, while others lament the heat death of the universe [54, 55, 56, 84]. The following are exemplars.

George Gordon (Lord Byron), from *Darkness* [85]

I had a dream, which was not a dream.
The bright sun was extinguish'd, and the stars
Did wander darkling in the eternal space,
Rayless, and pathless, and the icy earth
Swung blind and blackening in the moonless air;
Morn came and went — and came, and brought no day...
... The world was void,
The populous and the powerful was a lump
Seasonless, herbless, treeless, manless, lifeless,
A lump of death — a chaos of hard clay...
The waves were dead; the tides were in their grave,
The Moon, their mistress, had expired before;

The winds were wither'd in the stagnant air,
And the clouds perish'd; Darkness had no need
Of aid from them — She was the Universe.

D.P. Patrick, from *Helena Lost*

... Though Time must carve into our flesh his name,
Erase the precious ledgers of our minds,
Corrode our bones, corrupt our breaths and maim,
And, thus, to us Oblivion consign ...

Robert Frost, *Fire and Ice* [85]

Some say the world will end in fire,
Some say in ice.
From what I've tasted of desire
I hold with those who favor fire.
But if it had to perish twice,
I think I know enough of hate
To say that for destruction ice
Is also great
And would suffice.

Archibald Macleish, from *The End of the World* [85]

... And there, there overhead, there, there, hung over
Those thousands of white faces, those dazed eyes,
There in the starless dark the poise, the hover,
There with vast wings across the canceled skies,
There in the sudden blackness the back pall
Of nothing, nothing, nothing — nothing at all.

In the literary sphere, resignation to the second law and heat death seems almost a romantic death wish; however, it is firmly rooted in scientific faith. In the scientific sphere this resignation, though resting on broad experimental and theoretical support, ultimately, is also rooted in faith, especially as it pertains to second law inviolability. It has been a goal of this volume to shake this faith in the hope of attaining something more illuminating.

References

[1] Sheehan, D.P., Phys. Plasma **2** 1893 (1995).

[2] Sheehan, D.P., Phys. Plasma **3** 104 (1996).

[3] Sheehan, D.P. and Means, J.D., Phys. Plasma **5** 2469 (1998).

[4] Sheehan, D.P., Phys. Rev. **E 57** 6660 (1998).

[5] Duncan, T., Phys. Rev. **E 61** 4661 (2000); Sheehan, D.P., Phys. Rev. **E 61** 4662 (2000).

[6] Sheehan, D.P., Phys. Lett. A **280** 185 (2001).

[7] Sheehan, D.P., Glick, J., and Means, J.D., Found. Phys. **30** 1227 (2000).

[8] Sheehan, D.P., Glick, J., Duncan, T., Langton, J.A., Gagliardi, M.J. and Tobe, R., Found. Phys. **32** 441 (2002).

[9] Sheehan, D.P., Putnam, A.R. and Wright, J.H., Found. Phys. **32** 1557 (2002).

[10] Zhang, K. and Zhang, K., Phys. Rev. A **46** 4598 (1992).

[11] Gross, D.H.E., *Microcanonical Thermodynamics, Phase Transitions in "Small" Systems*, (World Scientific, Singapore, 2001).

[12] Gross, D.H.E., Entropy **6** 158 (2004).

[13] Sheehan, D.P., J. Sci. Explor. **12** 303 (1998); Sheehan, D.P., Infin. Energy **9/49** 17 (2003).

[14] Duncan, T. and Semura, J.S., Entropy **6** 21 (2004).

[15] Sheehan, D.P. *Thermal Life and the Second Law*, USD Deep Six Seminar, Feb. 2004. The term *thermosynthetic* was later coined by A. Sturz, USD Science Lecture Series, Feb. 2004.

[16] Uwin, P.J.R., Webb, R.I., and Taylor, A.P., Amer. Minerol. **83** 1541 (1998).

[17] Gold, T., Proc. Natl. Acad. Sci. USA **89** 6045 (1992).

[18] Szevtzyk, U., et al., Proc. Natl. Acad. Sci. USA **91** 1810 (1994).

[19] Kashefi, K. and Lovley, D.R., Science **301** 934 (2003).

[20] Brock, T.D., Science **230** 132 (1985).

[21] Stetter, K.O., Fiala, G., Huber, R., Huber, G., and Segerer, A., Experientia **42** 1187 (1986).

[22] Jaenicke, R., Eur. J. Biochem. **202** 715 (1991).

[23] Leibrock, E., Bayer, P., and Lüdemann, H.-D., Biophys. Chem **54** 175 (1995).

[24] Baross, J.A. and Holden, J.F., Advances Protein Chem. **48** 1 (1996).

[25] Gold, T., *The Deep Hot Biosphere* (Springer-Verlag, New York, 1999).

[26] Hochachka, P.W. and Somero, G.N., *Biochemical Adaptation* (Oxford University Press, New York, 2002).

[27] Vreeland, R.H., Rosenzweig, W.D., and Powers, D.W., Nature **407** 897 (2000).

[28] Ellenbogen, J.C. and Love, C.J. in *Handbook of Nanoscience, Engineering, and Technology*, Goddard, W.A. III, Brenner, D.W., Lyshevski, S.E., and Iafrate, G.J., Editors, (CRC Press, Boca Raton, 2003)

[29] Zhou, C., Deshpande, M.R., Reed, M.A., and Tour, J.M., Appl. Phys. Lett. **71** 611 (1997).

[30] Metzger, R.M., et al., J. Amer. Chem. Soc. **119** 10455 (1997).

[31] Campbell, N.A., Reece, J.B., and Mitchell, L.G., *Biology* 5th Ed., (Addison Wesley Longman, Inc., Menlo Park, 1999).

[32] Brillouin, L. Phys. Rev. **78** 627 (1950).

[33] Van Kampen, N.G., Physica **26** 585 (1960).

[34] McFee, R., Amer. J. Phys. **39** 814 (1971).

[35] Dyson, F.J., Rev. Mod. Phys. **51** 447 (1979).

[36] Adams, F.C. and Laughlin, G., Rev. Mod. Phys. **69** 337 (1997).

[37] Ćirković, M.M., Am. J. Phys. **71** 122 (2003).

[38] Rees, M.J., J. Roy. Astron. Soc. **22** 109 (1981).

[39] Davies, P.C.W., Mon. Not. R. Astron. Soc. **161** 1 (1973).

[40] Islam, J.N., Vistas Astron **23** 265 (1979).

[41] Burdyuzha, V. and Khozin, G., Editors *The Future of the Universe and the Future of Our Civilization*, (World Scientific, Singapore, 2000).

[42] Ellis, G.F.R., Editor *Far-Future Universe: Eschatology from a Cosmic Perspective* (Templeton, Radnor, 2002).

[43] Eddington, A.S., Nature **127** 447 (1931).

[44] Gold, T., Am. J. Phys. **30** 403 (1962).

[45] de Chardin, T., *The Phenomenon of Man*, translated by Wall, B., (Collins, London, 1959).

[46] Barrow, J.D., and Tipler, F.J., Nature **276** 453 (1978).

[47] Kutrovátz, G., Open Syst. Inf. Dyn. **8** 349 (2001).

[48] Popper, K.R., *The Poverty of Historicism* (Routledge and Kegan Paul, London, 1957; originally published in Economica, 1944/45).

[49] Popper, K.R., Brit. J. Philos. Sci **1** 117 (1950).

[50] Shakespeare, W., *Hamlet* (ca. 1600).

[51] von Helmholtz, H., On the Interaction of the Natural Forces (1854), in *Popular Scientific Lectures*, Kline, M., Editor (Dover, New York, 1961).

[52] Clausius, R., Ann. Phys. (Leipziz) **125** 353 (1865).

[53] Clausius, R., Philos. Mag. **35** 405 (1868).

[54] Russell, B., *Why I Am Not a Christian* (Allen and Unwin, New York, 1957) pg. 107.

[55] Bernal, J.D., *The World, the Flesh, and the Devil*, 2nd Ed. (Indiana University Press, Bloomington, 1969, 1st Ed. 1929) pg. 28.

[56] Atkins, P., in *The Nature of Time*, Flood, R. and Lockwood, M., Eds. (Basil Blackwell, Oxford, 1986) pg. 98.

[57] Dodelson, S., *Modern Cosmology* (Academic Press, Amsterdam, 2003).

[58] Rees, M.J., Observatory **89** 193 (1969).

[59] Bludman, S.A., Nature **308** 319 (1984).

[60] Ford, L.H., Gen. Relativ. Gravit. **19** 325 (1987).

[61] Krauss, L.M. and Turner, M.S., Gen. Relativ. Gravit. **31** 1453 (1999).

[62] Frautschi, S., Science **217** 593 (1982).

[63] Bekenstein, J.D., Phys. Rev. D **23** 287 (1981).

[64] Bekenstein, J.D., Phys. Rev. D **7** 2333 (1973).

[65] Hawking, S.W., Nature **274** 30 (1974).

[66] Hawking, S.W. Commun. Math. Phys. **43** 199 (1975).

[67] Sheehan, D.P., unpublished (1995).

[68] Hoyle, F., *The Black Cloud*, (Harper, New York, 1957).

[69] Čapek, K., *R.U.R.* trans. Selver, P. (Doubleday, Garden City, N.Y., 1923).

[70] Wheeler, J.A., IBM J. Res. Develop. **32** 4 (1988).

[71] Rauen, K., Infin. Energy **10/55** 29 (2004).

[72] Einstein, A. 'Autobiographical Notes', in Schilpp, P.A., Editor, *Albert Einstein: Philosopher-Scientist, Vol. 2* (Cambridge University Press, Cambridge, 1970).

[73] Eddington, A., *The Nature of the Physical World* (Everyman's Library, J.M. Dent, London).

[74] Fermi, E. *Thermodynamics* (Dover Publications, New York, 1936).

[75] Clausius, R., Phil. Mag. **4**, 12 (1856).

[76] Maxwell, J.C., in *Life of John William Strutt, Third Baron Rayleigh*, Strutt, R.J., Editor (E. Arnold, London, 1924) pp. 47-48.

[77] Çengel, Y.A. and Boles, M.A., *Thermodynamics: An Engineering Approach*, 2nd Ed. (McGraw-Hill, New York, 1994).

[78] Horgan, J., *The End of Science* (Broadway Books, New York, 1996).

[79] Maddox, J. *What Remains to be Discovered* (The Free Press, New York, 1998).

[80] Park, R.L., *Voodoo Science* (Oxford University Press, Oxford, 2000).

[81] Maddox, J., Nature **412**, 903 (2002).

[82] Lieb, E.H. and Yngvason, J., Phys. Today **53**, 32 (2000).

[83] Brillouin, L., Amer. Sci. **37**, 554 (1949).

[84] Zolina, P., *Heat Death of the Universe and Other Stories* (McPherson, Kingston, NY, 1988).

[85] Gordon, G. (Lord Byron) in *Poetry, An Introduction*, Lane, W.G. (D.C. Heath and Co., 1968); Frost, R., *ibid.*; Macleish, A., *ibid.*.

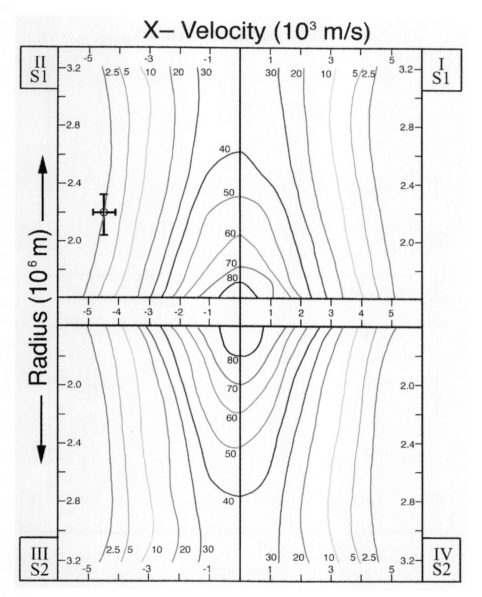

Plate I Phase space diagram v_x versus r for *standard gravitator*. (3000 patch crossings; 30° polar domes over S1 and S2); ($r_{patch} = 2.475 \times 10^6$m).

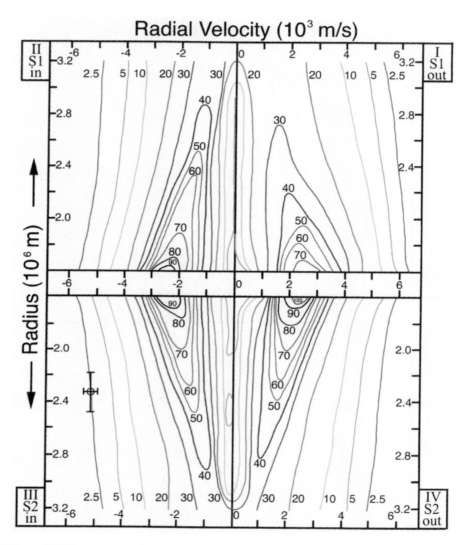

Plate II Phase space diagram v_r versus r for *standard gravitator*. (3000 patch crossings; 30° polar domes over S1 and S2; $r_G < r_{patch} < r_c$)

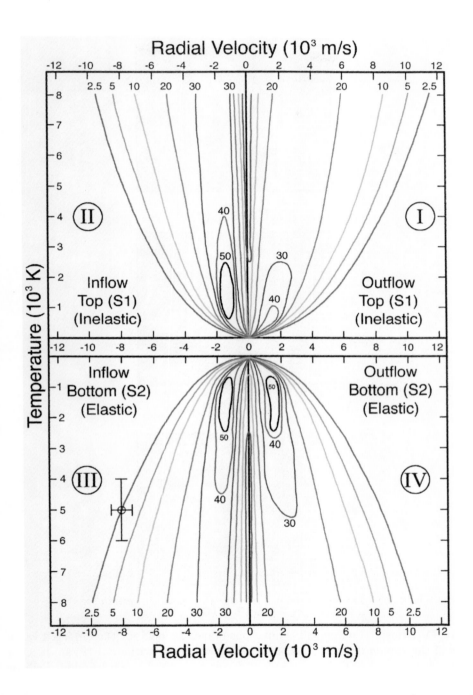

Plate III Phase space diagram v_r versus T for *standard gravitator*. (3000 patch crossings, 30° polar domes over S1 and S2, $r_{patch} = 2.405 \times 10^6$m)

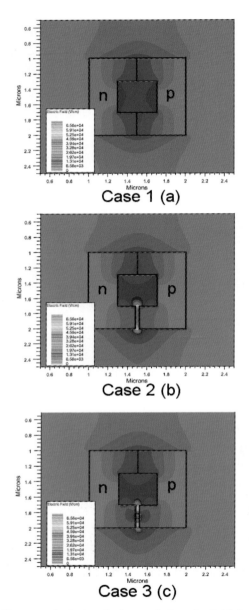

Plate IV Atlas 2-D numerical simulations of electric field for three related variations of the *standard device*. a) Case 1: *standard device* without J-II gap; b) Case 2: *standard device*; c) Case 3: *standard device* with 300Å x 600Å undoped silicon piston at gap center.

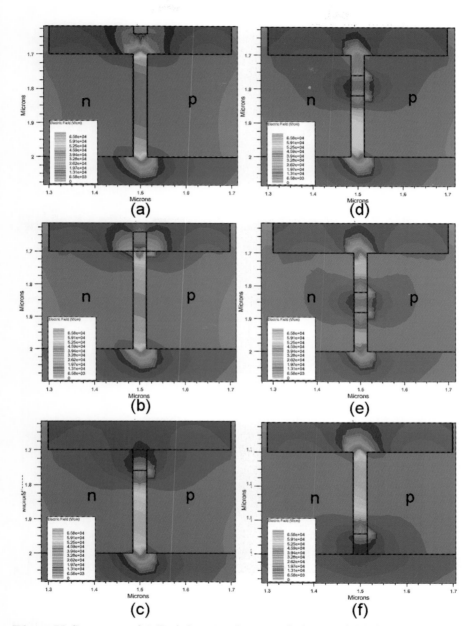

Plate V Sequence of 2-D Atlas simulations of electric field for *standard device* with static piston at various locations in J-II channel.

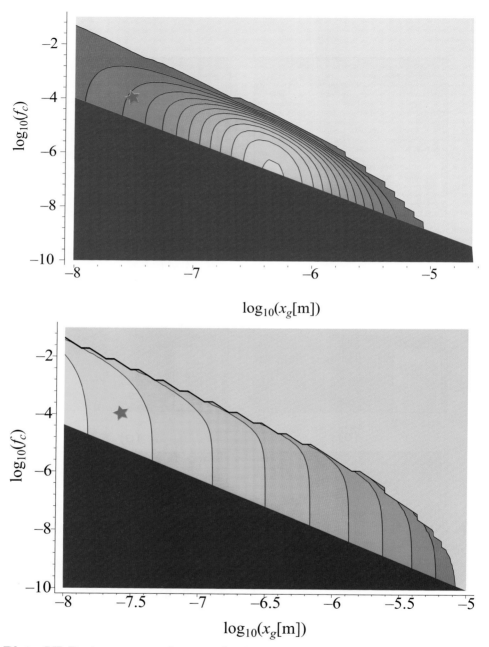

Plate VI Device power and power density over range of devices size-scaled to *standard device.*
(TOP): Maximum power output versus gap width (x_g) and contact fraction (f_c). Contours vary linearly from 10^{-9}W/device (red) to 1.2×10^{-8}W/device (yellow). (BOTTOM): Power density (Wm^{-3}) versus x_g and f_c. Contours vary logarithmically from 10^1 Wm^{-3} (red) to 10^{10}Wm^{-3} (yellow). Star indicates location of *standard device.*

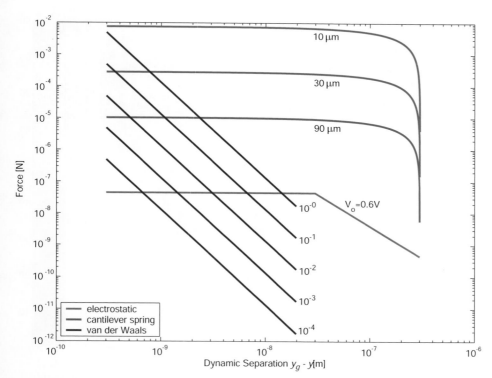

Plate VII Force (N) on hammer versus dynamic separation (m) with anvil (SUMMiT^{TM} case). Spring force (green); van der Waals force (blue); Electrostatic force (red). Device parameters: $l_z = l_h = 5\mu m$, $t_c = 4.5\mu m$, $t_h = 6\mu m$, $l_c = 10, 30, 90\mu m$; $10^{-4} \leq \eta \leq 1$, $V_o = V_{bi} = 0.6V$.

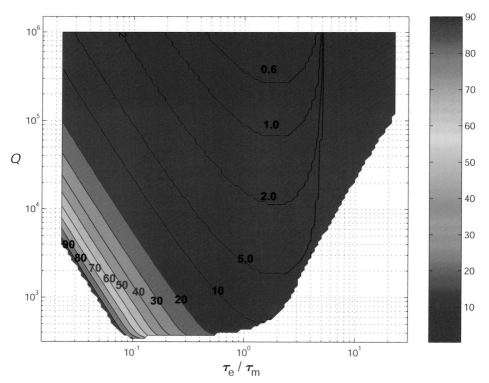

Plate VIII Minimum external bias voltage required for sustained oscillation for SUMMiT^{TM} device (Figure 9.13b), plotted versus Q and $\frac{\tau_e}{\tau_m}$. In sweet spot ($0.25 \leq \frac{\tau_e}{\tau_m} \leq 4$) and ($2 \times 10^3 \leq Q \leq 10^5$), low dc-voltage ($1V \leq V \leq 5V$) drives oscillation.

Index

Fundamental Theories of Physics

Series Editor: Alwyn van der Merwe, University of Denver, USA

Fundamental Theories of Physics

Fundamental Theories of Physics

46. P.P.J.M. Schram: *Kinetic Theory of Gases and Plasmas*. 1991   ISBN 0-7923-1392-5
47. A. Micali, R. Boudet and J. Helmstetter (eds.): *Clifford Algebras and their Applications in Mathematical Physics*. 1992   ISBN 0-7923-1623-1
48. E. Prugovečki: *Quantum Geometry*. A Framework for Quantum General Relativity. 1992   ISBN 0-7923-1640-1
49. M.H. Mac Gregor: *The Enigmatic Electron*. 1992   ISBN 0-7923-1982-6
50. C.R. Smith, G.J. Erickson and P.O. Neudorfer (eds.): *Maximum Entropy and Bayesian Methods*. Proceedings of the 11th International Workshop (Seattle, 1991). 1993   ISBN 0-7923-2031-X
51. D.J. Hoekzema: *The Quantum Labyrinth*. 1993   ISBN 0-7923-2066-2
52. Z. Oziewicz, B. Jancewicz and A. Borowiec (eds.): *Spinors, Twistors, Clifford Algebras and Quantum Deformations*. Proceedings of the Second Max Born Symposium (Wrocław, Poland, 1992). 1993   ISBN 0-7923-2251-7
53. A. Mohammad-Djafari and G. Demoment (eds.): *Maximum Entropy and Bayesian Methods*. Proceedings of the 12th International Workshop (Paris, France, 1992). 1993   ISBN 0-7923-2280-0
54. M. Riesz: *Clifford Numbers and Spinors* with Riesz' Private Lectures to E. Folke Bolinder and a Historical Review by Pertti Lounesto. E.F. Bolinder and P. Lounesto (eds.). 1993   ISBN 0-7923-2299-1
55. F. Brackx, R. Delanghe and H. Serras (eds.): *Clifford Algebras and their Applications in Mathematical Physics*. Proceedings of the Third Conference (Deinze, 1993) 1993   ISBN 0-7923-2347-5
56. J.R. Fanchi: *Parametrized Relativistic Quantum Theory*. 1993   ISBN 0-7923-2376-9
57. A. Peres: *Quantum Theory: Concepts and Methods*. 1993   ISBN 0-7923-2549-4
58. P.L. Antonelli, R.S. Ingarden and M. Matsumoto: *The Theory of Sprays and Finsler Spaces with Applications in Physics and Biology*. 1993   ISBN 0-7923-2577-X
59. R. Miron and M. Anastasiei: *The Geometry of Lagrange Spaces: Theory and Applications*. 1994   ISBN 0-7923-2591-5
60. G. Adomian: *Solving Frontier Problems of Physics: The Decomposition Method*. 1994   ISBN 0-7923-2644-X
61. B.S. Kerner and V.V. Osipov: *Autosolitons*. A New Approach to Problems of Self-Organization and Turbulence. 1994   ISBN 0-7923-2816-7
62. G.R. Heidbreder (ed.): *Maximum Entropy and Bayesian Methods*. Proceedings of the 13th International Workshop (Santa Barbara, USA, 1993) 1996   ISBN 0-7923-2851-5
63. J. Peřina, Z. Hradil and B. Jurčo: *Quantum Optics and Fundamentals of Physics*. 1994   ISBN 0-7923-3000-5
64. M. Evans and J.-P. Vigier: *The Enigmatic Photon*. Volume 1: The Field $B^{(3)}$. 1994   ISBN 0-7923-3049-8
65. C.K. Raju: *Time: Towards a Consistent Theory*. 1994   ISBN 0-7923-3103-6
66. A.K.T. Assis: *Weber's Electrodynamics*. 1994   ISBN 0-7923-3137-0
67. Yu. L. Klimontovich: *Statistical Theory of Open Systems*. Volume 1: A Unified Approach to Kinetic Description of Processes in Active Systems. 1995   ISBN 0-7923-3199-0; Pb: ISBN 0-7923-3242-3
68. M. Evans and J.-P. Vigier: *The Enigmatic Photon*. Volume 2: Non-Abelian Electrodynamics. 1995   ISBN 0-7923-3288-1
69. G. Esposito: *Complex General Relativity*. 1995   ISBN 0-7923-3340-3

Fundamental Theories of Physics

Fundamental Theories of Physics

Fundamental Theories of Physics

Fundamental Theories of Physics